PERFORMANCE ANALYSIS OF TELECOMMUNICATIONS AND LOCAL AREA NETWORKS

THE KLUWER INTERNATIONAL SERIES
IN ENGINEERING AND COMPUTER SCIENCE

PERFORMANCE ANALYSIS OF TELECOMMUNICATIONS AND LOCAL AREA NETWORKS

by

Wah Chun Chan
The University of Calgary

Springer Science+Business Media, LLC

Library of Congress Cataloging-in-Publication

Chan, Wah Chun.
  Performance analysis of telecommunications and local area networks / Wah Chun Chan.
    p. cm. --- (Kluwer international series in engineering and computer science ; SECS 533)
  Includes bibliographical references and index.
  ISBN 978-1-4757-8431-2   ISBN 978-0-306-47312-8 (eBook)
  DOI 10.1007/978-0-306-47312-8
    1. Telecommunication systems. 2. Local area networks (Computer networks) 3.
  Queuing theory. 4. High performance computing. I. Title. II. Series.

  TK5101 .C43 2000
  621.382'1--dc21

                                                                        99-047410

To my wife Yu-Chih and our children

Eileen, Jean, Vivian and An-Wen

Wah-Chun Chan

ABOUT THE AUTHOR

Wah-Chun Chan is Professor of Electrical and Computer Engineering at the University of Calgary at Calgary, Alberta, Canada. He received his B.Sc. degree from National Taiwan University, M.Sc. degree from the University of New Brunswick and Ph.D. degree from the University of British Columbia. Prior to his appointment at the University of Calgary in 1967, he was a Systems Engineer for Northern Electric Co. (now Nortel Technology) in Ottawa, Ontario.

Dr. Chan has published extensively in professional journals. In addition to telecommunication systems, his research interests include optimal control systems, variable structure control systems, queueing theory and reliability analysis.

Dr. Chan and his co-workers were awarded the IEE Ambrose Fleming Premium in 1974 for the papers: (a) "Waiting time in common-control queueing system", (b) "Transient in a single-server queueing system", and (c) "Multiserver computer controlled queueing system with preemptive priorities and feedback".

PREFACE

A telecommunication network conveys information by means of transmission links. These links are connected by switching systems and controlled by a signaling system. The network provides many services for exchange of information over distance. Thus, public-switched telephone networks as well as computer networks have become an integral part of modern society's infrastructure. Today, these networks are used extensively in business, in social life, in education and in entertainment. In particular, virtually all engineers and computer scientists need to understand the basic principles governing the operation and performance of telecommunications and computer networks.

SCOPE

The book is concerned with performance analysis in telecommunications and local area networks. It is designed to provide an understanding of the fundamental principles of teletraffic engineering. Emphasis is placed on the modeling techniques using queueing theory for the public-switched telephone network and local area networks.

The Telephone Network. The telephone network interconnects millions of telephones around the world. It offers a two-way, circuit-switched voice service and achieves a quality of service by setting up a communication path between two or more users.

Local Area Networks. A local area network (LAN) provides interconnection of a variety of data communicating devices within a small area. It is generally privately owned by a single organization. A typical example of the LAN technology is the Ethernet.

WHY WRITE SUCH A BOOK

These exist several books on the performance analysis of data networks or computer networks. Most of them are at a higher mathematical level.

This book attempts to present the essentials of queueing theory at a level that undergraduate students and practicing engineers can understand. After presenting the theory, applications to the analysis of practical networks follow.

WHO ARE THE INTENDED READERS

The book should appeal to undergraduate students. It can be used as a textbook for a course on telecommunications and local area networks. With some supplementary material the book can also be used for a graduate course. The practicing engineers may find the book useful for self-study, because of the emphasis on practical applications.

The material covered in this book is based on a series of lecture notes developed by the author in the Department of Electrical and Computer Engineering at the University of Calgary, Alberta, Canada.

The text is organized as follows. After an introductory chapter presenting some basic concepts and terminology of teletraffic engineering, Chapter two provides some basic theory on transmission systems. Chapter three discusses the congestion problem in switching systems, while Chapter four introduces the basic principles of queueing theory. Emphasis is placed on modelling techniques and the physical significance of the theory. The rest of the text, Chapter five to Chapter ten, is devoted to applications of queueing theory to performance analysis of the public-switched telephone networks and local area networks. There are numerous examples which illustrate the applications of the theory. The text assumes that the reader has some background in elementary probability theory and random processes.

ACKNOWLEDGEMENTS

The author owes a special thanks to Mrs. Ella Gee for her skillful typing of the manuscript several times. In addition, the author wishes to thank his wife, Jane (Yu-Chih Liu). Without her support and understanding this text could not have been written.

Contents

CHAPTER 1

INTRODUCTION TO TELECOMMUNICATION SYSTEMS

1-1. INTRODUCTION

Telecommunications has played a vital role in the advancement of engineering and science. In addition to its importance in public switched telephone network (PSTN), radio and television networks, the Internet, and ATM networks, telecommunications has become an important and integral part of modern society. For example, telecommunications is essential in worldwide airline reservation systems. It is also essential in commercial data processing industries, in remote accessing of data bases, in information and financial services, in inventory control systems, in automatic teller systems, and in automated order systems.

Since advancement in the theory and practice of telecommunications provide the means for attaining good quality of service in voice and data communications, sharing data bases and computer resources, etc., most engineers and scientists should have a good understanding of the fundamental concepts of telecommunications.

Historical Overview. The first significant work in telecommunications was F.B. Morse's code for telegraphy. Other significant works in the early stages of the development of telecommunications were due to A.G. Bell, G. Marconi and C.E. Shannon, among many others. In 1876, Bell invented a telephone system. In 1897, Marconi patented a wireless telegraph system. Teletypewriter service was initiated in 1931. Satellite communication was proposed by J. Pierce in 1953 and began service in 1962. During the decade of 1960, TV systems and cable TV services were offered commercially. Starting in 1969, the ARPA net created the world's first network for computer communications. The fundamental basis of the ARPA net is that messages are transmitted in a store-and-forward manner.

Store-and-forward packet switching was invented by P. Baran in 1961 in the United States. He proposed that if messages are divided into smaller units called packets for transmission, it would be more reliable, less susceptible to nuclear annihilation, and more economical than the PSTN. By the early 1980s, computer communication networks, such as the ARPA net, the Systems Network Architecture (SNA) of IBM, and the DEC net of Digital Equipment Corporation, etc. were interconnected to form the Internet.

In 1968, A.G. Fraser proposed the concepts of virtual circuits and fixed-size packets for Spider, the first asynchronous time-division multiplexing (ATDM) network. Fraser's work laid the foundation for asynchronous transfer mode (ATM) networking. By the late 1980s, CCITT agreed in principle that the broad-band integrated services digital network (B-ISDN) would be built using ATDM, and the international standard was named ATM (asynchronous transfer mode).

Unlike the PSTN and the Internet, ATM technology is still under development. ATM networks try to combine the ideas of the telephone network, such as connection-oriented service and end-to-end quality of service, and those of the Internet, such as the virtual circuits, fixed-small-size packets and integrated services.

Definitions. In order to provide the service which permits people or machines to communicate at a distance, a telecommunication system must provide the means and facilities for connecting the user terminals or telephones at the beginning of the service and disconnecting them when the service is completed.

In telephone systems, each subscriber telephone is connected by a subscriber line to a central switching point, called a switching center, the main function of which is to provide immediate connection between pairs of subscribers. The term telephone call, or simply call, means the demand to set up a connection. A call can be regarded as a series of dialing attempts to the same number, where the last attempt is either abandoned or results in a successful connection. The subscriber line, also known as the subscriber loop, is the pair of wires connecting the subscriber to the local switching center. The subscriber loop provides a path for the two-way speech signals, ringing, switching and supervisory signals, and is permanently associated with a particular subscriber. The end office is the local central switching center typically serving only subscriber lines. To set up a connection, the end office must perform switching, signaling, and transmitting functions.

In automatic switching centers, the connections are performed by means of switching elements, generally referred to as switches. A telephone switch actually has two parts: the switching hardware carrying voice, and the switch controller handling call set-up and tear-down signals.

In studying telecommunication engineering, we need to define additional terms that are necessary to describe telecommunication systems:

Trunks. A trunk in telephone systems is a communication path that contains shared circuits that are used to interconnect central offices; in general, 2-wire lines are used for interlocal centers and 4-wire lines for intertoll centers.

Trunk Groups. A trunk group is a group of trunks between two points, both of which are switching centers and/or individual message distribution points, and which employ the same multiplex terminal equipment.

Junctors. A junctor is a connection between networks in the same central office. Its function is equivalent to an interoffice trunk.

Links. A link is a transmission path used to make a connection between successive stages of a switching network.

Network. A network is a set of communication links and switches which, when activated, is capable of supporting a multiplicity of distinct transmission paths for voice or data transmissions.

Central Offices. A central office is a central switching center that comprises a switching network and its control and support equipment.

Toll Points. A toll point is a class 4 (switching) office not offering operator assistance for incoming calls.

Toll Centers. A toll center is a class 4 office offering operator assistance for incoming calls.

Tandem Offices. A tandem office is a switching center between local offices, reducing the number of direct trunks required.

Blocking. Blocking is the state of a group of paths between two points in a network in which all the paths are occupied and hence no further interconnections are possible.

Blocking Network. A blocking network is a network that, under certain conditions, may be unable to set up a connection from one end of the network to the other. In general, all networks used in telecommunications are of the blocking type.

BORSCHT Circuit. A BORSCHT circuit is a line circuit in the central office. It is used as a mnemonic for the functions that must be performed by the circuit: Battery, Overvoltage, Ringing, Supervision, Coding (in a digital office), Hybrid and Testing.

Modems. A modem (modulator/demodulator) is a device which modulates and demodulates signals transmitted over communication facilities. The modulator is for transmission and the demodulator is for reception. A modem is used to permit digital signals to be sent over analog lines.

Multiplexing. Multiplexing is a means for the division of a transmission facility into a number of channels either by splitting the frequency band of the channel into narrower bands, each of which is used as a distinct channel (FDM, frequency-division-multiplexing), or by allotting the entire channel to a number of users, one at a time on a time-shared basis (TDM, time-division-multiplexing).

Packets. A packet is a group of binary digits including data and control information arranged in a specified format which is transmitted as a basic unit.

Protocols. A protocol is a well-defined set of rules or conventions which governs the format and control of information that is transmitted through a network or that is stored in a data base. The information may be voice, text, data or image.

Telecommunications. The term telecommunications is a service that permits people and machines to communicate at a distance. It includes telephony, video-telephony, data transmission, teleconferencing, e-mail, etc.

Stations. A station in a telecommunication system is the device used as a means of communication by the user to the system and to other users. It includes telephone sets, terminals, printers, computers, or other types of data-communicating and data-handling devices.

1-2. EXAMPLES OF TELECOMMUNICATION SYSTEMS

In this section, we shall present several examples of telecommunication systems.

Step-by-Step $(S \times S)$ **Systems.** The basic principle of a step-by-step system is illustrated in the schematic diagram in Figure 1-1. The $S \times S$ system makes use of two-dimensional stepping switches. These stepping or selection switches are designed for both vertical and horizontal movements, which are called the selection and hunting stages, respectively. They can step a maximum of ten levels vertically in unison with dial pulses and then, in the hunting stage, automatically rotate horizontally over ten positions to search for an idle path to the next selector or other circuit. They are directly controlled by the dialed pulses.

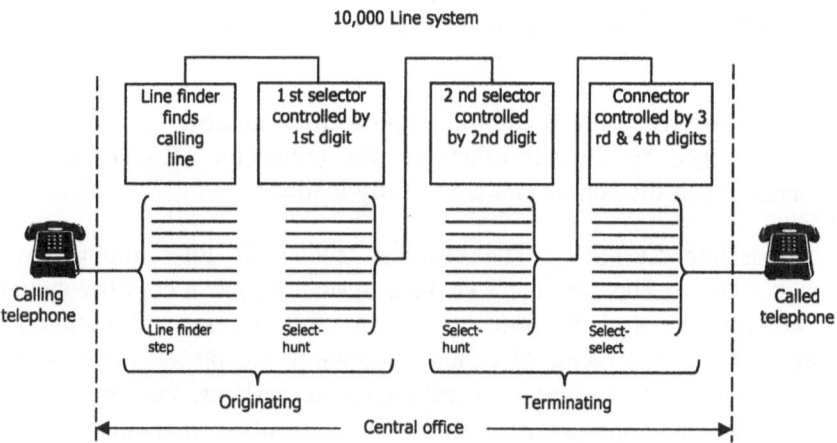

Figure 1-1 Direct control $S \times S$ switching system.

The line finders are simpler switches, one for each group of incoming subscriber lines. The line finder, once connected, supplies a dial tone to the calling subscriber.

To illustrate the processing of an intraoffice call within the $S \times S$ system, consider the four-digit line numbers 2345 to be called from a subscriber. When the calling subscriber initiates a call (handset off-hook), an idle line finder (always associated with a selector) finds and connects to the calling line. A dial tone is sent to the calling party by this first selector. Dialing the first digit, 2, steps the first selector vertically to its second level. This first selector then rotates horizontally to select an idle path to a second selector that serves the 2000 - 2999 number series. The next digit, 3, will step the second selector to its third level where it will hunt for an idle path to connectors associated with lines in the 2300 - 2399 number sequence. The digit, 4, will cause the connector switch to move to its fourth level. Since the connector has no hunting feature, the last digit, 5, will move this connector horizontally 5 positions, thereby connecting to the called line number 2 3 4 5. The connector next tests the called line. If it is idle, ringing current is applied. If it is busy, a busy tone is returned to the caller.

The $S \times S$ system has the advantage of being inexpensive for small systems and highly reliable due to the distributed nature of equipment. However, the system has several drawbacks:

1.   Since the office code digits specify the exact levels to which selectors are stepped, it is not feasible to select an alternate route for interoffice calls if all the trunks are busy.

2.   The last two digits of the called line number are specifically determined by their locations on the connector. Congestion could arise when the connectors are heavily loaded.

   Crossbar Systems. The crossbar switch is the simplest space-division switch. The crossbar switching system employs crossbar switches and common control in order to obtain full access and nonblocking capabilities. Active elements called crosspoints are placed between input and output lines. In common control switching systems, the switching and the control operations are separated. This permits a particular group of common control circuits to route connections through the switching network for many calls at the same time on a shared basis.

   The principal control device of the crossbar system consists of two types of controllers known as originating markers and terminating markers (see Figure 1-2). Each marker is capable of advancing the state of a call, but only remains with a call for about one-half second. Each intraoffice call is served by one originating marker and one terminating marker during its setup.

The philosophy of design of the No. 1 Crossbar system is based on the assumption that the majority of traffic handled will be interoffice. It is often used for telephone service in large metropolitan areas.

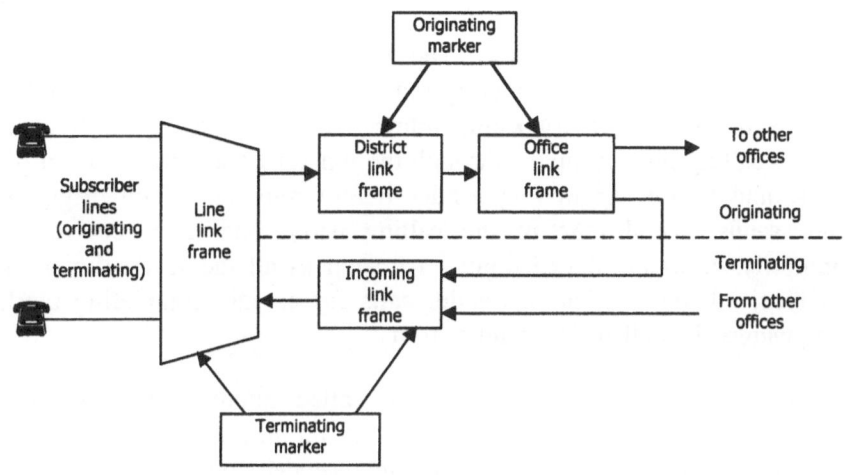

Figure 1-2 No. 1 Crossbar System

The originating calls are served by originating senders (devices for receiving and buffering dialed digits) and markers (controllers capable of translating dialed digits and locating idle paths and trunks for routing calls). After the originating marker sets up the connection, the originating sender sends the dialed digits to the terminating sender, which calls in a terminating marker to complete the connection.

After having performed their operations, the originating and terminating senders and markers drop off. The originating sender disengages after transmitting the last four digits, and the terminating sender releases after signaling the incoming trunk circuit to ring the called station or to return a busy tone to the calling station.

For intermediate size central offices with low and primarily intraoffice traffic, the No. 5 Crossbar system shown in Figure 1-3 has been widely used in North America. Compared to the No. 1 Crossbar system, the markers of a No. 5 Crossbar system are given more control functions and the sender functions are reduced and modified. Originating registers accept dialed digits and need only transmit information to a sender for an outgoing call. Two types of markers are employed, one for dial tone control and the other for completing call connections.

The processing of an interoffice local call is as follows. A subscriber lifting the handset to initiate a call will activate the line relay on a particular line-link frame, which in turn activates the associated connector to seize an idle dial-tone marker. The dial-tone marker locates the calling line, and secures an idle originating register (a storage element capable of storing a number) on a trunk-line frame. It then closes a selected path from the calling line terminals, through the junctor group and the trunk-link frames, and to the originating sender before releasing. The originating sender sends a dial tone to the calling party, and then receives and temporarily stores the dialed digits. As soon as all the seven digits have been received, the originating sender calls for an idle completing marker and transmits the called line number to it.

Upon receiving the digits of the called number, the completing marker identifies the office code. If it is an interoffice call, then the completing marker seizes an idle outgoing sender and transmits to it all seven digits received from the originating register. At the same time, the

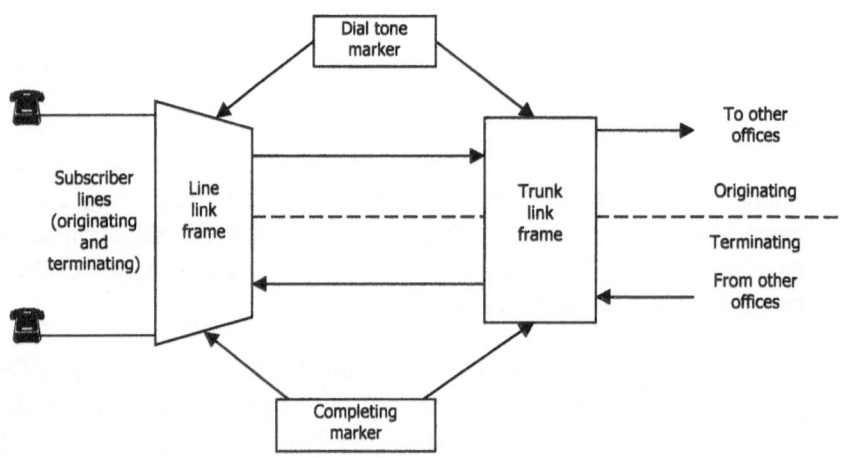

Figure 1-3 No. 5 Crossbar System.

completing marker selects an idle outgoing trunk circuit on the trunk-link frame to the distant central office. If the call is an intraoffice one, the completing marker directs the outgoing sender to prepare to transmit only the last four digits corresponding to the called line number and then closes a path to connect the calling line through the trunk-link frame to the called line on the line-link frame. Note that for an interoffice call, the incoming trunk in the distant central office is the terminating end of the previously selected outgoing trunk in the calling office.

When an incoming trunk is seized by the outgoing trunk circuit in the calling office, it selects an incoming register. This register then signals the outgoing sender in the calling office to send the four digits of the called line number. It also selects an idle completing marker and forwards the four digits and other related data to the marker. The marker then selects the particular line-link frame and locates and tests the called line. If it is idle, the marker selects a path to connect the called line on the line-link frame with the incoming trunk on the trunk-link frame, and then releases. If the called line is busy, the marker instructs the incoming trunk circuit to cause its associated outgoing trunk circuit in the calling office to send a busy signal to the calling party. The outgoing trunk circuit supervises the call and also controls the disconnection of the operated crosspoints of the crossbar switches on the line-link and trunk-link frames when the calling party hangs up the phone.

Stored-Program Control Electronic Switching System. The stored-program control (SPC) electronic switching system (ESS) offers much greater flexibility, maintainability, lower cost and larger capacity than crossbar switching systems. The former two advantages are the result of the stored-program control approach used, and the latter two are due to solid-state electronic technology.

The major impact of electronic switching has been in the speeds of switching and control. Both the $S \times S$ and crossbar switching systems are electromechanical systems. Their switching and control functions may require several milli-seconds. With electronic switching, they require only a few nano-seconds. The high switching speed makes it possible to design a single common control for the entire system.

Other advantages of electronic switching systems include the long service life of the equipment and its capacity for adapting new features and services without changing the system design. This design flexibility is the result of utilizing the stored program and changeable memory concepts for processing calls in electronic switching systems.

Figure 1-4 shows the schematic diagram for the organization of a space-division ESS. The principal elements of the system are grouped into two main parts: the central processor and the switching network. These two parts are functioning separately. The central processor consists of a set of program stores (semipermanent memory), a set of call stores (temporary memory), and a central control. The principal peripheral equipment associated with the central processor are the subscriber lines and line scanners, junctors and junctor scanners, trunks and trunk scanners, signal distributors, and the master control center.

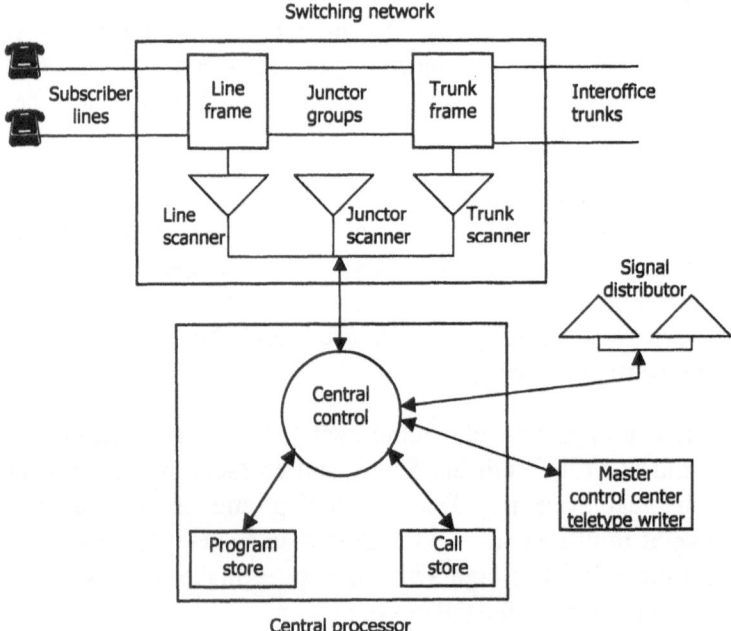

Figure 1-4 Principal elements of space-division ESS.

There is a continuous exchange of information between the central control and the call store as well as the program store. This information may concern the state of subscriber lines, paths in the switching network and outgoing trunks. Instructions for the call-processing operation are stored in the (semipermanent memory) program store. Information concerning the progress of a call is stored in the (temporary memory) call store.

The switching network, comprised of the line-link frame, junctor groups, and trunk-link frame, interconnects calling subscribers' lines with selected outgoing or intraoffice trunk circuits.

The line scanners, junctor scanners and trunk scanners are used to constantly monitor subscriber lines, junctors, and trunk circuits for changes in their status. For example, a subscriber line may change its status from on-hook to off-hook, indicating that a call (request for service) has been initiated. On the other hand, a subscriber line may change its status from off-hook to on-hook, indicating that a call has been terminated or disconnected. A junctor circuit or trunk circuit may be idle or busy. Information concerning the status of lines, junctor circuits and trunk circuits is detected by the corresponding scanners and periodically sent back to the central control.

The master control center administers the operation and maintenance of the entire central office. It provides information for accounting, system operations and trouble reports.

The ARPA Net. The ARPA (Advanced Research Project Agency) net was developed to interconnect the various types of large computers throughout the United States, allowing users at one computer center to access facilities at other computer centers at a distance. It is a resource-sharing communication network in which messages are first segmented into smaller fixed-size units called packets. The packets traverse the network independently in a stored-and-forward manner until they reach the destination node, where they are reassembled into the original message. Thus many packets of the same message may be transmitted simultaneously in the network using different paths, thereby providing the pipelining effect in reducing the message transfer delay.

The primary functions of the ARPA net comprise message storage, coding, routing, flow control and error control, among others. These functions are performed by communication processors known as IMPs (Interface Message Processors).

When a complete message is received at the destination IMP, an acknowledgement will be sent back to the sender node, indicating the correct delivery of the message.

Message routing in the ARPA net uses the locally determined routing technique. Each node in the ARPA net makes its own decision as to which node a given packet will be forwarded next.

The design objective for the ARPA net is to minimize the cost subject to the constraint that the average message delay between any two nodes be 0.2 second.

The SITA Network. The SITA (Societe Internationale de Telecommunications Aeronautiques) network is a worldwide airline reservations network. It also handles airline teletype messages using circuit-switching techniques.

There are two types of traffic handled by the SITA network: the inquiry/response traffic labeled as type A and the telegraphic traffic labeled as type B.

The type A traffic consists of inquiries and responses between the agents, terminals in airline offices, and their associated reservation computers located at some geographic distance. The type B traffic includes telegraphic messages destined to and generated by airline teleprinters and computers, and messages to and from local Telex networks. In general, type A messages are much shorter than type B messages. Furthermore, type A messages receive priority over type B messages when traversing the network. For the type A messages, the average response time is 3 seconds, while that for type B messages is in the order of several minutes.

Agents' terminals are polled cyclically by the processor asking them to transmit (if any) waiting inquiry messages. Upon receipt of a polling message, the agent will transmit his waiting inquiries to the processor.

Ethernet. The Ethernet is a baseband mode, broadcast bus-type local area network designed at the Xerox Palo Alto Research Center, whose transmission medium is a coaxial cable named the Ether. The access method used is the carrier-sense multiple access with collision detection

(CSMA/CD).

Since there is no centralized control structure superimposed onto the Ether, it is purely a passive communication medium. Control of the network is therefore distributed throughout the system.

The Ethernet specifications and the IEEE 802 standards specify the baseband buses, the use of 50 Ω coaxial cable and terminators, and a data rate of 10 Mbps. The terminators absorb signals, preventing reflection from the ends of the bus. The main components of an Ethernet are: a coaxial cable with terminators, taps, transceivers, interfaces, and controllers. The transceiver makes a physical connection to the Ether by a tap. The interface is used to serialize and deserialize the bit streams between the controller and the transceiver. The controller is responsible for the correct transmission and reception of packets across the network. Figure 1-5 shows the main components of an Ethernet connection.

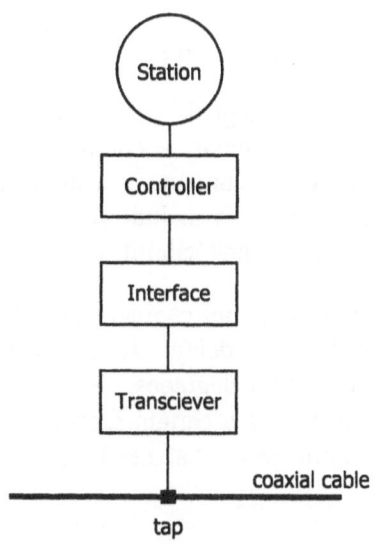

Figure 1-5 An Ethernet connection for a node.

1-3. ELEMENTS OF A TELECOMMUNICATION SYSTEM

Messages or data may be defined as entities that convey meaning. They can be transmitted in either an analog or a digital form of signals.

An analog signal is a physical quantity that varies with time in a continuous fashion. Examples of analog signals are the acoustic pressure produced by speech on the telephone, the sound waves generated from a musical instrument and the light intensity at a certain point in a television image. A digital signal is an ordered sequence of symbols produced by a message source. Examples of digital signals are the messages produced by teleprinters or computer terminals, and computer data.

A telecommunication system may be considered as a network consisting of the following four components:

1.   End systems or stations

2   Transmission system

3.   Switching system

4.   Signaling system

End Systems. End systems in the telephone network have evolved from analog telephones to digital handsets and cellular phones.

Transmission System. Signals generated by a station are carried over transmission links. In general, a communication path between two distant points can be set up by connecting a number of transmission links in tandem. The transmission links include two-wire lines, co-axial cables, microwave radio, optical fibers, and satellites.

A transmission link can be characterized by its bandwidth, link attenuation, and the propagation delay. As the length of a link increases, the quality of the carried signal degrades. To maintain signal quality, the signal must be regenerated after a certain distance. However, with optical fibers it is possible to build links that need regeneration only once every 5000 km or so.

Switching System. A switching system is a collection of switching elements arranged and controlled in such a way as to set up a communication path between any two distant points. A switching center of a telephone network comprising a switching network and its control and support equipment is called a central office. In computer communication networks, switching is performed in a store-and-forward manner. Each node on the path of the message must store the message in a buffer, check for any errors, and route it to the next node. The switching technique used in computer communication networks is known as packet switching or message switching, while the switching method used in telephone networks is called circuit switching.

Signaling System. A signaling system is a collection of facilities that supply and interpret the control and supervisory signals needed to perform the supervisory, routing, and management operations in a telecommunication network. It links the various switching centers or nodes in a network to enable the network to function as a whole.

Modern telephone networks use a separate common channel interoffice signaling network to interconnect switch controllers. Because control information is separated from the voice channel, common channel interoffice signaling is called out-of-band signaling.

1-4. TOPOLOGICAL STRUCTURE OF TELECOMMUNICATION NETWORKS

In a public switched telephone network, central offices may be interconnected by direct trunk groups or by an intermediate office known as a tandem, transit, toll, or gateway office. In the telephone network, the interconnection of central offices may have the structure as shown in Figure 1-6. Note that a star connection utilizes a tandem office, such that every central office is interconnected via this tandem office. Star connections may be used where the traffic level is comparatively low. Mesh connections are used when there are relatively high traffic levels between offices such as in metropolitan networks.

In telecommunication networks using broadcasting, cable, radios or satellites may be used. In all these networks, at any instant only one user is allowed to transmit. A conflicting situation arises when two or more users transmit simultaneously. To resolve this conflict we must have a rule or protocol to specify which user may have permission to send messages at a given time.

The AT&T and ITU-T (or CCITT) Hierarchical Networks. In a large telecommunication network, such as the public-switched telephone network, the network structure is subject to continued changes as the usage grows, subscribers' behavior alters, and equipment costs and characteristics change. In selecting the structure and layout, we must give some

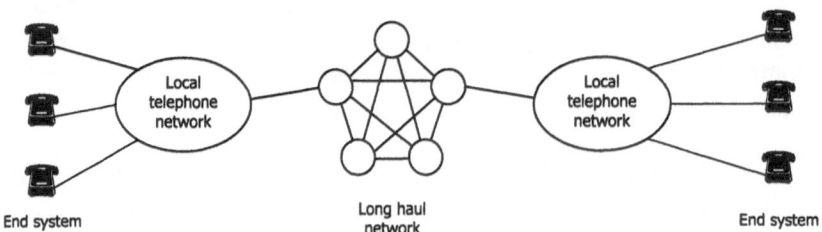

Figure 1-6 Simplified telephone network.

consideration to the trade-off between the cost of the network and the grade of service. A popular concept in networking is to provide alternate routing when some parts of the network are congested. Furthermore, the network must be adaptable to various traffic patterns, such as telephone calls from resort areas in summer time, weekend traffic, Christmas traffic, etc.

Alternate routing introduces switching complexities. When alternate routings are sought over a long distance on many transmission links, a number of difficulties could arise. A call might be routed in a circle or over such a complex path that quality of the call suffers severely. In order to avoid unnecessary complications, a logical scheme is to employ a hierarchy of switching offices.

To route traffic effectively and economically, national telephone networks employ some form of hierarchy, giving orders of importance of the exchanges (central offices) making up the network with certain restrictions on traffic flow. There exist two types of hierarchical networks today, each serving about 50% of the world's telephones: (1) the AT&T (American Telephone and Telegraph in the United States) network, generally used in North America, and (2) the ITU-T (International Telecommunications Union-Telecommunications sector) (or the CCITT (Consultative Committee on International Telegraph and Telephone)) network, typically used in Europe and areas of the world under European influence.

The hierarchical structures of the AT&T and the CCITT networks are shown in Figure 1-7 and Figure 1-8, respectively. In a hierarchical network, traffic is always routed through the lowest available level of the network. This approach not only uses fewer facilities but also provides better circuit quality because of shorter paths and fewer switching points.

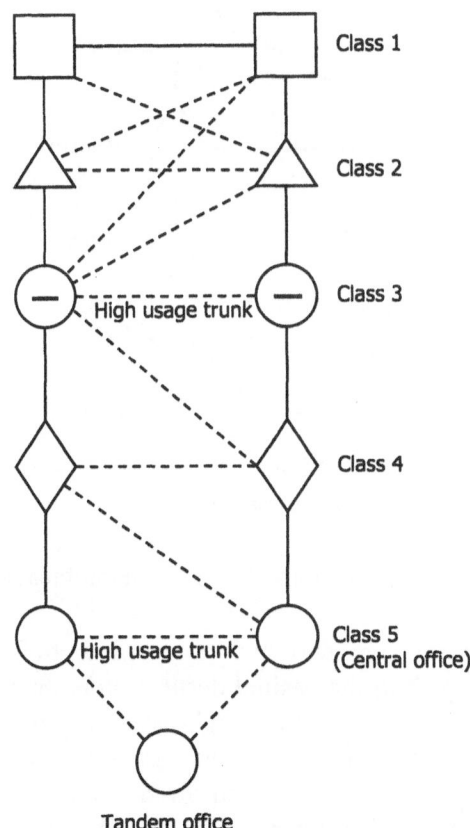

Figure 1-7 The North American (AT&T) hierarchical network (dashed lines show high-usage trunks). Note that the two highest ranks are connected in mesh.

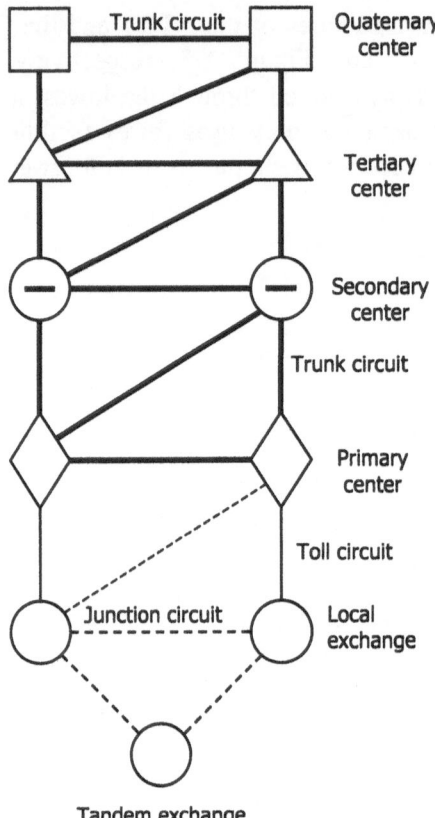

Figure 1-8 The ITU-T (or CCITT) hierarchical structure.

Trunk groups that are designed to overflow traffic to an alternate route to establish a path to the desired destination are known as high-usage groups. When the offered load to a high-usage group includes first-route traffic only (no overflow traffic), the trunk group is called a primary high-usage group. When the offered load to a high-usage group includes overflow traffic, it is called an intermediate high-usage group.

In Figure 1-7 the direct interoffice (high-usage) trunks are depicted as dashed lines, while the backbone trunks are shown with solid lines. The highest rank in the hierarchy is the class 1 center and the lowest, the class 5 office (central office). In general, a high-usage trunk group may be established between any two switching centers regardless of location and rank, whenever the traffic volume and distance are considered to be economical. When routing is through the highest level in the hierarchy, the route is called the final route. The nomenclature of the AT&T and the ITU-T (or CCITT) hierarchy is shown in Table 1-1.

Table 1-1. North America (AT&T) and ITU-T (CCITT) hierarchical structure

North American	ITU-T (or CCITT)
Class 1. Regional center	Quaternary center
Class 2. Sectional center	Tertiary center
Class 3. Primary center	Secondary center
Class 4. Toll center (toll point)	Primary center
Class 5. End office	Local exchange

Routing Rules for Hierarchical Networks. The use of hierarchical structure in telephone networks can simplify network administration and switch design, for only the information regarding the order of each switching center in the hierarchy and the high usage routes need be known. The CCITT (Rec. Q. 13) suggests the far-to-near rule for advancing calls, whereby the first choice in advancing a call is to advance the call as far as possible using the backbone route to measure distances. The second choice is the next best and so on. Thus the use of high-usage routes in the hierarchical network will reduce the number of links in a communication path and hence improve signal quality on long distance calls.

One weakness in hierarchical networks is the low network reliability and security in case of the loss of one or several links. The current trend is to reduce the number of ranks in the hierarchy and offer more alternate routes. Large national telecommunication networks with computer switching will be designed with only three levels of hierarchy.

For reasons of transmission quality and signaling, it is desirable to limit the number of circuits for the connection of a call. According to CCITT Rec. Q. 13, section on Basic Rules for Routing, a specification on routing design states that the maximum number of circuits to be used for an international call is 12, with up to a maximum of 6 circuits being international. In a few exceptional cases, the total number of circuits used by a call may be 14, of which the number of international circuits remains at the maximum of 6. Therefore, the absolute maximum number of circuits to be used for the national portion is 8. Thus, 4 circuits are used for each national network.

Routing Methods. There are three methods of routing calls in a telecommunication network: (1) right-through routing, (2) own-exchange routing, and (3) computer-controlled routing (with common-channel signaling).

In right-through routing, the originating exchange determines the route from source to sink and no alternate routing is allowed at intermediate switching points. However, the initial outgoing circuit group may have alternate routes. Because of its limitations in alternate routing, right-through routing is limited almost exclusively to the local area network.

Own-exchange routing allows for changes in routing as the call proceeds to its destination. This routing method is particularly suited to networks with alternate routing and changes in routing pattern in response to load configuration. Another advantage in own-exchange routing is its flexibility. When new exchanges are added or the network is modified, minimal switch modifications are required in the network. One disadvantage is the possibility of forming a closed routing loop in the network. However, the use of a hierarchical routing system would ensure that such closed loops cannot be generated.

Conventional telephone networks have signaling information for a particular call carried on the same path that carries the speech. This path is called the conversation path. Here signaling is the generation and transmission of information that sets up a desired call and routes it through the network to its destination. Modern computer-controlled networks often use a separate path to carry the signaling information. In this case, the computer in the originating exchange can optimally route the call through the network on a separate signaling path. The originating computer would have a map of the network in the memory with updated details of network conditions. The necessary adaptive information is broadcasted on a separate path that connects the various computers in the network. This method is called computer-controlled routing. Such routing is termed routing with common-channel signaling and with adaptive network management signals.

1-5. SIGNALS AND THEIR CHARACTERISTICS

Telecommunications engineering is mainly concerned with the transmission of messages between two distant points. The signal that contains the message is usually converted into electrical waves before transmission. The most commonly used parameter that characterizes an electrical signal is its bandwidth if it is an analogue signal, or its bit-rate if it is a digital signal as shown in Table 1-2.

Table 1-2. Signals and their bandwidth or bit-rate.

Type	bandwidth, Hz, or bit-rate, bits/s
Telegraph	50 bps
Telephone (speech)	300-3400 Hz
Music	50 Hz-16 kHz
Facsimile	40 kHz
Broadcast television	0-55 MHz
Data	up to 100 M bps

When a signal is transmitted via the transmission medium or channel, the signal will be degraded by the channel due to the bandwidth limitation, attenuation, distortion, and delay. Because of the presence of noise, the maximum theoretical capacity of a channel of bandwidth W Hz and a signal-to-noise power ratio (SNR) is given by Shannon's law:

$$C = W\log_2 (1 + \frac{S}{N}), \quad bits \ per \ second \qquad (1-1)$$

1-6. TRANSMISSION MEDIA AND THEIR CHARACTERISTICS

The main transmission media for telecommunications and their characteristics are shown in Table 1-3.

Table 1-3. Transmission media and their characteristics.

Medium	Characteristics
Wire-pair cables	With repeater and balancing, wire-pair cables can be used up to several hundred kHz and are susceptible to cross-talk.
Coaxial cables	Coaxial cables can be useful from 50 kHz to hundreds of MHz. Cross-talk is negligible.
Microwave radio	Widely used for television transmission, susceptible to radio interference.
Tropospheric scatter circuits	Usable for distances up to 600 km, experiences multiple reflections and bad fading.
H.F. radio	Usable for shortwave broadcasting in range 3-30 MHz, experiences bad fading.
V.H.F. radio	Usable for point-to-point and mobile communication in the range 150-180 MHz.
Other media	Microwave waveguides and optical fibers. Optical carrier frequency is in the range 10^{13}-10^{16} Hz.

1-7. QUALITY OF SERVICE IN TELEPHONE NETWORKS

In this section, we shall present the criteria for the design of telecommunication systems. Since equipment requirements in a telecommunication system are determined on the basis of the traffic intensity of a busy hour, it is necessary to precisely identify the criteria for the measures of quality of service.

In order to quantitatively define the quality of service, we shall introduce the degree of congestion as the primary measure of inconvenience that a call will frequently encounter.

Grade of Service. Grade of service is a measure of congestion expressed as the probability that a call will be blocked or delayed. This probability represents the percentage of the offered traffic which will be blocked or delayed under busy hour conditions. Thus grade of service involves not only the ability of a system to set up a connection when requested, but also the rapidity with which the connections are made.

Grade of service is also commonly expressed as the fraction of calls or demands that fail to receive immediate service (blocked calls), or the fraction of calls that are forced to wait longer than a given length of time for service (delayed calls).

Blocking Criteria. If the design of a system is based on the fraction of calls blocked (the blocking probability), then the system is said to be engineered on a blocking basis. Blocking can occur if all devices are occupied when a demand for service is initiated. Blocking criteria are often used for the dimensioning of switching networks and interoffice trunk groups.

Delay Criteria. If the design of a system is based on the fraction of calls delayed longer than a specified length of time (the delay probability), the system is said to be engineered on a delay basis. Delay criteria are used in telephone systems for the dimensioning of registers where the delay in providing a dial tone is specified. In computer communication networks, the average transfer delay is commonly used, where the average transfer delay of a message shall not exceed a given value.

Note that when dealing with the grade of service in teletraffic engineering, it is important to differentiate between the concepts of time congestion and call congestion. Congestion is the condition in a switching center when a subscriber cannot obtain a connection to the wanted subscriber immediately.

Time Congestion. Time congestion is the fraction of time during which all the devices are busy (occupied) during the busy hour. It is the probability that an outside observer who observes the system at a certain time during the busy hour and finds all the devices busy. It is also called the probability of blocking.

Call Congestion. Call congestion is the fraction of calls that fail to receive service at first attempt during the busy hour. It is the probability that all facilities in the system are busy just prior to an arrival epoch of a call. Call congestion in a loss system, is also known as the probability of loss while in a delay system it is referred to as the probability of waiting.

When the input process of calls is Poisson, the time congestion and call congestion are equal. In general, the time congestion and call congestion are different; however, in most practical cases the discrepancies are found to be small.

1-8. FUNDAMENTALS OF VOICE TRAFFIC AND DATA TRAFFIC

The traffic intensity in a telecommunication system varies from hour to hour, day to day, and year to year. There is usually more traffic from 9-11 a.m. and 4-5 p.m. on a work day, more traffic on Mondays and Fridays, and less traffic on Wednesdays.

In telecommunication systems, there are two types of traffic: the voice traffic and the data traffic. Since traffic is a random quantity, the probability distributions of traffic may be divided into three categories according to the peakedness factor z which is defined as the variance-to-mean ratio of a traffic distribution,

$$z = peakedness\ factor$$

$$= \frac{Variance\ of\ N(t)}{Mean\ of\ N(t)}$$

$$(1\text{-}2)$$

where $N(t)$ denotes the number of calls arriving in a time interval of length t.

A traffic distribution is said to be smooth if its peakedness factor is less than unity. If a traffic distribution has a unity peakedness factor, then it is said to be random or Poisson. Note that the mean and variance of the Poisson distribution are equal. For example, the telephone industries in North America assume the (voice) traffic distribution originating from the subscribers to be Poisson. Moreover, if a traffic distribution has a peakedness factor greater than unity, then the traffic is said to be rough or peaked. For example, in computer communication networks, users tend to use the network intermittently, with the interdata arrival time in general much greater than the user data transmission time. Thus the data traffic in computer networks is usually very rough and often said to be bursty.

A fundamental problem in the design of telecommunication networks concerns the dimensioning of a route. That is, for a given volume of offered traffic and a given grade of service, how many trunk circuits are required to interconnect two end offices? What capacity (in bits per second) should be made available on each link of the network to provide a specified grade of service for a given volume of offered traffic?

In order to answer the above questions, some statistical information about the nature of the offered traffic and the usage of the trunks or links is needed. In general, the traffic offered to a group of devices (lines, circuits, links, trunks or traffic paths) can be specified by two parameters:

1. The average arrival rate λ; that is, the average calling rate in calls per hour for voice traffic, or the average message arrival rate in messages per second for data traffic.

2. The average service time τ; that is, the average holding time per call in hours or 100 seconds for voice traffic, or the average transmission time per message in seconds for data traffic.

For voice traffic, the calling rate is usually defined as the number of calls per traffic path during the busy hour. The average holding time is the average duration of occupancy of a traffic path by a call. For data traffic, the average message arrival rate is the average number of messages per second arriving at a node for processing. The reciprocal of the average holding or service time, $\mu = \dfrac{1}{\tau}$, in calls per hour, is referred to as the service rate. Moreover, the dimensionless quantity which is the ratio of average arrival rate to the average service rate

$$a = \frac{\lambda}{\mu} = \lambda \, \tau \qquad (1\text{-}3)$$

is internationally called the erlang, named after A.K. Erlang, the father of teletraffic theory. This quantity is also called the offered traffic or traffic intensity. One erlang represents a circuit occupied for one hour.

If a group of s trunk lines carried a' erlangs of traffic, then the carried traffic per line

$$\rho = \frac{a'}{s} \qquad (1\text{-}4)$$

is called the occupancy of a line, where

$$a' = \text{carried traffic} = \text{offered traffic} - \text{lost traffic} \qquad (1\text{-}5)$$

Example 1-1. Consider the voice traffic originating from a large number of subscriber lines with calling rate, $\lambda = 100$ calls/hour. The average holding time of a call is $\tau = 3$ minutes. Calculate (a) the traffic intensity or offered traffic, (b) the occupancy of a circuit if the traffic calculated in (a) is offered to 10 circuits which encounter a 1% loss, and (c) the peakedness factor.

Solution. (a) The calling rate is $\lambda = 100$ calls/hour and the average holding time is $\tau = 3$ minutes. Thus the traffic intensity or offered traffic is given by

$$a = \lambda\tau = 100 \times \frac{3}{60} = 5 \ erlangs$$

(b) The carried traffic is

$$a' = \text{offered traffic} - \text{lost traffic} = 5 - 5 \times 0.01 = 4.95 \ erlangs$$

Hence occupancy of a circuit is

$$\rho = \frac{a'}{s} = \frac{4.95}{10} = 0.495$$

(c) Since the number of subscriber lines is large, we may assume that the originating traffic is Poisson with calling rate λ. According to the theory of probability (see Example 4-1), it is known that the Poisson distribution has equal mean and variance. This implies that if $N(t)$ denotes the number of calls originating in a time interval of t time units long, then

$$E[N(t)] = Var[N(t)]$$

Therefore, using (1-2) we find the peakedness factor

$$z = 1$$

In the analysis and design of telecommunication systems, busy-hour traffic data are commonly used. Traditionally, the design of telephone switching centers is based on the traffic intensity during the busy hour in the busy season. In teletraffic analysis, the busy hour is defined as that continuous sixty-minute period during which the traffic intensity is highest.

The concept of the busy hour can be defined in several ways. The four most commonly used definitions of busy hour traffic are as follows:

Definition 1. The average of the busy hour traffic on the 10 busiest days of the year (North American standard).

Definition 2. The average of the busy hour traffic on the 30 busiest days of the year (defined as the mean busy hour traffic).

Definition 3. The average of the busy hour traffic on the 5 busiest days of the year (referred to as the traffic on exceptionally busy days.)

Definition 4. The average weekday traffic over one or two weeks in a given busy season (normal practice for manual or operator switched traffic).

From the above definitions, we see that the busy hour traffic based on definition 1 is greater than the one of definition 2. As a result the design based on definition 1 would require more equipment than definition 2.

Teletraffic may be defined as the occupancy of certain devices in the network during the process of setting up a connection. Note that traffic is generated from the moment the calling subscriber lifts the handset (the telephone goes off hook), until the call is released by replacing the handset back on its cradle (the telephone goes on hook).

Traffic Units. The international unit of telephone traffic is the erlang. One erlang represents the occupancy of a device for one hour. Thus one erlang equals one call-hour per hour. Note that the traffic intensity of a channel represents the occupancy or efficiency of the channel.

Traffic Intensity. Traffic intensity is defined as the product of the calling rate and the average holding time.

When the average call holding time is expressed in hundred seconds, the resulting traffic unit is called hundred-call-seconds or centum-call seconds (CCS). In this case, 1 erlang equals 36 CCS, for there are 36 hundred-seconds in 1 hour. The traffic unit CCS is only used in North America.

Example 1-2. Consider a group of 1000 subscribers which generate 500 calls during the busy hour. The average call holding time is 200 seconds. What is the offered traffic in erlang and in CCS?

Solution. The offered traffic of the group is

$$a = \lambda\tau = 500\frac{calls}{hour}\times\frac{200\ seconds}{3600\ seconds/hour} = 27.78\ erlangs$$

or

$$a = \lambda\tau = 500\frac{calls}{hour}\times\frac{200\ seconds}{100} = 1000CCS$$

The offered load or offered traffic is also referred to as the traffic intensity.

1-9. OUTLINE OF THE BOOK

In this book, we present the basic concepts and fundamental methods for analysis of telecommunication networks and local area networks. For switching networks the primary performance measure is the blocking probability between a pair of inlet and outlet. For telephone networks with Poisson input traffic, such as the Erlang loss and Erlang delay systems, the performance measures are the blocking probability and the delay probability, respectively. For local area networks, the main performance measures are the average transfer delay for a message traversing the network and the throughput.

Chapter 1 covers a brief historical overview of the development of telecommunications. After defining some important terms needed for the description of telecommunication systems, we describe several typical telecommunication networks such as the public telephone network, the ARPA network for computer communication, the SITA network for world wide airline reservation service, and the Ethernet for local area networks. In addition, we introduce the concepts of time congestion and call congestion, the peakedness factor as a basic parameter for the characterization of the input traffic in a telecommunication network, and the busy hour traffic.

Chapter 2 provides the background knowledge of transmission systems for telecommunication networks. We present methods for subscriber loop design and digital transmission, and techniques for signal multiplexing, and discuss their differences, advantages and disadvantages.

Chapter 3 introduces the basic principles of switching systems, switching techniques and methods of calculation of congestion. We use network graphs and channel graphs to simplify the calculation. Also we investigate both the blocking and nonblocking networks.

Chapter 4 provides a detailed study of the modeling techniques of traffic flows. We point out the important properties of the Poisson traffic flow and the Markov property of the exponential service time distribution. Using the methods of Markov chain and imbedded Markov chain, we investigate the $M/M/1$ queue, the $M/G/1$ queue, and the $GI/M/1$ queue in detail. We determine the state probability distributions for the $M/M/1$ and $GI/M/1$ queues and for the birth and death process, which forms a basic model for the study of congestion in telephone systems. Furthermore, we derive the Pollaczek-Khinchin formula for mean waiting time for the $M/G/1$ queue and also formulas for the average transfer delay in the $M/G/1$ queue with vacation and with priority discipline.

Chapter 5 investigates the Erlang loss and Erlang delay systems as the central models for the design of telephone systems in North America. Using the result obtained from the birth and death process, we derive two formulas for the calculation of congestion in telephone systems. These formulas are known as the Erlang B and the Erlang C formulas. In addition, we determine the waiting time distribution for the Erlang delay system with service in order of arrival. We also study the overflow in alternate routing networks and present the equivalent random method for analysis and design.

Chapter 6 treats the Engset loss and Engset delay systems with quasi-random input. These systems form the primary models for the design of telephone systems in Europe. By means of the result obtained from the birth and death process, we determine two formulas for the calculation of congestion in telephone systems. These formulas are known as the Engset loss and Engset delay formulas.

Chapter 7 introduces the fundamentals of local area networks with a bus structure. For simplicity of analysis we assume the data traffic to be Poisson. We discuss three multiaccess techniques for controlling the access to the transmission medium on the network: fixed assignment, random assignment, and demand assignment. We then calculate the average transfer delay and throughput for all three multiaccess methods.

Chapter 8 presents the basic features and the two modes of operation in polling networks as well as their configurations. Having described the operation of a bus polling network with either roll-call polling or hub-polling, we examine the polling cycle and the average waiting time. Then we develop an expression for the average waiting time of the packet for an exhaustive network, partially gated network, and gated network. Finally, we obtain an expression for the average transfer delay in terms of the average waiting time.

Chapter 9 provides an investigation of the performance of token ring networks for multiple-token operation, single-token operation, and single-packet operation. After introducing the concepts of ring latency and station latency, we develop expressions for the average transfer delay for the above three operations.

A basic problem in the design of slotted ring networks is to adjust the number of slots so that the gap is minimized. If the station latencies are not sufficient for the required ring latency, artificial delays can be added in the station latency.

We show that the throughput can be increased by allowing multiple packets on the register insertion ring network. For the performance analysis of register insertion ring networks, a simplified model is employed. In this case, the maximum throughput can be greater than unity. The average transfer delay is shown to be the same for both the ring priority and station priority schemes.

Chapter 10 investigates the performance of slotted ALOHA networks for cases of Poisson input and a finite number of stations. For the latter case, we present a Markov chain analysis for the study of the dynamic behaviour of the slotted ALOHA network. Moreover, we discuss an analytical technique called equilibrium point analysis for the investigation of more complicated protocols.

We also carry out the performance analysis for both the carrier-sense multiple access protocols and carrier-sense multiple access with collision detection protocols under the assumption that propagation delays are much smaller than the packet transmission time. By means of carrier-sense, the throughput can be increased.

1-10. SUMMARY

This chapter gives a perspective on telephone networks, local area networks and computer communication networks. We define a number of technical terms pertinent to the understanding of the structure and operation of telephone and computer communication networks.

A telecommunication network can be viewed as a communication network which consists of a transmission subnetwork, a switching subnetwork, and a signaling subnetwork. These three subnetworks interact and function in a specific and cooperative way to provide good quality of service for communications.

To simplify network operation and management, two types of hierarchical telephone networks, the AT & T and the ITU-T (or CCITT) networks, were established to serve all the countries in the world. Modern trends in telecommunications are to establish a universal communication network known as the broad band integrated services digital network (B-ISDN) which can handle voice, data, video and image traffic. The international standard for this network is named asynchronous transfer mode (ATM).

It is well known that the call arrival process to a telephone network is a Poisson process with a constant arrival rate and that the data traffic is bursty. However, the bursty nature of data traffic is not well understood and requires much research.

REFERENCES

[1] Keshav, S., An Engineering Approach to Computer Networking, Reading, Mass.: Addison-Wesley, 1997.

[2] Flood, J.E., Telecommunication Networks, editor, Herts, England: Peter Peregrinus Ltd., 1975.

[3] Talley, D., Basic Electronic Switching for Telephone Systems, Rochelle Park, N.J.: Hayden Book, 1975.

[4] Martin, J., Telecommunications and the Computer, 2nd ed., Englewood Cliffs, N.J.: Prentice Hall, 1976.

[5] Briley, B.E., Introduction to Telephone Switching, Reading, Mass.: Addison-Wesley, 1983.

[6] Boucher, J.R., Voice Teletraffic Systems Engineering, Norwood, M.A.: Artech House, 1988.

[7] Grunn, H.J., Principles of Traffic and Network Design, Geneva, Il.: abc TeleTraining Inc., 1986.

PROBLEMS

1-1. What is the difference between a primary high usage trunk group and an intermediate high-usage trunk group?

1-2. What is the grade of service for a telephone network?

1-3. What is the difference between the concepts of time congestion and call congestion?

1-4. Suppose that a data traffic distribution has a mean of 10 messages per second and a standard deviation of 7 messages per second. What is the peakedness factor of this data traffic?

1-5. Poisson traffic originates from 10,000 subscribers with a rate of 500 calls per hour. The average holding time of the calls is 2.5 minutes. (a) What is the offered load? (b) If the Poisson traffic is served by 25 circuits with a 2% loss, what is the carried load? (c) What is the occupancy of a circuit?

CHAPTER 2

TRANSMISSION SYSTEMS

2-1. INTRODUCTION

The primary function of a transmission system is to provide circuits having the capability of accepting electrical signals at one point and delivering them to a distant point with good quality. A transmission system in its simplest form may be a pair of wires connecting two telephones. In a telephone network, a call between two distant points can be set up by connecting a number of transmission systems in series to form a communication path between the two points. The telephone set converts the acoustic signal (voice) into an electrical signal. It also converts a received electrical signal to its corresponding acoustic form. In addition, it generates supervising signals (on-hook and off-hook) and the address information for the switching system to establish connections.

A Simple Telephone Connection. A connection between two telephones requires two wires as shown in Figure 2-1(a). If the distance between the two parties is substantial, it may be necessary to use amplifiers (repeaters) to compensate for signal attenuation. Since amplifiers are uni-directional devices, for two-way communications it is necessary to use a four-wire transmission, as shown in Figure 2-1(b). However, the switching equipment in the local office and the local loops on the premises are designed for only two-wire operation. Thus, we must convert the locally used two-wire circuits to the four-wire circuits used for long distance transmission. This conversion is accomplished by a hybrid coil (or hybrid transformer), as shown in Figure 2-1(c). The rules of operation for the hybrid transformer are that signals entering at A only go to C but not to B, while signals entering at C only go to B but not to A.

Twisted Pair Cable. A twisted pair is made by twisting two insulated conductors together. The wire pairs are stranded in units, and the units are then cabled into cores. Common wire sizes used are 19-, 22-, 24-, and 26-gauge.

The characteristics of a pair cable are determined by the primary constants: R, the series resistance in ohms per unit length; L, the series inductance in henries per unit length; G, the shunt conductance in siemens

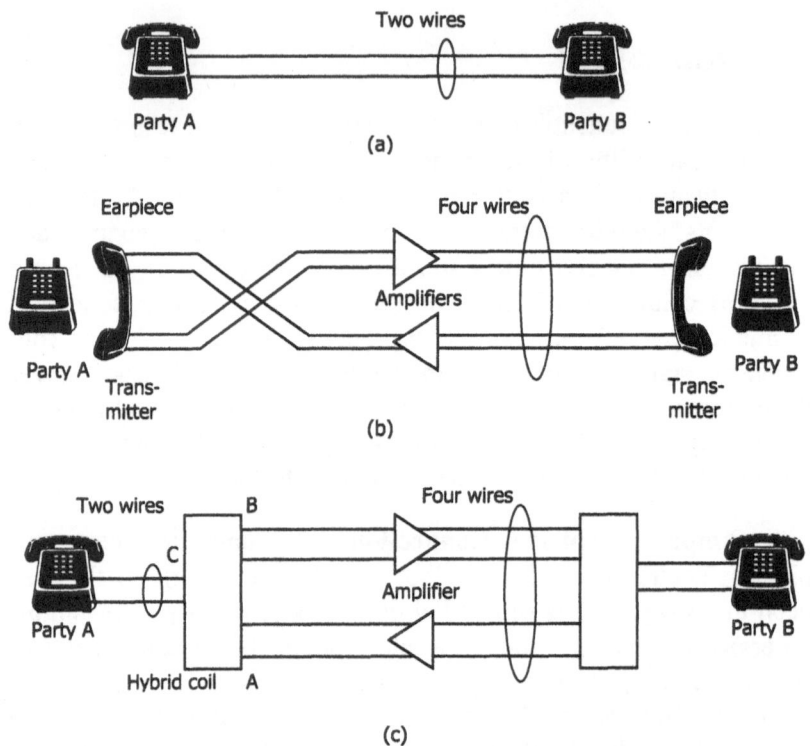

Figure 2-1.   (a) Simple two-wire connection between two telephones.
               (b) Four-wire connection between two telephones.
               (c) Connection between two two-wire telephones over a
                    four-wire transmission facility.

per unit length; and C, the shunt capacitance in farads per unit length. These primary constants are functions of the frequency: only C is relatively independent of the frequency; G is generally small; L decreases to about 70 percent of its initial value as the frequency increases from 50 kHz to 1 MHz and is stable beyond; and R, relatively constant over the voice band, is proportional to the square root of the frequency at higher frequencies due to the skin effect.

Wire pairs may be used to transmit both analog and digital signals. The most common use of wire pairs is for the transmission of voice. If a circuit uses a separate transmission path for each direction, the circuits are called channels. Digital signals may be transmitted over an analog voice channel by using a modem. Analog signals may be sent over a digital channel by using analog-to-digital conversion.

The Transmission Line. Consider a transmission line consisting of a pair cable as shown in Figure 2-2. Analog signals may be sent over a digital channel by using analog-to-digital conversion.

Since $dV = Izdx$ and $dI = Vydx$, we can write the differential equations

$$\frac{d^2V}{dx^2} = yzV \qquad \text{and} \qquad \frac{d^2I}{dx^2} = yzI \qquad (2\text{-}1)$$

where

$$z = R + j\omega L = \text{series impedance per unit length}$$

and

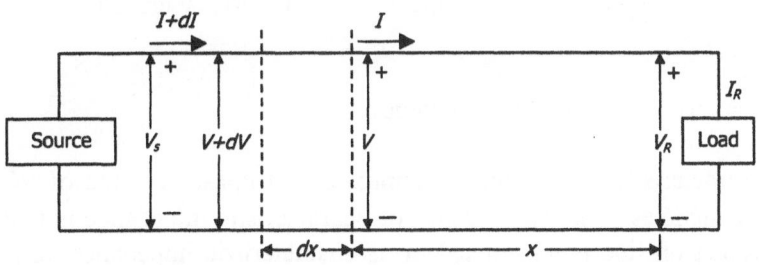

Figure 2-2. Schematic diagram of a transmission line.

$$y = G + j\omega C = \text{shunt admittance per unit length}$$

Using the conditions at the receiving end of the line, namely, when $x = 0$, $V = V_R$ and $I = I_R$, the solutions to the above differential equations are given by

$$V = \frac{V_R + I_R Z_c}{2} e^{\gamma x} + \frac{V_R - I_R Z_c}{2} e^{-\gamma x} \qquad (2\text{-}2)$$

and

$$I = \frac{V_R/Z_c + I_R}{2} e^{\gamma x} - \frac{V_R/Z_c - I_R}{2} e^{-\gamma x} \qquad (2\text{-}3)$$

where

$$Z_c = \sqrt{z/y} = \textit{the characteristic impedance of the line}$$

$$\gamma = \sqrt{yz} = \alpha + j\beta = \textit{the propagation constant}$$

$$\alpha = \frac{1}{\sqrt{2}} \left[\sqrt{(R^2 + \omega^2 L^2)(G^2 + \omega^2 C^2)} + RG - \omega^2 LC \right]^{1/2}$$

$$\beta = \frac{1}{\sqrt{2}} \left[\sqrt{(R^2 + \omega^2 L^2)(G^2 + \omega^2 C^2)} - RG + \omega^2 LC \right]^{1/2}$$

and $\omega = 2\pi f = $ the angular frequency.

The characteristic impedance Z_c approaches a constant value of $\sqrt{L/C}$ at high frequencies. The significance of characteristic impedance is that when a transmission line is terminated in its characteristic impedance, any signal on the line is absorbed when it reaches the terminating resistance. Thus there are no reflections. The propagation constant γ represents the attenuation α, in nepers per unit length, and the phase shift β, in radians per unit length.

By examining the propagation constant at various frequencies, we shall consider the approximate value for α in four different cases: the frequencies $\omega << R/L$ and $\omega << G/C$, the frequencies $G/C << \omega << R/L$, the frequencies $\omega >> R/L$ and $\omega \gg G/C$, and higher frequencies.

Case 1. If $\omega \ll R/L$ and $\omega \ll G/C$, then

$$\alpha \approx \sqrt{RG} \ , \ nepers/unit \ length$$

Case 2. If $G/C << \omega << R/L$, then

$$\alpha \approx \sqrt{\omega RC/2} \ , \ nepers/unit \ length$$

Case 3. If $\omega >> R/L$ and $\omega >> G/C$, then neglecting the terms with G and using the approximation

$$\sqrt{R^2 + \omega^2 L^2} \approx \omega L \left[1 + \frac{1}{2} \frac{R^2}{\omega^2 L^2} \right]$$

the value of attenuation reduces to

$$\alpha \approx \frac{R}{2} \sqrt{\frac{C}{L}} \ , \ nepers/unit \ length \tag{2-5}$$

Case 4. If the frequency is higher than that of Case 3, then R is proportional to $\sqrt{\omega}$ due to the skin effect and hence α is proportional to $\sqrt{\omega}$.

Note that the attenuation α is proportional to $\sqrt{\omega}$ for both cases 2 and 4. For most practical applications, twisted pair cables are used in the frequency range of Case 2. Typical values of the primary constants for a 0.63 mm diameter copper twisted pair are:

$$R = 100 \ \Omega/km$$

$$L = 1 \ mH/km$$

$$G = 10^{-5}S/km$$

$$C = 0.05 \ \mu F/km$$

However, attenuation is not the dominant effect of the operating frequencies of twisted pairs. The most important effect is the crosstalk, caused by the shunt capacitance between pairs in the same cable. The effect of crosstalk can be reduced by complicated balancing networks of resistors and capacitors at each end of the cable. Nevertheless, crosstalk imposes a practical upper limit of 500 kHz on operating frequencies. At frequencies of 1 MHz or higher, coaxial cables may be used in order to overcome the crosstalk limitations of twisted pairs.

A Simple Transmission Scheme. Since the internal impedance of a microphone is much less than that of the line, greater power transfer into the telephone line may be achieved by the use of a transformer, as shown in Figure 2-3(b). Because of the necessity of service and replacement, it is much more economical to use a large central battery of -48V at the local exchange. Figure 2-3(c) shows a method of feeding the direct current into the line. An auto-transformer is used in the instrument to match the impedances and a capacitor is used to prevent the direct current flowing through the receiver. For signaling purposes, two equal inductors are used to prevent the battery from producing a shunt on the signaling path. These inductors are part of a relay for signaling use.

Local Telephone Networks. A local telephone network includes the following components:

1. The subscriber plant;

2. The local exchanges (central offices);

3. The trunk plant connecting local exchanges and those trunks connecting local exchanges to the next level exchange of network hierarchy (Class 4 exchanges or toll centers);

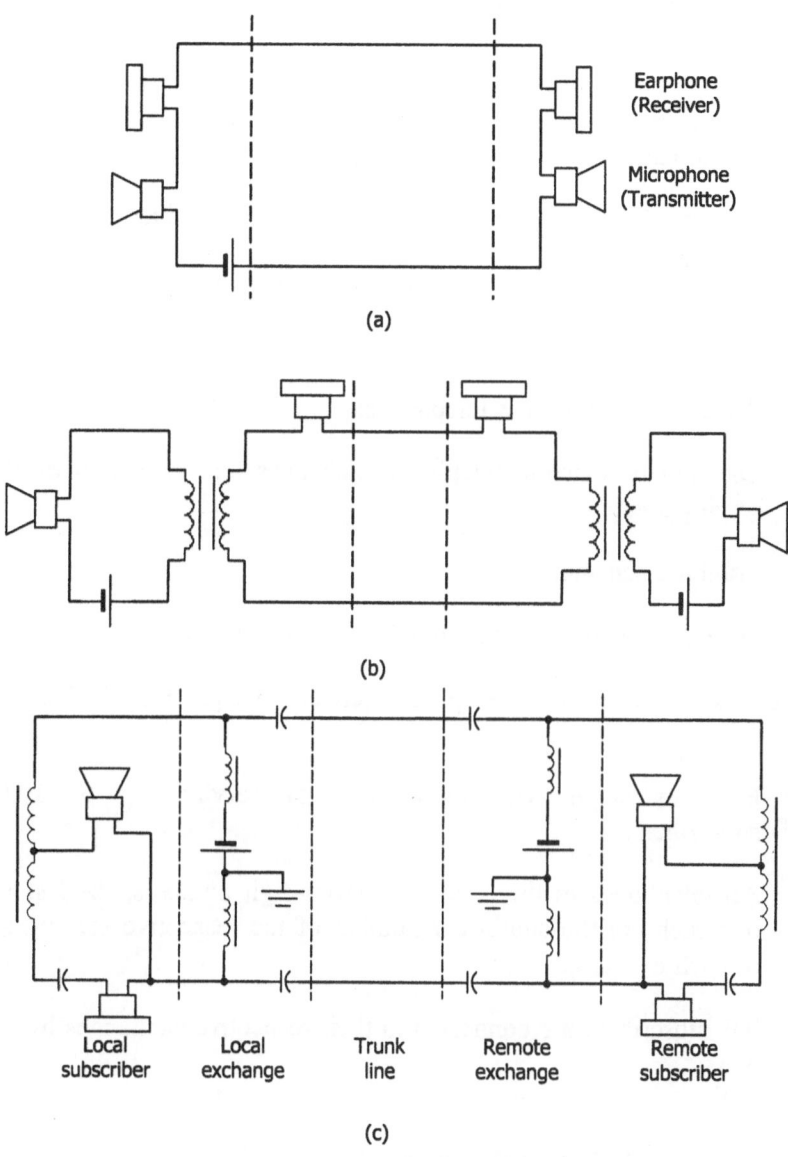

(a)

(b)

| Local subscriber | Local exchange | Trunk line | Remote exchange | Remote subscriber |

(c)

Figure 2-3. Simple transmission schemes.
      (a) Simplest form of two-wire speech link.
      (b) Use of transformer to increase power matching.
      (c) Use of central battery with capacitor feeding bridge.

4.   The class 4 exchanges (USA) or the primary centers (CCITT).

For a given quality of service, in order to build the most economical local network, certain constraints will have to be placed on the design. Important information required for the design includes:

- Geographic extension of the local area under consideration;

- Number of inhabitants and existing telephone density;

- Calling habits;

- Percentage of business telephones;

- Location of existing telephone exchanges and extension of their serving area;

- Trunking scheme;

- Present signaling and transmission characteristics.

For the study of local telephone network design, we shall assume that:

- Each exchange will be capable of serving up to 10,000 subscribers;

- All telephones in the area have seven-digit numbers, the last four of which are the subscriber number of the respective serving area of each exchange;

- All subscribers are connected to their respective exchanges by wire pairs.

2-2. SUBSCRIBER LOOP DESIGN

The subscriber loop (the pair of wires connecting the subscriber telephone set to the local end office) is a dc loop powered by a battery of -48 V. It provides a metallic path for:

- The talk battery for the telephone transmitter;

- An ac ringing voltage for the bell on the telephone;

- A current to flow through the loop when the telephone instrument is taken out of its cradle (off hook), informing the switch of the need for service and thus causing a line seizure at that switch;

- The telephone dial that, when operated, makes and breaks the dc current on the closed loop, indicating to the switching equipment the called party number with which communication is desired.

When designing subscriber loops, we must deal with the two limiting factors on their length: *attenuation* and *IR drop*. Here attenuation refers to energy loss in the line at a reference frequency, measured in decibels. The reference frequency is 1000 Hz in North America and 800 Hz in Europe. As a telephone loop is extended in length, its loss at the reference frequency increases. Thus, at some point as the loop is extended, the subscriber cannot hear sufficiently well because of attenuation. Likewise, as the loop is extended in length while the battery (supply) voltage is constant, the effectiveness of signaling is decreased. This limit is due to the IR drop of the line. When a telephone is taken off hook, the telephone loop is closed and a current flows, closing a relay at the local exchange. If the current flow is insufficient, the relay will not close or will close and open intermittently (chatter) such that line seizure cannot be effected.

Suppose that the following limits are specified for a subscriber loop:

Attenuation limit = 6 *dB*   (*AC loss limit*)

Loop resistance limit = 1250Ω   (*IR drop limit*)

2-2-1. Basic Resistance Design. The formula for calculating the dc loop resistance for a copper conductor is

$$R_{dc} = \frac{43.7691}{d^2} \ \Omega/km \tag{2-6}$$

where R_{dc} is the dc loop resistance in Ω per kilometer and d is the diameter of the conductor in millimeters.

For a given loop length of, say, 10 km, the dc loop resistance is calculated by the formula

$$R_{dc} = \frac{loop\ resistance\ limit}{loop\ length} \qquad (2\text{-}7)$$

or for a given dc loop resistance of a wire, the maximum subscriber loop length can be expressed as

$$l = \frac{loop\ resistance\ limit}{R_{dc}} \qquad (2\text{-}8)$$

where the loop length is defined as the distance from the subscriber to the central office.

Thus,

$$R_{dc} = \frac{1250}{10} = 125\ \Omega/km$$

Table 2-1 lists the diameter and dc loop resistance for several subscriber cables of which the sizes of 19-, 22-, 24-, and 26-gauge are the most commonly used.

The diameter of copper wire required is then equal to

$$d = \sqrt{\frac{43.7691}{R_{dc}}} = \sqrt{\frac{43.7691}{125}} = 0.592\ mm.$$

From the American wire gauge (Table 2-1), we see that a 22-gauge wire is required. Note that the wire resistance in Ω/km (Table 2-1) must be doubled for the loop resistance R_{dc} in Ω/km, since there is a go and return path in a loop.

Table 2-1. American Wire Gauge (B & S) Versus Wire Diameter and Resistance.

American Wire Gauge	Diameter (mm)	Resistance (Ω/Km) at 20° C
19	0.91	26.39
22	0.64	52.95
24	0.51	84.22
26	0.40	133.9

If the size of the gauge wire is given, we can determine the maximum loop length for a given signaling resistance from a loss and resistance table typified in Table 2-2.

Table 2-2. Loss and Resistance per 1000 m of Subscriber Cable

Cable Gauge	Loss/km (dB)	R_{dc} Ω/km
26	1.68	274.0
24	1.35	170.3
22	1.05	106.3
19	0.69	52.8

For a 26-gauge loop and a 1250 Ω loop resistance limit, the maximum permissible loop length is

$$l = \frac{1250\Omega}{274 \ \Omega/km} = 4.56 \ km$$

This method of determining the maximum subscriber loop length using the signal resistance limit as a basis is called *the basic resistance design*.

Example 2-1. Consider a subscriber loop 10 km long. The dc loop resistance limit is assumed to be 1250 Ω. Calculate the dc loop resistance and determine the cable gauge for the loop.

Solution. The dc loop resistance is given by

$$R_{dc} = \frac{loop\ resistance\ limit}{loop\ length}$$

$$= \frac{1250\ \Omega}{10\ km} = 125\ \Omega/km$$

The diameter of the conductor is given by

$$d = \sqrt{\frac{43.7691}{R_{dc}}}$$

$$= \sqrt{\frac{43.7691}{125}}$$

$$= 0.5917\ mm$$

Using Table 2-1, we find that a 22-gauge cable should be used.

2-2-2. Basic Transmission Design. The maximum permissible loop length may also be determined using the attenuation or loop loss as a basis. The formula for calculating the maximum permissible loop length is given by

$$l = \frac{Attenuation\ limit}{Loss\ /km} \tag{2-9}$$

For a 26-gauge loop and a 6 dB loss, the maximum permissible loop length is

$$l = \frac{6 \; dB}{1.68 \; dB/km} = 3.57 \; km$$

This method of determining the maximum subscriber loop length using the attenuation limit (loop loss) is called *the basic transmission design.*

Example 2-2. Calculate the maximum subscriber loop length for a 22-gauge cable with an attenuation limit of 6 dB.

Solution. For a 22-gauge cable, the loss per kilometer is 1.05 dB/km (see Table 2-2). The maximum permissible loop length is

$$l = \frac{Attenuation \; limit}{Loss/km} = \frac{6 \; dB}{1.05 \; dB/km} = 5.71 \; km$$

Comparing this result with that of Example 2-1, we see that in this particular example the attenuation limit is more severe than the signal resistance limit.

If a 22-gauge cable with H-66 type of loading is used, then we see from Table 2-4 that the attenuation is reduced to 0.56 dB/km. In this case, the maximum permissible loop length is extended to

$$l = \frac{6 \; dB}{0.56 \; dB/km} = 10.71 \; km$$

However, the diameter of the 22-H-66 cable has a loop resistance of 110 Ω/km and only has a dc loop resistance of 1178 Ω. Therefore, using a 22-H-66 cable will satisfy both the AC loss limit and the IR drop limit.

In many situations, it is desirable to extend subscriber loop lengths beyond the permissible limit without increasing the conductor diameter. Common methods involve the use of amplifiers and/or loop extenders, as well as inductive loading coils.

A loop extender is a device that increases the battery voltage on a loop and hence extends its signaling range. It may also contain an amplifier, thereby extending the transmission loss limit as well.

Inductive loading consists of inserting series inductances (loading coils) into the loop at fixed intervals. This will reduce the transmission loss on subscriber loops.

Loaded cables are coded according to the spacing of the loading coils. The standard code for loading coils regarding spacing is shown in Table 2-3. The most commonly used spacings are B, D, and H.

Table 2-3. Code for Load-coil Spacing

Code Letter	Spacing (ft)	Spacing (m)
A	700	213.5
B	3000	914.4
C	929	283.2
D	4500	1371.6
E	5575	1699.3
F	2787	850.0
H	6000	1830.0
X	680	207.3
Y	2130	649.2

When determining signaling limits in subscriber loop design, approximately 15 Ω per loading coil should be added, as if the coils were series resistors.

Table 2-2 is useful for the calculation of attenuation of loaded loops for a given length. For example, for a 6 *dB* attenuation limit, the maximum permissible loop length using a 19-gauge wire is

$$l = \frac{6\ dB}{0.69\ dB\,/km} = 8.6956\ km$$

However, using a 19-H-88 loaded cable, the maximum permissible loop length is extended to

$$l = \frac{6 \ dB}{0.26 \ dB \ /km} = 23 \ km$$

Table 2-4 shows more properties of cable conductors, where loading arrangement code 19-H-88 indicates: the number 19 denotes the 19-gauge wire; H indicates the spacing between coils; the number 88 refers to the inductance (in milli-henries) of the coils.

Both the transmission limit and the signaling limit can be extended at a cost. If the pairs to be extended are few, they should be extended. However, if the pairs to be extended are many, it may be more economical to set up a new exchange area. The current tendency is to reduce the wire gauge where possible.

The size of an exchange area depends on the subscriber density, the subscriber traffic, and its distribution. Exchange sizes are often in units of 10,000 lines. Ten thousand is a convenient number which has important significance in telephone numbering. Consider a seven-digit number, say, 284-5000. The first three digits identify the local exchange, and the last four digits identify the individual subscriber line and is called the subscriber number. For a four-digit subscriber number, there are 10,000 numbers, from 0000 to 9999. Also 10,000 lends itself to crossbar switch unit size and is an average unit for subscriber densities in suburban areas and midsized towns in fairly well-developed countries. The term wire center is often used to denote a single location housing one or more 10,000-line exchanges. Some wire centers house up to 100,000 lines with a specific local serving area.

Various gauges of cables may be combined to provide good transmission with minimum plant cost. Figure 2-4 shows that customers within 15 kft of the central office are served with all 26-gauge nonloaded cables. Those between 15 and 18 kft are served with loops made up of partly 26-gauge and partly 24-gauge. Beyond 18 kft, inductive loading must be added. Customers between 40 and 50 kft are served with a combination of 19- and 22-gauge. In general, the longer the loop, the coarser the gauge that must be used.

Loaded Cable Circuits. Loading coils are used in telephone cable circuits. They have a resistance of a few ohms and an inductance of 10-100 milli-henries. They introduce lumps of inductance into the line at fixed intervals. These loading coils tend to enhance voice transmission

Table 2-4. Some Properties of Cable Conductors.

Diameter (mm)	AWG No.	Mutual Capacitance (nF/km)	Type of Loading	Loop Resistance (Ω/km)	Attenuation at 1000 Hz (dB/km)
0.405	26	40	None	270	1.61
		50	None		1.79
		40	H-66	273	1.25
		50	H-66		1.39
		40	H-88	274	1.09
		50	H-88		1.21
0.50		40	None	177	1.30
		50	H-66	180	0.92
		50	H-88	181	0.80
0.511	24	40	None	170	1.27
		50	None		1.42
		40	H-66	173	0.79
		50	H-66		0.88
		40	H-88	174	0.69
		50	H-88		0.77
0.60		40	None	123	1.08
		50	None		1.21
		40	H-66	126	0.58
		50	H-88	127	0.56
0.644	22	40	None	107	1.01
		50	None		1.12
		40	H-66	110	0.50
		50	H-66		0.56
		40	H-88	111	0.44
0.70		40	None	90	0.92
		50	H-66		0.48
		40	H-88	94	0.37
0.80		40	None	69	0.81
		50	H-66	72	0.38
		40	H-88	73	0.29
0.90		40	None	55	0.72
0.91	19	40	None	53	0.71
		50	None		0.79
		40	H-44	55	0.31
		50	H-66	56	0.29
		50	H-88	57	0.26

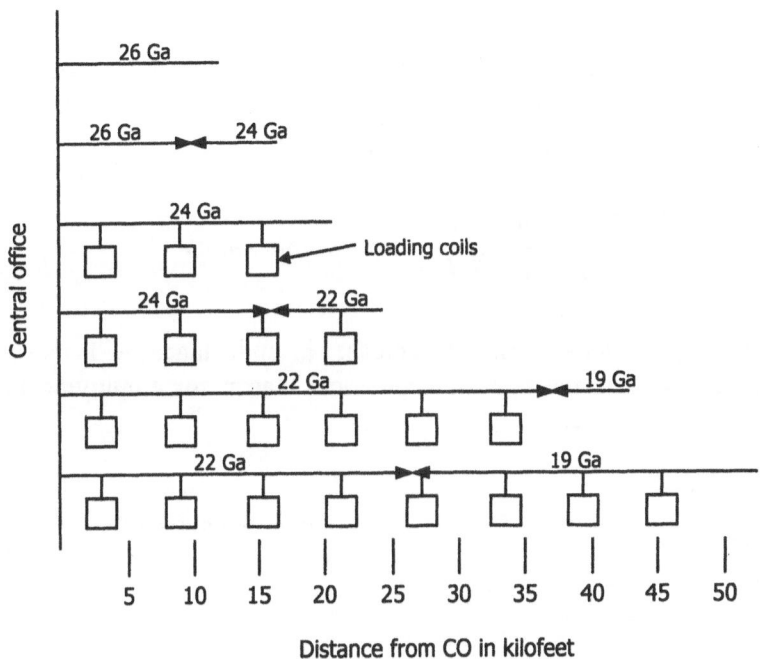

Figure 2-4. Basic resistance design.

quality within the 0 to 4 kHz voice band.

Loading coils provide a means of introducing a fairly large amount of inductance into a line. The coils are placed on the line at distances not much greater than a quarter-wavelength of the higher frequency transmitted so that the practical effect is much the same as if the distributed inductance were increased. Loading coils, acting with the capacitance and inductance of the line, constitute a low-pass filter. Thus, a line with loading coils will not transmit the frequencies required for carrier-current transmission. An important practical effect of loading coils on the line is that the characteristic impedance is increased, and becomes more nearly resistive and more nearly uniform over the range of frequencies transmitted. This helps with impedance matching. Loading also tends to equalize the velocity of propagation at different frequencies, which tends to avoid delay distortion.

Figure 2-5 shows the marked reduction of attenuation of transmitted frequencies that results from loading and the increase of attenuation as the cut-off frequency is approached.

To show that attenuation is reduced by increasing the inductance of a cable, consider an approximation of the attenuation constant at radio frequencies in (2-5); that is,

$$\alpha = \frac{R}{2} \sqrt{\frac{C}{L}} , \; nepers\,/unit \; length$$

This expression shows that an increase in inductance will reduce the attenuation. This is the important practical reason for employing loading coil on telephone lines.

Note that if x is the distance between the sending end and the receiving end of the line, an expression for attenuation on a line that is

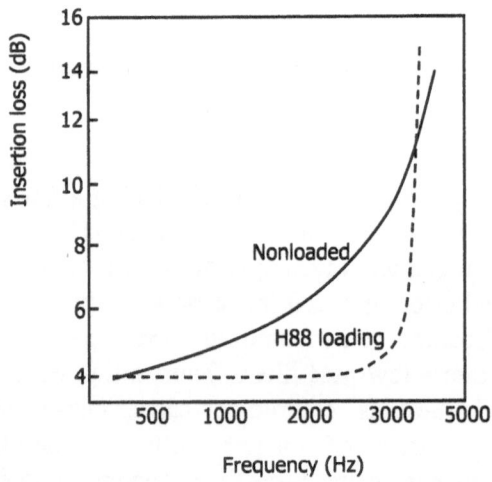

Figure 2-5. Attenuation in a loaded cable circuit.

terminated in its characteristic impedance is

$$\alpha \, x = \frac{1}{2} \, \ln \frac{P_s}{P_r} \tag{2-10}$$

where P_s and P_r represent the sending power and receiving power, respectively. In terms of dB, attenuation is defined by

$$Attenuation \ in \ dB = 10 \, \log_{10} \frac{P_s}{P_r} = 8.686 \, \alpha \, x \tag{2-11}$$

2-3. UNIGAUGE DESIGN FOR TELEPHONE CUSTOMER LOOP PLANTS

Using the basic resistance design method, customer loops are developed with a maximum conductor resistance of 1300 ohms. In the Bell system of the United States of America, in order to use minimum copper while keeping within the 1300 Ω limit, four standard cable conductor gauges (AWG 19, 22, 24 and 26) are used. For urban and suburban loops up to 52 kft in length, a systematic approach known as *the uniform gauge* or *unigauge design* has been developed. This approach treats the entire loop as one system including the local switching equipment, the outside plant, and the telephone set. It is capable of increasing the transmission, signaling, and supervisory range of local switching equipment from 1300 to 2500 Ω, thus permitting the exclusive use of 26-gauge conductors for all loops within 30 kft of the wire center.

Consider a typical layout of a subscriber plant based on the unigauge design (Bell system). It can be seen from Figure 2-6 that subscribers within 15 kft (5 km) of the switching center are connected over loops made up of 26-gauge nonloaded cable fed with the standard -48V battery. These loops are referred to as *short unigauge loops*. This length range includes 80% of all Bell system main stations. All loops in the 15 to 30 kft range are referred to as *long unigauge loops,* which are also served over 26-gauge nonloaded cable but require a range extender in the central office. This range extender provides 72V for signaling, an adequate transmitter current, and a 5 dB midband gain for speech. To extend the loop to 30 kft, 88 mH loading coils are added at 15 kft and 21 kft

points. Loops measuring more than 30 kft long may also use the unigauge principle and are referred to as *extended unigauge*.

It should be noted that the long 15 kft nonloaded sections, into which the switch faces, provide a uniform impedance for all conditions when an active amplifier is switched in. This is an important factor with regards to stability.

The unigauge application is particularly attractive in wire center serving areas that have the following characteristics:

- Extensive requirements for additional cable pairs in the 15-30 kft range;

- Low requirement for party-line service;

- Low requirement for PBX and special service lines.

2-4. SIGNAL MULTIPLEXING

In telecommunication systems, it is almost always the case that the capacity of the transmission medium exceeds that required for the transmission of a single signal. To make cost-effective use of the transmission system, it is desirable to use the medium by having it carry several signals simultaneously.

Multiplexing in transmission systems is a means of utilizing the same transmission medium for many different users concurrently. There are several ways in which signals can be multiplexed, the most important ones being space-division multiplexing, frequency-division multiplexing, and time-division multiplexing.

2-4-1. Space-Division Multiplexing. Space-division multiplexing is the grouping of many physically separate transmission paths into a common cable. A telephone cable consisting of hundreds (or thousands) of twisted pairs constitutes a space-division multiplexed system since many conversations can be carried on the single cable.

It should be emphasized that space-division multiplexing is not limited to voice-frequency circuits. In fact, many high-capacity transmission systems using either frequency or time-division multiplexing

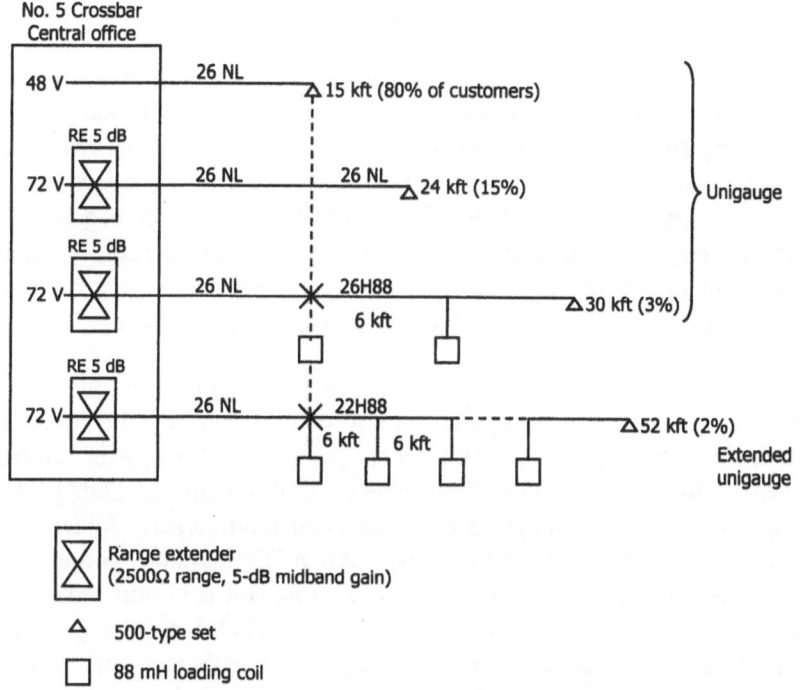

Figure 2-6. Layout of unigauge design.

can also be space-division multiplexed. For example, a carrier system may put thousands of message channels on a single coaxial line that is in turn part of a large cable containing many coaxial lines.

2-4-2. Frequency-Division Multiplexing. In frequency-division multiplexing (FDM), each channel of the system is assigned a fixed separate segment of the total bandwidth of the transmission medium. The signal spectrum for each user's signal is shifted into the proper bandwidth slot through modulation. Thus, many narrow bandwidth channels can be accommodated by a single wide bandwidth transmission system.

To separate the signals at the receiving end, band-pass filters, one for each channel, are required so that each channel will be shifted back to the original frequency spectrum.

2-4-3. Time-Division Multiplexing. Time-division multiplexing (TDM) is the sharing of a common transmission medium in time. TDM has often been used in telephone communications. For example, most telephones are in use only a small portion of the time. Thus, several telephones can time share a common line to the nearest central office.

If the sharing of the transmission medium or multiplexing is carried out by a deterministic, sequential allocation of time slots to each user, the multiplexer, transmitter, and receiver operate like a pair of synchronized switches. This type of multiplexing is also called synchronous time-division multiplexing (STDM). The time slots are often organized into frames so that each user is allocated a time slot in each frame. As a result of the fixed allocation of time slots, channel capacity is wasted if a user does not have a message to transmit in the allocated time slot.

Utilization of the shared channel can be significantly increased by combining a number of signals into a central buffer and transmitting through the shared channel. This technique of multiplexing is referred to as asynchronous or statistical time-division multiplexing (ATDM). Figure 2-7 shows a schematic diagram of a statistical multiplexer. ATDM allows time slots to be allocated dynamically. An ATDM system assigns a time slot only when a source becomes active. A time slot is eliminated when the respective source becomes inactive. Thus an ATDM system periodically redefines the length of its frame to change the number of time slots, and hence the number of channels. Circuit-switched telephone networks use synchronous TDM whereas data networks use asynchronous TDM techniques.

Each user has a short-term buffer to store input messages. These buffers are polled by a scanner that transfers messages from the line buffers into a central buffer. Since messages from all user inputs are mixed on the shared channel, each message must be given an identification for its destination.

Now it is appropriate to distinguish between multiplexing and concentration. Multiplexing generally refers to static channel allocation schemes in which given frequency bands or time slots on a shared channel

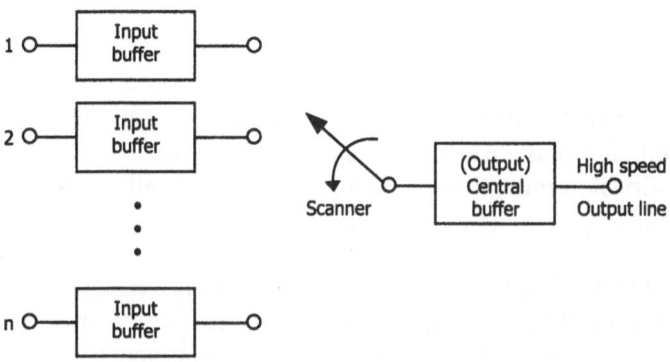

Figure 2-7. Schematic diagram of a statistical multiplexer

are assigned on a fixed, predetermined basis. Concentration is a scheme in which a number of users dynamically share a smaller number of output sub-channels on a demand basis.

2-5. DIGITAL TRANSMISSION SYSTEMS

In digital transmission systems, the message signals are converted into digital form which produces a rugged signal with a high degree of immunity to transmission distortion, interference, and noise. Digital transmission also allows the use of regenerative repeaters for long distance communication. However, the quantization process associated with analog-to-digital conversion results in quantization noise which becomes the fundamental limitation on waveform reconstruction. To keep the quantization noise small, a coded pulse modulation system generally requires a much larger bandwidth than a comparable analog transmission system.

Transmission Media. There is a variety of transmission media that can be used for digital links. For lower speed transmission up to 10 Mbps, twisted pair cables are the most commonly used. For higher speeds up to 1,000 Mbps, coaxial cables, microwaves, waveguides, and optical fibers may be employed.

Twisted pair cables and coaxial cables have larger attenuations at high frequencies. In addition to attenuation, twisted pair cables suffer from cross-talk caused by electromagnetic coupling between physically isolated circuits.

A circular waveguide provides good transmission bandwidth and low attenuation. Optical fibers provide still larger transmission capacity and low transmission loss. Microwave systems have been extensively used for frequency-division-multiplexed analog signals as well as for transmitting digital signals by modulating a carrier.

The block diagram of a digital baseband transmission system is shown in Figure 2-8. With the exception of local area networks limited to a few kilometers at maximum, most data transmission links make use of lines of the public switched telephone networks. These lines have relatively well-defined characteristics.

Table 2-5 summarizes typical capacities for transmission link technology.

Table 2-5. Characteristics of Transmission Links

Transmission Link	Bandwidth
Twisted pair	200 Mbps over short distances
Coaxial cable	1 Gbps up to 1 km
Terrestrial microwave	2.4 Gbps
Satellite microwave	4 - 6 Gbps
Optical fiber	10 - 100 Gbps

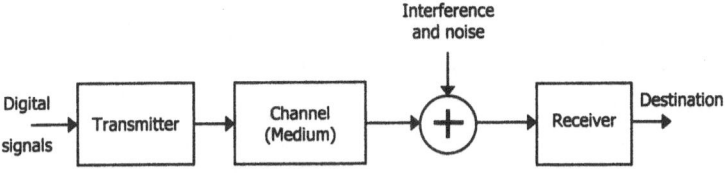

Figure 2-8. Block diagram of a digital baseband transmission system

A digital transmission system may consist of a transmitter, a transmission channel (medium), and a receiver which contains an equalizer, a detector, and a clock (timing) recovery device. The main task of a digital transmission system is to transfer a digital signal from the source to the destination. A digital signal is an ordered sequence of symbols produced by a discrete information source. Because of the finite bandwidth of the system, there is an upper limit for the signaling rate. In addition, distortion and noise can cause errors to the transmitted digital signal. Thus, signaling rate and error probability play important roles in the performance of digital transmission systems.

The function of the transmitter is to provide a transmittable signal that accurately represents the source information and is matched to the characteristics of the transmission channel, the interference, and the receiver. The transmitter also supplies the required transmitted power.

In general, the characteristics of the transmission channel are time, frequency, amplitude, and temperature dependent. Therefore the transmitted signal is always distorted during transmission via the channel. Also, interference such as cross-talk and noise are imposed on the transmitted signal.

For theoretical analysis, the noise amplitude X is often assumed to be Gaussian with zero mean and variance σ_n^2; that is, the noise amplitude is represented by a Gaussian (or normal) random variable with the probability density function

$$p(x) = \frac{1}{\sqrt{2\pi}\sigma_n} e^{-x^2/2\sigma_n^2}, \quad -\infty < x < \infty \tag{2-12}$$

For polar binary signals with positive or negative pulses of amplitude $+V_p$ or $-V_p$ with equal probability, an error occurs if the noise is more negative than $-V_p$ when the pulse is positive, or if the noise is more positive than $+V_p$ when the pulse is negative. Thus the error probability can be expressed as

$$P_E = \frac{1}{2} P\{X > V_p\} + \frac{1}{2} P\{X < -V_p\}$$

$$= \frac{1}{2}(1 - P\{X \leq V_p\}) + \frac{1}{2} P\{X \leq -V_p\}$$

$$= \frac{1}{2}(1 - 2P\{X \leq V_p\}) \tag{2-13}$$

$$= \frac{1}{2}(1 - \frac{2}{\sqrt{2\pi}\sigma_n} \int_0^{V_p} e^{-x^2/2\sigma_n^2} dx)$$

By making a change of variable, $y = x/\sqrt{2}\sigma_n$, the integral term can be written as

$$\frac{2}{\sqrt{\pi}} \int_0^{V_p/\sqrt{2}\sigma_n} e^{-y^2} dy$$

Hence

$$P_E = \frac{1}{2}(1 - \frac{2}{\sqrt{\pi}} \int_0^{V_p/\sqrt{2}\sigma_n} e^{-y^2} dy), \quad -\infty < y < \infty \tag{2-14}$$

The integral term in (2-14) is known as the error function and is often denoted by

$$erf(v) = \frac{2}{\sqrt{\pi}}\int_0^v e^{-y^2}dy \tag{2-15}$$

Furthermore, the quantity $1 - erf(v)$ is called the complementary error function and is denoted by

$$erfc(v) = 1 - erf(v)$$

Thus, in terms of the complementary error function, (2-14) can be expressed as

$$P_E = \frac{1}{2}erfc(V_p/\sqrt{2}\sigma_n) \tag{2-16}$$

This formula is useful in determining the probability of decision error in digital transmission.

If the transmission of pulses can take on any of m amplitude levels with equal probability, then the separation between levels is $2V_p/(m-1)$ rather than $2V_p$ for polar binary signals. Consequently, an error will occur if the magnitude of the noise at the decision time is greater than $V_p/(m-1)$. In this case, the formula for determining the error probability for polar m-ary signals is given by

$$P_E = \frac{m-1}{m}erfc\left[\frac{V_p}{(m-1)\sqrt{2}\sigma_n}\right] \tag{2-17}$$

Clearly, for $m = 2$, formula (2-17) reduces to (2-16).

In digital data communication (analog or digital signal), a fundamental requirement is that the receiver knows the starting time and duration of each bit that it receives.

There are two schemes for meeting this requirement: asynchronous transmission and synchronous transmission. Asynchronous transmission is the earliest and simplest scheme for transmitting messages in separate groups. Within an individual group, a specific predefined time interval is used for each message. However, the durations of transmission of the groups are unrelated to each other. Thus the sample clock in the receiver is reestablished for reception of each message. Synchronous transmission is a more efficient scheme of transmitting digital data continually at a constant rate. Hence the receiver must establish and maintain a sample clock that is always synchronized to the incoming data.

2-5-1. Asynchronous Transmission. An asynchronous line can be in an idle state or in an active state. The idle state of the line is logical 1. In this scheme, data are transmitted one character at a time. Each character is preceded by a start bit (a 0) and followed by a stop bit (a 1), as shown in Figure 2-9(a). When there is no data to send, the transmitter sends stop bits continuously. The receiver detects the beginning of a new character by the transition from 1 to 0 and uses the start bit to synchronize itself to the bit timing of the incoming signal. To establish synchronization, a clock running at 16 times the bit rate is used and the alignment is made to one sixteenth of a bit time.

This scheme works well at low speeds, but at higher rates, providing a 16 times clock can be a difficult problem. Also, transmitting the start and stop bits with every character means that the line utilization is only 80% of its maximum. The main attraction of asynchronous transmission is the ease of determining the sample times in the receiver. Asynchronous transmission is used only in voiceband data sets (modems) for transmission rates up to 1200 bps. The major drawback of asynchronous transmission is its poor performance in terms of error rates on noisy lines.

2-5-2. Synchronous Transmission. With synchronous transmission, bits or characters are transmitted without start and stop bits. To allow the receiver to determine the beginning and end of a block of data, synchronization is required. Here a clock is maintained in the receiver with the same frequency as the transmitter's clock and the same phase as the incoming data. Each block of data begins with a preamble bit pattern and ends with a postamble bit pattern. Together, the preamble and postamble plus the data are called a frame. Depending on whether the block of data is character-oriented or bit-oriented, there are two types of preamble and postamble.

With the character-oriented type, each block of data is preceded by one or more synchronization characters, as shown in Figure 2-9(b). The synchronization character, called SYNC, is chosen such that its bit pattern is completely different from any of the regular characters being transmitted. The postamble is a different character. The receiver detects the beginning of an incoming block of data by the SYNC character and the end by the postamble character.

The bit-oriented type treats the block of data as a bit stream. The preamble-postamble principle is the same, with one difference in that the preamble or postamble pattern may possibly appear in the data. To overcome this problem, the procedure known as bit stuffing may be used. For example, the high-level data link control and synchronous data link control bit-oriented schemes make use of the pattern 01111110 (called a flag) as both preamble and postamble. To avoid the appearance of this pattern in the data stream, the transmitter will always insert an extra 0 bit after each occurrence of five 1's in the data to be transmitted. When the receiver receives a sequence of five 1's, it examines the next bit. If the bit is 0, the receiver deletes it.

Advantages and Disadvantages of Digital Transmission

An important property of digital transmission is that the noise generated in the terminal and the noise encountered in the transmission line can be separated. Quantization noise generated by the terminal equipment can be made smaller by proper choice of the codec (coder-decoder). Because of the regenerative repeater of the digital transmission system, line noise has little effect on the message signal.

Because the signal-to-line noise ratio requirement is lower in digital than in analog transmission, digital transmission often results in better utilization of noisy media. In cables, digital transmission can provide more channels at a lower cost than can analog transmission. In radio media, digital transmission can tolerate greater external interference.

In digital transmission, because all signals are in digital form, there is complete freedom to intermix them in a common facility. This flexibility makes a digital system more useful than an analog system. A major disadvantage of digital transmission is that much greater bandwidth is required.

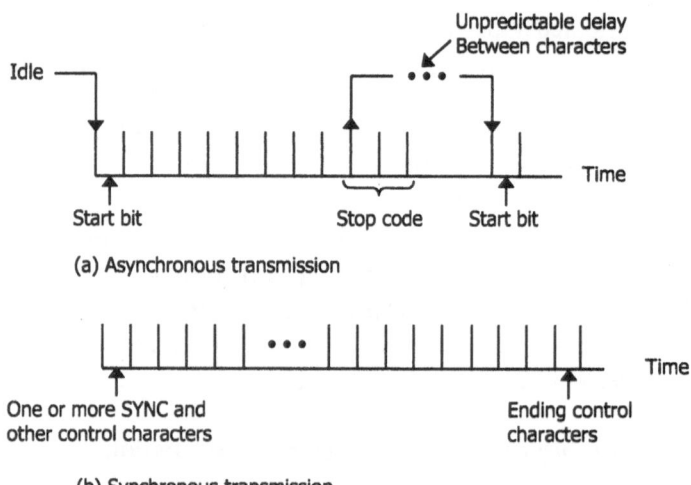

Figure 2-9. Asynchronous and synchronous transmission

2-6. OPTICAL FIBER TRANSMISSION SYSTEMS

Optical fiber as a transmission medium possesses many special characteristics such as low loss, high bandwidth, small physical cross section, light weight, electromagnetic interference immunity, and security, which make it very useful for telecommunication systems. When compared with coaxial cable, a major advantage of optical fiber is that no equalization is necessary. Also, because of its low attenuation properties, optical fibers can have repeater separation in the order of 10 to 100 times that of coaxial cable for equal transmission bandwidth. Optical fiber links need regeneration only once every 5000 km or so.

The basic elements of an optical fiber transmission system, as shown in Figure 2-10, are the optical transmitter, the optical fiber, and the optical receiver.

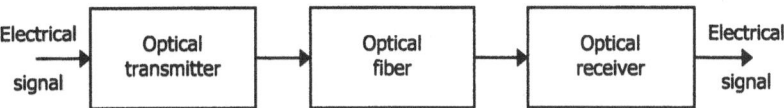

Figure 2-10. Block diagram of an optical transmission system

The optical transmitter consists of a semiconductor light source which may be a laser diode or a light-emitting diode. The primary function of the light source is to convert electrical signals (currents) into optical signals (lights). Optical fibers are grouped into two classes: single mode fibers and multimode fibers. Single mode fibers are designed such that only one mode is propagated. Multimode fibers allow the propagation of more than one mode. In a multimode fiber, each mode travels with a slightly different group velocity. Thus it causes a light pulse to have more distortion than in a single mode fiber. Single mode fibers have two distinct advantages. First, single mode fibers have smaller diameter cores. Second, single mode fibers have less internal (Rayleigh) scattering, which reduces attenuation.

As a transmission medium for telecommunication systems, optical fibers are characterized by two important parameters: attenuation and dispersion. Attenuation, expressed in decibels per kilometer, determines how far a light signal can travel in a fiber before it must be regenerated. Dispersion, expressed in megahertz per kilometer, is a measure of distortion of a light pulse. Dispersion at the receiver end is caused by two factors: material dispersion and modal dispersion. Material dispersion is caused by the change in the refractive index of the material with frequency. Modal dispersion arises in multimode fibers due to the fact that different modes have different phase and group velocities so that different modes arrive at the receiver at different times. Thus delay distortion (dispersion) will result.

Attenuation. The low attenuation or low transmission loss of optical fibers has proven to be one of the most important factors in bringing about their wide acceptance in telecommunications.

In optical fiber communications, the attenuation is usually expressed in decibels per unit length

$$\alpha l = 10 \log_{10} \frac{P_i}{P_o}$$ (2-18)

where α is the signal attenuation in decibels per unit length, l is the fiber length, P_i is the input (transmitted) optical power, and P_o is the output (received) optical power.

Dispersion. Dispersion of the transmitted optical signal causes distortion for transmission along optical fibers. When optical fiber transmission involves some form of digital modulation, dispersion in the fiber causes broadening of the transmitted light pulses as they travel along the optical fiber channel.

The optical receiver is a light detector, which is a photodiode. The light detector is used to convert the received optical signal to an electrical signal. The ratio of the root mean square (rms) value of the output current or voltage of a photodetector to the rms value of the incident optical power is called responsivity. Thus responsivity is a measure of the amount of electrical output power of a photodiode for a given incident light input power. Mathematically, the responsivity R of a photodiode is defined as the output current I_{ph} in amperes produced per unit of incident optical power P_i in watts. That is,

$$R = \frac{I_{ph}}{P_i} , A/W$$ (2-19)

Since the photocurrent I_{ph} is given by

$$I_{ph} = \frac{q \, \eta \, P_i}{hf} \quad amperes$$ (2-20)

where

$q = 1.602 \times 10^{-9}$ coulomb is the charge of an electron,

η = the quantum efficiency which is the ratio of the average number of electrons excited into the conduction band to the number of photons,

h = 6.626 x 10^{-34} J - s is Planck's constant, and

c = 2.998 x 10^8 m/s is the speed of light.

Using the relationship between the wave length λ and the frequency f , $c = \lambda f$, the responsivity can be expressed as

$$R = \frac{q \, \eta \, \lambda}{hc} = \frac{\eta \, \lambda}{1.24} \, , A/W \qquad (2\text{-}21)$$

where the unit of λ is in micrometers.

2-7. SUMMARY

The transmission system is the building block of a communication network. It provides the physical communication links to interconnect the nodes (or transmitting and receiving devices) in a network. The most commonly used transmission media for telephone and local area networks are the twisted pair cables and coaxial cables. Because of the relatively high cost, at the current state of technology optical fiber links are suited for point-to-point long distance communications.

In the design of subscriber loops in telephone networks, the attenuation due to ac energy loss and IR drop in dc voltage must be taken into account. These two factors limit the length of a subscriber loop. We have presented the basic resistance design and the basic transmission design to determine the maximum permissible length of subscriber loops. To enhance voice transmission quality in telephone networks, loading coils may be introduced into the transmission lines at fixed intervals. For urban and suburban loops, the unigauge design may be employed. This method permits the exclusive use of 26-gauge conductors for all loops within 30 kft or 9 km of the wire center.

To increase the cost-effectiveness of the transmission system, signal multiplexing techniques such as space-division multiplexing, frequency-division multiplexing and time-division multiplexing may be employed.

Because of low loss, high bandwidth, and electromagnetic interference immunity, optical fibers, have great potential to be used as a transmission medium for future telecommunication networks.

REFERENCES

[1] McNamara, J.E., Technical Aspects of Data Communication, 3rd ed., Bedford, Mass.: Digital Press, 1988.

[2] Hills, M.T. and Evans, B.G., Transmission Systems, London: George Allen & Unwin, 1973.

[3] Freeman, R.L., Telecommunication System Engineering, 2nd ed., New York: John Wiley & Sons, 1989.

[4] Bell Telephone Laboratories, Transmission Systems for Communications, Revised 4th ed., Winston-Salem, N.C.: Bell Telephone Laboratories, 1971.

[5] Bellamy, J., Digital Telephony, 2nd ed., New York; John Wiley & Sons, 1991.

[6] Inose, H., An Introduction to Digital Integrated Communications Systems, New York: Peter Peregrinus, 1981.

[7] Jones, W.B. Jr., Introduction to Optical Fiber Communication Systems, New York: Holt, Reinehart and Winston, 1988.

PROBLEMS

2-1.   Let $y = G + j\omega C$ and $z = R + j\omega L$ be the shunt admittance per unit length and series impedance per unit length of a transmission line, respectively. The propagation constant of the line is defined by

$$\gamma = \sqrt{yz} = \alpha + j\beta$$

Show that

$$\alpha = \frac{1}{\sqrt{2}}\left[\sqrt{(R^2 + \omega^2 L^2)(G^2 + \omega^2 C^2)} + RG - \omega^2 LC\right]^{\frac{1}{2}}$$

and

$$\beta = \frac{1}{\sqrt{2}}\left[\sqrt{(R^2 + \omega^2 L^2)(G^2 + \omega^2 C^2)} - RG + \omega^2 LC\right]^{\frac{1}{2}}$$

2-2. For low-loss lines, the quantities α^2 and RG are small and can be neglected. Calculate (a) the attenuation constant α, (b) the characteristic impedance Z_c if the dielectric losses are also small, and (c) the attenuation constant α with the additional condition as in (b).

2-3. Determine (a) the minimum value of α by varying L, and (b) the corresponding phase shift β.

2-4. Given the values for the primary constants at 100 MHz, $R = 100\ \Omega/km$, $L = 1\ mH/km$, $G = 10^{-5} S/km$, and $C = 0.05\ \mu F/km$ of a transmission line, can we assume that this is a low-loss line?

(a) What is the value of the attenuation constant α?

(b) What is the value of the characteristic impedance Z_c?

2-5. A low-loss transmission line which has a capacitance $C = 0.051\ \mu F/km$ and an inductance $L = 0.54\ mH/km$, is operated at radio frequencies. Calculate its characteristic impedance Z_c.

2-6. A particular cable operated at high frequencies has a capacitance of $0.052\ \mu F/km$ and a characteristic impedance of $296\ \Omega$. What is the inductance per kilometer of this cable?

2-7. What is the capacitance per kilometer of a transmission line having a characteristic impedance of $417\ \Omega$ and an inductance of $0.54\ mH/km$?

2-8. What is the maximum subscriber loop length for a 24 AWG loaded cable of diameter 0.511 mm with $H-66$ loading?

2-9. If the loop resistance limit is $1250\ \Omega$, calculate the maximum permissible loop length for a 26 AWG copper wire.

CHAPTER 3

SWITCHING SYSTEMS

3-1. INTRODUCTION

The purpose of a switching system is to provide a means to pass information from one terminal device to any other terminal device selected by the originator. In telephony, switching is used to establish a communication path between two subscribers. Thus, a switch sets up a communication path on demand and takes it down when a communication path is no longer needed. It performs logical operations to establish the path and automatically charges the subscriber for usage.

User requirements for a telecommunication switching system may be summarized as follows:

1. The capability of providing a communication path between any two users;

2. The connection time should be relatively small compared to the holding time or conversation time;

3. The probability of completion of a call should be high, usually 99%;

4. The availability to the user at any time;

5. The primary mode of communication for most users will be voice; and

6. The privacy of user conversation is assumed, but it is neither specifically requested nor guaranteed.

Basic Switching Functions. To set up a communication path, a local switching center must respond to a calling signal originating from a terminal. The switching functions are remotely controlled by the calling subscriber by off hook, on hook, and dial information. We shall give a brief explanation of each of the basic functions, according to the order of occurrence during the progress of a call in a switching system.

Attending - When a call is originated from a station or through another office, a central office or switching center receives the signal of request for service.

Signal Reception - After receiving the request for service, the central office sends a dial-tone signal to the calling station. Upon receiving the dial information, usually numerical, it responds to address the desired called station.

Signal Interpretation - Based on the received signal information, the switching office interprets and carries out the required action.

Path Selection - The switching office finds an idle link or series of links or channels through the switching center network.

Route Selection - The switching office determines the trunk group to which a path is to be established; includes intra-office calling.

Busy Testing - This function is to test whether a link or trunk is in use or reserved for use on another call. When a tested link or trunk is found busy, successive testing of trunks or links is known as hunting.

Path Establishment - This is the control of the elements of the switching center network to establish a path for communication. In circuit switching this function requires some form of memory to retain the connection for the duration of the call.

Signal Transmission - For inter-office calls, this is the transmission of the address of the called station to the distant office. Address signaling in station-office communication consists of rotary dial and touch tone. Signaling between telephone offices can be inband or out-of-band using a signal frequency. Moreover, a scheme for communication between the controls of central offices is used. This scheme is known as common channel inter-office signaling in which the controls of central offices are sent via a private channel.

Alerting - This informs the called station or office that a call is being sent to it. For intra-office calls, this is known as ringing. For inter-office calls, it is the transmission of the attending signal.

Supervision - This is to detect when the connection is no longer needed and to effect its release. Supervision is also required for other purposes such as for call service features, which include call abandonment, no address transmitted, partial address received, unassigned address received, and unequipped address received.

Types of Switching Systems - In telecommunication switching, there are two principal types of switching known as circuit or line switching and message switching or store-and-forward switching.

Circuit Switching - Circuit switching is a form of switching which sets up a dedicated communication path between two or more stations through one or more intermediate switching nodes. It is also known as line switching. Circuit switching can also be applied to a system with a switching center network that is used for two-way transmission with no chance of transmission delay due to passage through the network. Circuit switching is widely used in telephone networks. The information may be voice, data (including telex and teletypewriter), or video.

Store-and-Forward Switching - In this form of switching, messages are received at intermediate routing points and recorded (stored), checked for any errors, and then retransmitted to a further routing point or to the ultimate destination. This is also known as message switching. Generally, it implies that no switching center network is employed, but a signal access network is required. With message switching, it is not necessary to establish a dedicated path between two stations. Line utilization is increased. A substantial delay may be produced and may relate to one of the following factors:

1. The lack of a common user group of idle transmission facilities at the moment the message reaches the switching office;

2. The need for multiple transmissions of the same message and the need for storage;

3. The called station or terminal being busy;

4. The speed of message transmission facilities or medium may be less than the received rate; or

5. The formatting or coding of the transmitted message being different from the way it was received.

The length of delay depends upon the memory capacity provided in a switching center and the transmission capacity available for use in routing the message to its destination. In these networks, it is desirable to employ switching memory at the expense of more adequate transmission facilities.

The message may be divided into smaller, usually uniform size segments called packets. This form of store-and-forward switching is known as *packet switching*. In packet switching, each packet contains data plus a destination address. If a station has a message to send that is of a length greater than the maximum packet size, the message is broken into packets for transmission. There are two approaches for the network to handle the packets: datagram and virtual circuit.

In the datagram approach, each packet is treated independently. Packets with the same destination address do not follow the same route. Thus, it is more flexible and more reliable. It has the advantage that the delays are shared by all users rather than serving complete messages in the order of arrival. Thus, many packets of the same message may be in transmission simultaneously, thereby giving one of the main advantages of packet switching, namely, the pipelining effect. As a result, the transmission delay may be reduced considerably (over message switching). At the receiving node, the packets are reassembled to reconstruct the original message.

In the virtual circuit approach, a logical connection is established before any packets are sent. Each packet now contains a virtual circuit identifier as well as data. No routing decisions are required at each node on the preestablished route. Thus, it differs from the datagram approach. One disadvantage of the virtual circuit approach is that if a node fails, all virtual circuits that pass through that node are lost.

3-2. CENTRALIZED SWITCHING

In a telephone system serving n subscriber stations, $n(n-1)/2$ line connections or two-way transmission facilities are required for interconnection. When every station is interconnected with every other station in a telecommunication system, the system is called a mesh connection or a full mesh. An eight-subscriber system, in which each subscriber station is directly connected to every other station, is shown in Figure 3-1(a).

The switches per station used to select the required connection of paths at both ends of the transmission facilities are called crosspoints. Crosspoint is the term more commonly used to denote an electronic circuit that can electrically connect or disconnect two conductors in response to some external control signals. A telephone system consisting of n stations would require $n - 1$ switches per station and hence in total $n(n - 1)$ crosspoints. Full mesh systems have only limited applications, generally where n is small and the traffic levels are comparatively high. These fully connected systems are also known as noncentralized switching systems.

From the inception of telephony, the concept of centralized switching was realized to be more efficient. A centralized switching network with remote control is shown in Figure 3-1(b). With centralized switching, only one two-way line is required per station, but the number of crosspoints remains the same as that of noncentralized switching. The difference is that for centralized switching the crosspoints must be remotely operated. Centralized switching can be made more efficient in reducing the number of crosspoints, as shown in Figure 3-1(c), where only $n(n - 1)/2$ (or half as many) crosspoints are required. This is known as a single stage nonblocking switching network, which can provide for all stations to be connected simultaneously without any blocking at all. From Figure 3-1(c), we see that the saving in crosspoints is achieved at the expense of introducing the $n - 1$ links (or lines within the centralized switching system or central office). If fewer than $n/2$ simultaneous connections are required at a given time, then fewer than $n(n - 1)/2$ crosspoints may be used. The number of links required is determined by the desired grade of service for a given offered load. In Figure 3-1(d), we may use $L < (n - 1)/2$ links to further reduce the number of crosspoints. When all the L links are used for connecting $2L$ stations simultaneously, the remaining $n - 2L$ stations cannot be connected until one of the L links becomes available. Thus, congestion or blocking is always possible.

For a subscriber of one central office to communicate with a subscriber of another central office, call control information (signaling) between central offices is necessary. When more central offices and trunk groups that connect them are added, it becomes clear that a centralized switching office, known as a tandem office in telephony, for central offices is needed (see Figure 3-2).

Typically, an exchange area defined for telephone service will be made up of a number of local central offices and one or more tandem

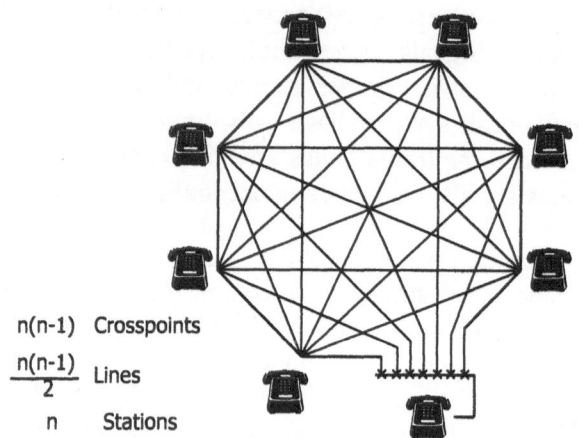

(a) Noncentralized switching system.

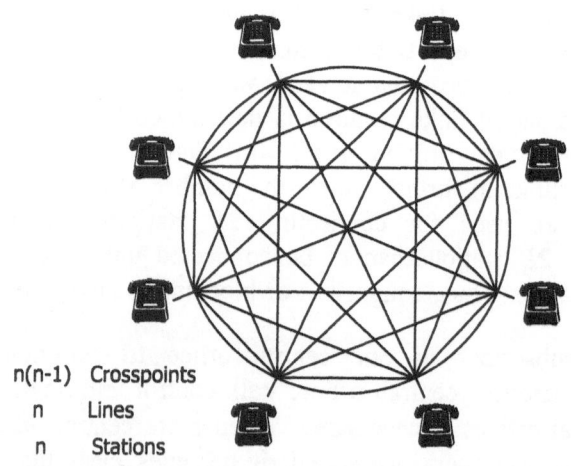

(b) Centralized switching system with remote control.

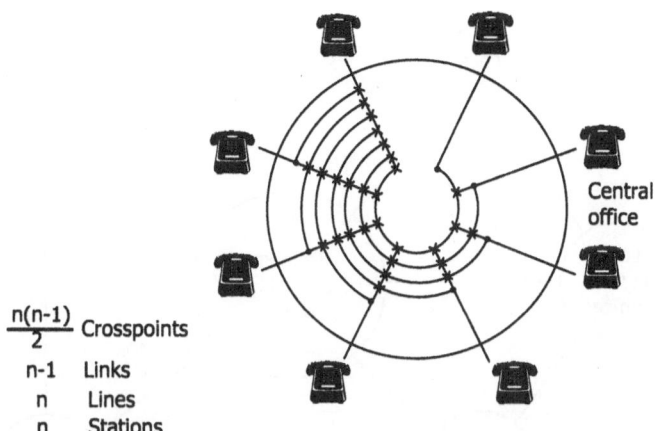

$\dfrac{n(n-1)}{2}$ Crosspoints

n-1   Links

n   Lines

n   Stations

(c) Centralized switching system with no blocking.

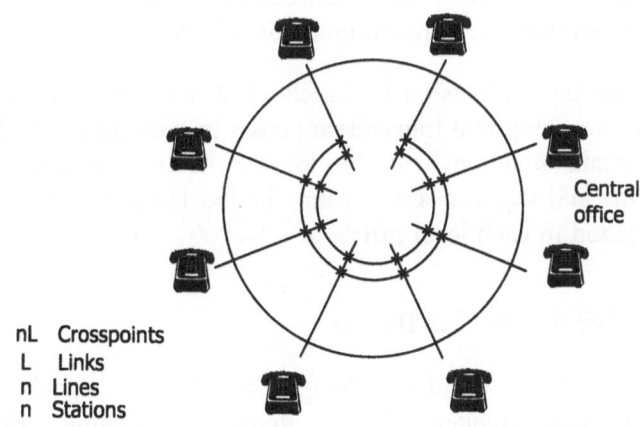

nL   Crosspoints

L   Links

n   Lines

n   Stations

(d) Centralized switching system with possible blocking.

Figure 3-1. Centralized switching system.

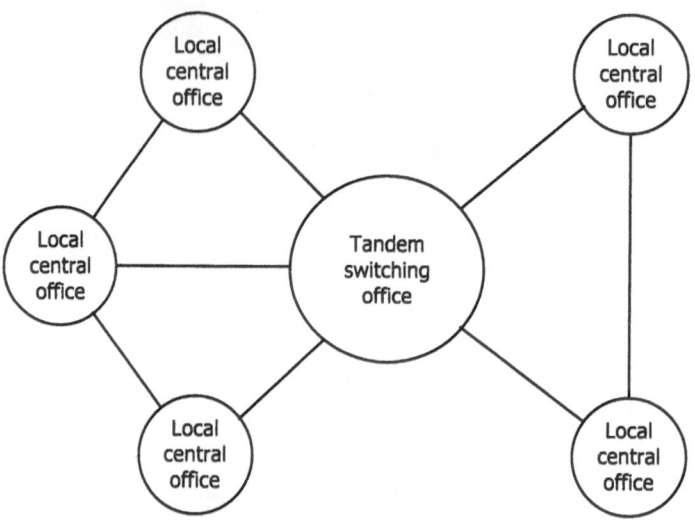

Figure 3-2. Tandem switching office.

offices. The number of trunks between the offices will be based on economics to provide the required grade of service.

What we have discussed is the classical network problem of trading off between switching and transmission costs to minimize overall costs. In addition, certain functions can be located in the tandem office. For example, time and weather services may be located in the tandem office to avoid replication in each local office.

3-3. SWITCHING TECHNIQUES

There are three switching techniques based on the division of calls in space, time and frequency: space-division switching, time-division switching and frequency-division switching.

3-3-1. Space-Division Switching. A network consisting of metallic contacts is called a space-division network because the communication paths set up by the closure of metallic contacts in series are physically separated from each other. The metallic contacts are called crosspoints. They are grouped in a two-dimensional array to form a switch unit. In space-division switching, different calls are connected through different physical paths separated in space for the duration of a call. A 4 × 4 space-division switching is shown in Figure 3-3.

The advantage of space-division switching is its simplicity and good transmission properties. For example, the bandwidth of the signal is not unduly limited by the switch mechanism. The disadvantage of space-division switching is that the switches can be relatively slow to operate, can be bulky, and can involve a large amount of wiring because each connection needs a separate set of wires.

3-3-2. Time-Division Switching. In time-division switching, all the inlets and outlets are connected to a common transmission path via high

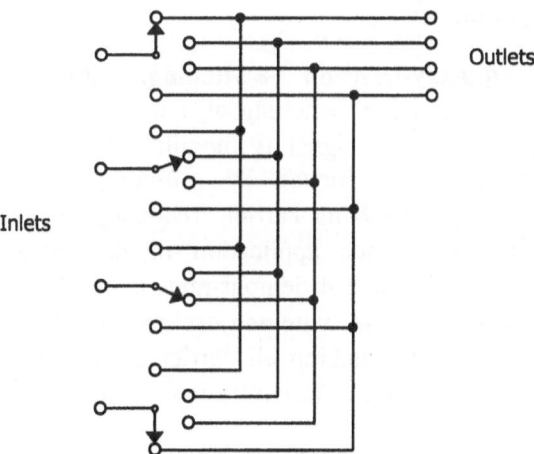

Figure 3-3. Space-division switching.

speed switches. These switches connect the required inlet and outlet onto the same path for a short time. Time-division switchings are applicable to digital switching only. Modern large switching systems often consist of combinations of time-division and space-division switches.

The signals are time-multiplexed by the switch onto a fast bus, and the resulting signal is demultiplexed by accurately timed gates in accordance with the user's requirements. It uses much less switch points than space-division switching. A typical example using time-division switching in telephony is the pulse-code modulation (PCM) transmission of voice samples. Each voice signal is sampled at a rate of 8000 samples/second with a sampling period of 125 μs. These samples are digitized to 8 bits/sample, so that a typical PCM voice channel requires 64 kbps of transmission capacity. If 24 8-bit voice channels are multiplexed with 1 frame bit for each sample, the bit rate is $C = (24 \times 8 + 1)$ bits/125 μs = 1.544 Mbps. This is known as the T1 system. Note that the 24-channel frame can also be used for data transmission.

To carry out time switching, each frame arriving at the switch is first written into a memory. Switching is then accomplished by reading out the stored data in a desired order. Such a device is called a time-slot interchange (TSI) unit. Time switching may be integrated into digital transmission, so that modulation and demodulation take place only at the far end of the system.

3-3-3. Frequency-Division Switching. As in time-division switching, frequency-division switching also uses a common transmission path, but in this case, each signal is modulated onto a different carrier frequency. Switching is achieved by providing each outlet with a demodulator which can have its carrier frequency changed. Frequency-division switching now finds application in demand-assigned satellite communication links. Figure 3-4 demonstrates a simple frequency-division switching. Frequency-division switching can be used to increase the traffic capacity of the satellite by making all carrier frequencies available to all ground stations. Note that the toll telephone network also uses demand assignment.

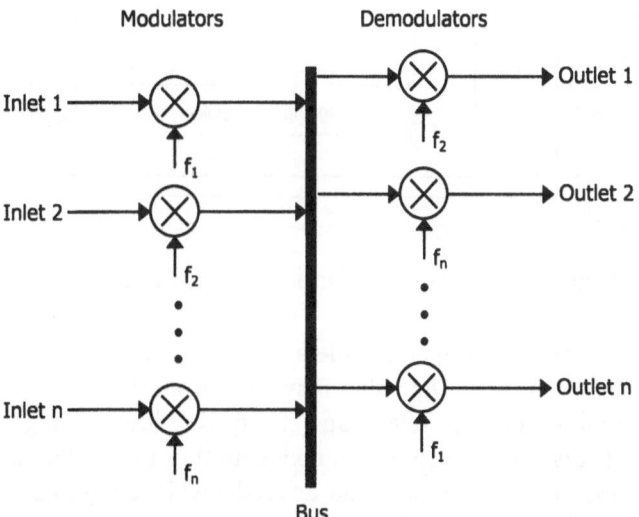

Figure 3-4. Frequency-division switching.

3-4. CONGESTION IN SPACE-DIVISION SWITCHING NETWORKS

Switching networks for connecting the subscriber lines are usually arranged with concentration. Figure 3-5 shows a connecting network using concentration, distribution and expansion. With concentration, the number of outlets is less than the number of inlets, while with expansion, the reverse is true.

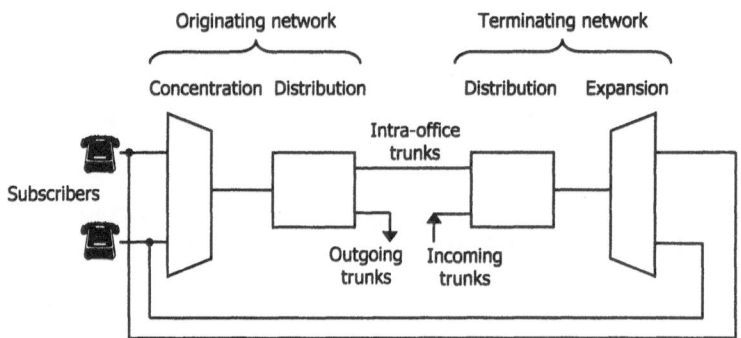

Figure 3-5. Standard form of connection network.

The ratio of the number of inlets to the number of outlets is known as the concentration ratio. Its value may vary from 2 for high calling rates, to 4 for medium calling rates, and to 6 for low calling rates. The expansion ratio is the inverse of the concentration ratio. Its value depends on the loading of the inlets and the desired level of internal blocking. A typical design for local switching networks is based on an average inlet loading of 0.6 erlang and total internal blocking not exceeding 0.002. Toll switching network dimensions are based on an average inlet loading of 0.7 erlang and first trial blocking of 0.005. Distribution (or intermediate) stages are usually arranged without concentration or expansion.

A basic requirement of switching network design is to minimize the total number of crosspoints needed to switch a given offered traffic load at a given grade of service. Other factors include complexity of control, flexibility for expansion, and overload performance.

The determination of offered traffic per line is complicated by the fact that an intra-office call occupies two lines, while an outgoing and incoming call occupies only one line. Similarly, calculations of busy-hour traffic per line must take incoming calls into account. This situation can be represented in a traffic H diagram as shown in Figure 3-6.

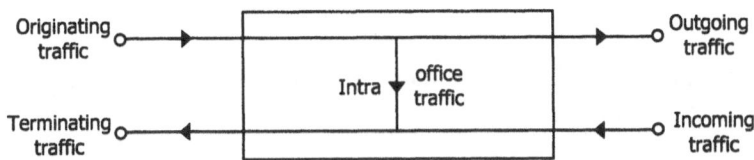

Figure 3-6. Traffic H diagram.

In the traffic H diagram, the intra-office traffic is considered as a part of the originating traffic. The traffic H diagram is drawn with the assumption that the sum of the originating traffic and the terminating traffic is equal to the sum of the incoming traffic and the outgoing traffic. In North America the offered traffic in CCS per line is calculated as the sum of the originating traffic and the terminating traffic (which means that the intra-office traffic is counted twice) divided by the number of lines. However, the busy-hour call (BHC) traffic in erlangs per line is calculated as the sum of the originating traffic and the incoming traffic (counting intra-office traffic only once) divided by the number of lines.

Example 3-1. A switching network in a central office serving 10,000 lines has an average origination rate of three calls per hour per line and an average holding time of 3 minutes. If the intra-office traffic is 10% of the originating traffic and the incoming traffic is 200 erlangs, what are the BHC and CCS traffic per line?

Solution. Since the calling rate per line is 3 calls per hour, the total originating traffic is given by

$$a = 10,000 \times \frac{3 \; calls}{hour} \times \frac{3 \; minutes}{60 \; minutes \, / hour} = 1,500 \; erlangs$$

or

$$a = 1{,}500 \times 36 = 54{,}000 \; CCS$$

The intra-office traffic is equal to

$$10\% \times 1{,}500 = 150 \; erlangs$$

The terminating traffic is the sum of the intra-office traffic and the incoming traffic, that is,

$$150 + 200 = 350 \; erlangs$$

Hence, the BHC traffic per line is given by

$$\frac{1{,}500 + 200}{10{,}000} = 0.17 \; erlangs \; per \; line$$

and the CCS traffic per line is given by

$$\frac{1{,}500 + 350}{10{,}000} \times 36 = 6.66 \; CCS \; per \; line$$

Note that the BHC traffic per line is equivalent to $0.17 \times 36 = 6.12$ CCS per line.

3-4-1. Switching Matrices. Switching matrices are used to inter-connect inlet and outlet terminals in a switching network. In the design of switching matrices, the number of inlets and outlets, the specified blocking probability and traffic characteristics, the cost of manufacturing, and the cost of control should be taken into consideration.

For space-division switching matrices, the building block is the switch array. Figure 3-7 shows a typical switch array for the connection of n inlets and m outlets (full availability).

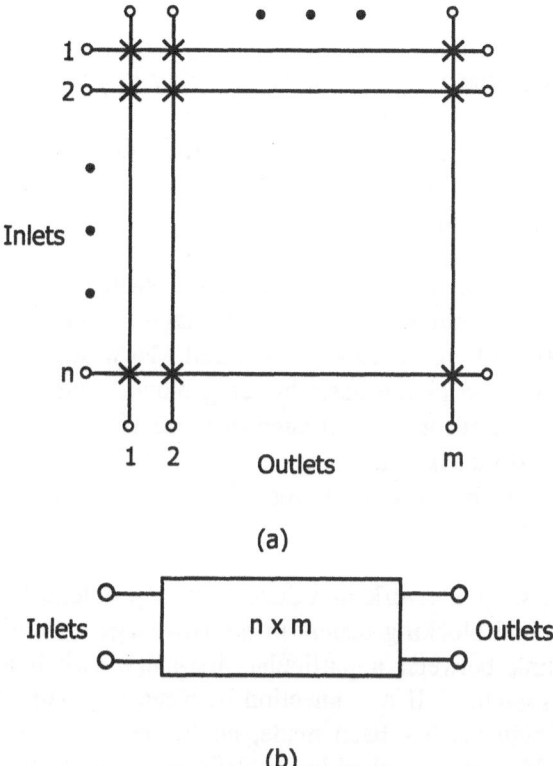

Figure 3-7. (a) Typical switching matrix.
(b) Representation of an $n \times m$ switching matrix.

The one-stage matrix in Figure 3-7 has nm crosspoints. If the traffic offered per inlet is a erlangs, then the total traffic offered to the switch is na erlangs and the traffic carried per outlet is na/m erlangs which is also the occupancy per outlet.

A one-stage square matrix with an equal number of inlets and outlets has the property of nonblocking, because a connection between a free inlet and a free outlet can always be made, irrespective of other conditions in the switch. However, nonblocking switching systems are unnecessarily costly because terminals are used for only some proportion of the time. Nonblocking switches provide unnecessary switching capacity.

3-4-2. Multistage Networks. It is possible to reduce the total number of crosspoints if the switching network is constructed from a number of stages consisting of smaller matrices. Generally, a switching network may consist of distribution stages, routing stages, and collection stages. A distribution stage increases the number of paths from an inlet to an outlet through the network. Corresponding to the distribution stage, a collection stage must be used to collect the distributed paths. A routing stage also increases the number of paths within the network, which has no corresponding collection stage, so that paths from a routing stage are directed to different outlets.

To illustrate the idea, let us consider a switching network with 100 inlets and 100 outlets. If a one-stage matrix is used, then $100 \times 100 = 10,000$ crosspoints are required. Now suppose that a 10×10 square matrix is used as the basic building block. One matrix can connect any of 10 inlets to 10 outlets. If each of these 10 outlets is connected to a different 10×10 matrix, each original inlet can then have access to a total of 100 outlets. In this case, two crosspoints have to be operated to make a connection.

This two-stage network introduces a new problem, known as internal blocking. Internal blocking occurs in the two-stage network, because there is only one link between a particular first-stage switch and each of the second-stage switches. If a connection between any pair of first-stage and second-stage switches has been made, no further access between that pair will be possible. Internal blocking is defined as the probability that a free inlet fails to be connected to a free outlet because of the absence of a free path (link) between them. The complete arrangement of this two-stage network is shown in Figure 3-8.

The two-stage network in Figure 3-8 consists of a total of twenty 10×10 basic switches, so that the total number of crosspoints is $20 \times 10 \times 10 = 2,000$ instead of 10,000 as in the case of one-stage network.

A larger network can be constructed by adding a further stage. Figure 3-9 shows a construction of a three-stage network with 1000 inlets and 1000 outlets. The first ten 100×100 two-stage sub-networks are connected to 100 separate third-stage switches. These then provide access to a total of 1000 outlets. The extra inlets on the third-stage switches can now be used by nine other 100×100 two-stage sub-networks.

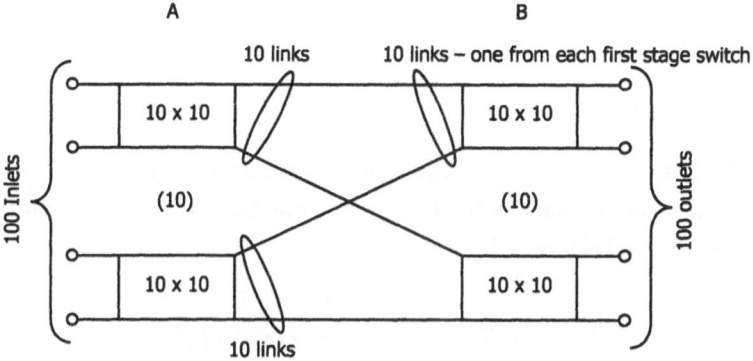

Figure 3-8. Two-stage network.

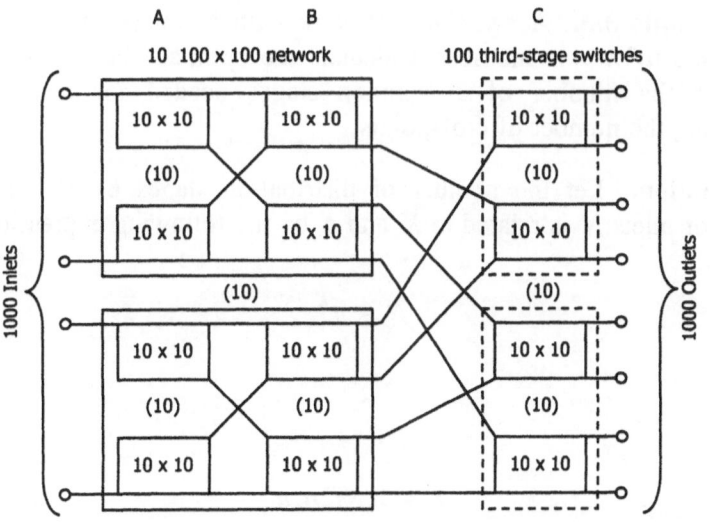

Figure 3-9. Three-stage network.

Continuing this arrangement for K stages, each stage consisting of $n \times n$ matrices as basic switches, gives a network with N inlets and N outlets, where $N = n^K$. In this general arrangement, each stage of switches is called a distribution stage because outlets are distributed over matrices in the next stage. Note that each stage consists of the same number of matrices, N/n.

Example 3-2. Consider the two-stage switching network of N inlets and N outlets. If $n \times n$ basic matrices are used, calculate the number of crosspoints in the network.

Solution. Since each matrix has n^2 crosspoints and there are two stages, each stage consisting of N/n matrices, the total number of crosspoints in the network is

$$X = 2 \times \frac{N}{n} \times n^2 = 2nN \ crosspoints$$

For a two-stage network, $N = n^2$; thus, the network requires a total of $2n^3$ crosspoints.

Example 3-3. A switching network with N inlets and N outlets is constructed by $n \times n$ matrices. Calculate the optimal size of the matrices used and the number of distribution stages needed for $N = 1000$ by minimizing the number of crosspoints.

Solution. Let the number of distribution stages be K. Then the number of inlets N is related to K and n by the following expression

$$N = n^K \qquad\qquad (3\text{-}1)$$

Thus

$$K = \ln N / \ln n \qquad\qquad (3\text{-}2)$$

Since the number of crosspoints X in a K-stage network is equal to the product of the number of stages, K, the number of matrices per stage, N/n, and the number of crosspoints per matrix, then

$$X = K \frac{N}{n}(n \times n) = nN \ln N/\ln n \tag{3-3}$$

Differentiating X with respect to n and equating to zero gives

$$\frac{dX}{dn} = N \ln N \left[\frac{\ln n - 1}{(\ln n)^2} \right] = 0$$

or $\ln n = 1$.

Hence, the optimal size n of the matrix is

$$n = e = 2.718$$

or the optimal size of the basic matrix is 3×3.

For a 1000-inlet and 1000-outlet network, the required number of stages will be

$$K = \ln N/\ln n = \ln 1000/\ln 3$$

$$= 6.28 \; or \; 7 \; stages$$

For $K = 7$ and $n = 3$, this arrangement yields a network of $N = n^K = 3^7 = 2187$ inlets and 2187 outlets. This network uses a total of $X = KnN = 7 \times 3 \times 2187 = 45,927$ crosspoints. Note that in this analysis the problem of internal blocking has been ignored. When internal blocking is taken into consideration the optimal arrangement will be based on a larger matrix.

3-4-3. Link Systems. The method of connecting the inlets and outlets via multistage of matrices is called the link system. A formal definition of the operation of the link system is as follows.

1. A connection between an inlet and an outlet in a link system is formed by one or more links;

2.   The link or links are seized at the same time as the chosen outlet;

3.   Only the link or links which connect to the chosen outlet are seized (condition selection).

A switching network may consist of several stages and many links and paths. In order to calculate the blocking probability, it is convenient to use a diagram known as *the network graph* to show the pattern of links. In the design and analysis of multi-stage networks, an even simpler graph is drawn for a particular inlet and a particular outlet. This simpler graph is known as the *channel graph*.

Network Graph. A network graph of a switching network is a graph in which each switch is represented by a dot and each link by a line joining appropriate dots in each stage.

Channel Graph. A channel graph of a switching network for a particular pair of inlet and outlet is a graph in which only the links and intermediate stages that could be used are drawn.

Internal blocking occurs in a link system when there are no idle links available for connection to an idle outlet. A convenient method for calculating internal blocking in a link system is the linear graph method developed by C.Y. Lee in 1955. A linear graph, also known as a channel graph, is a simplified network graph. In order to simplify the calculation of internal blocking in switching networks by use of channel graphs, we make the following assumptions:

1.   The occupancy of each link (the fraction of time that a link is in use) of the switching network is independent of other links;

2.   The blockings in the basic switches are negligible; and

3.   The traffic in the network is conserved.

3-4-3-1. Two-Stage Link Systems. Figure 3-10(a) shows a two-stage link system which uses $n \times m$ matrices. Each stage consists of m matrices and is thus a distribution stage. The corresponding network and channel graphs are shown in Figure 3-10(b) and (c), respectively.

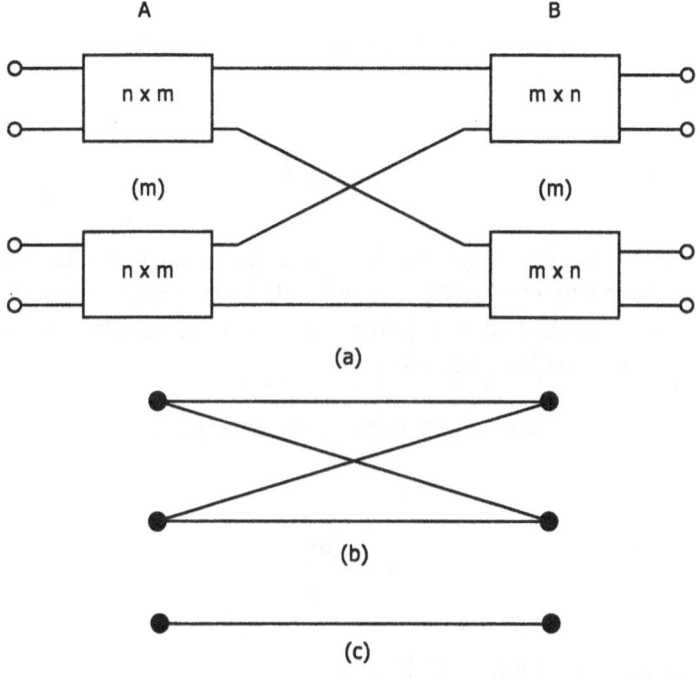

Figure 3-10. (a) Two-stage network.
                 (b) Network graph.
                 (c) Channel graph.

To calculate the internal blocking between stage A and stage B, we assume that the traffic offered per inlet is a_1 erlangs. Then the link occupancy, which is the traffic carried by a link in the network, is simply

$$a = \frac{na_1}{m} \; erlangs$$

Since there is only one possible path between any A-stage inlet and any B-stage outlet, the blocking probability is given by

$$B = a = na_1/m \qquad (3\text{-}4)$$

If the level of congestion is unacceptable, larger matrices may be used. Suppose that $n \times (2m)$ matrices are used, where two parallel paths (interstage links) are available between any A-stage inlet and any B-stage outlet. In this case, the blocking probability is reduced to

$$B = \left[\frac{na_1}{2m} \right]^2 \qquad (3\text{-}5)$$

3-4-3-2. Three-Stage Link Systems. Consider the three-stage network depicted in Figure 3-11(a) with all three stages being distribution stages. The corresponding network and channel graphs are shown in Figure 3-11(b) and (c), respectively.

If the traffic offered per inlet is a_1 erlangs, the occupancy on an $A - B$ link is

$$a = \frac{na_1}{m}$$

and the occupancy on a $B - C$ link is

$$b = k \times \frac{na_1}{m} \times \frac{1}{k} = \frac{na_1}{m}$$

Therefore, the blocking probability in a network with three distribution stages is given by

$$B = 1 - (1 - a)(1 - b) = 1 - (1 - a)^2$$

In general, if a network consists of K distribution stages in series, the blocking probability is given by

$$B = 1 - (1 - a)^{K-1} \qquad (3\text{-}6)$$

because, for a particular pair of inlet and outlet, there are $K - 1$ links connected in series, each with an occupancy of a. In the derivation of formula (3-6), it has been assumed that the link occupancies in different stages are independent.

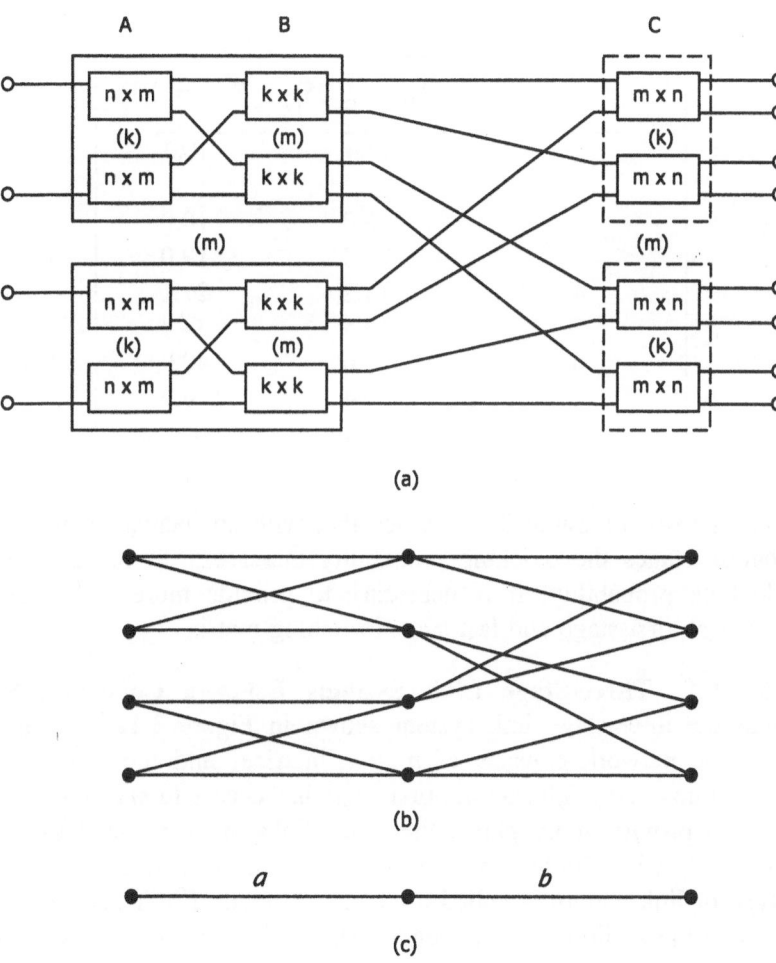

(a)

(b)

(c)

Figure 3-11.   (a) Three-stage network with distribution stages.
(b) Network graph.
(c) Channel graph.

Table 3-1 shows the blocking probabilities for networks with different numbers of stages and with link occupancies of 5% and 10%, that is, $a = 0.05$ and 0.1.

Table 3-1.   Blocking probabilities for simple distributed multi-stage networks.

No. of stages K	$B(a = 0.05)$	$B(a = 0.1)$
	(%)	(%)
2	5.0	10.0
3	9.7	19.0
4	14.2	27.1
5	18.5	34.4
6	22.6	40.9
7	26.4	46.9

From (3-6) or Table 3-1, we see that with an increasing number of distribution stages the blocking probability increases. In order to reduce the blocking probability, it is necessary to provide more than one path between each first-stage and last-stage switching matrix.

3-4-3-3.   Three-Stage Link Systems Using a Collection Stage. Consider the three-stage link system shown in Figure 3-12(a). The first stage of the network consists of $n \times m$ matrices and the second stage, $k \times k$ matrices. Any inlet on the first stage has access to mk $B - C$ links. In order to provide more paths, these mk links are connected to only k third-stage switches (rather than mk switches, as in a distribution stage). This type of link pattern is called a collection stage. From Figure 3-12(a), we see that the addition of a collection stage to a two-stage network does not increase the number of outlets accessible from a first-stage switch, but it does increase the number of paths between switches on the first and last stages.

There are now m possible paths, each consisting of two links connecting any inlet on the first stage and any outlet on the last stage. The corresponding network graph and channel graph are shown in Figure

3-12(b) and (c), respectively.

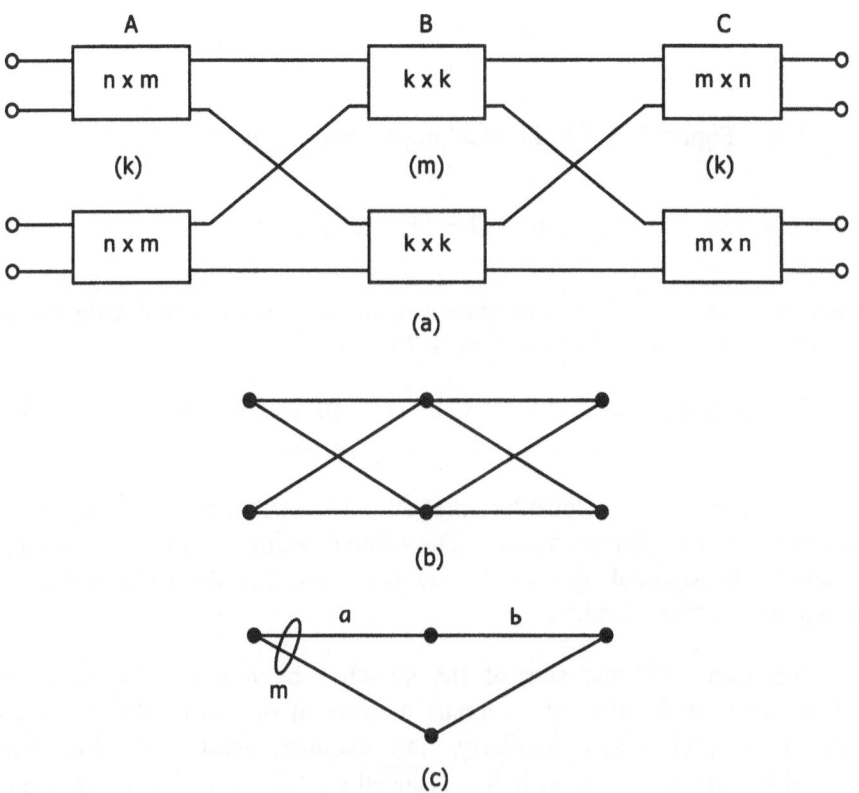

Figure 3-12. (a) Three-stage network.
                   (b) Network graph.
                   (c) Channel graph.

If the traffic offered per inlet is a_1, then the occupancy on an $A - B$ link is

$$a = \frac{na_1}{m}$$

and the occupancy on a $B - C$ link is

$$b = k \times \frac{na_1}{m} \times \frac{1}{k} = \frac{na_1}{m}$$

From Figure 3-12(c), the blocking probability is given by

$$B = [1 - (1 - a)^2]^m \qquad (3\text{-}7)$$

where $m < 2n - 1$. It will be shown in the next section that a three-stage network becomes nonblocking if $m \geq 2n - 1$.

For a three-stage network with $m = 10$ and $a = 0.1$, the blocking probability is $B = 6.13 \times 10^{-8}$, which is negligible.

Example 3-4. Consider the $N \times N$ switching network which consists of K distribution stages. The offered traffic per inlet is a erlangs. Calculate the optimal size of the switches so that the total number of crosspoints is minimized.

Solution. Let the size of the switches be $n \times n$. By taking the traffic into consideration, the average number of outlets of the network is reduced to $N(1 - a)$. Similarly, the average number of free links accessible from an $n \times n$ switch is reduced to $n(1 - a)$. Hence, by taking the busy links into account, the effective size of a switch appears to be $n(1 - a) \times n(1 - a)$. Therefore, the average number of outlets accessible from a first-stage switch becomes

$$N(1 - a) = [n(1 - a)]^K \qquad (3\text{-}8)$$

It follows that the number of stages is given by

$$K = \frac{\ln[N(1-a)]}{\ln[n(1-a)]} \tag{3-9}$$

and the number of crosspoints used in the network is

$$X = K(\frac{N}{n})(n \times n) = Kn \ N = Nn\frac{\ln[N(1-a)]}{\ln[n(1-a)]}$$

Differentiating X with respect to n and setting the result to zero yields

$$\ln[n(1-a)] = 1$$

or

$$n = \frac{e}{1-a} \tag{3-10}$$

Suppose that $a = 0.1$ erlangs and $N = 1000$. Using (3-10) we have

$$n = \frac{e}{1-0.1} = 3.02 \text{ or } 4$$

Thus the optimal size of switch is 4×4.

By (3-9), we obtain the number of stages

$$K = \frac{\ln[1000(1-0.1)]}{\ln[4(1-0.1)]} = 5.3125 \text{ or } 6$$

Thus the number of stages needed is 6. With $n = 4$ and $K = 6$, the actual number of outlets is

$$N = 4^6 = 4096$$

and the number of crosspoints in the network is given by

$$X = KnN = 6 \times 4 \times 4096 = 98{,}304$$

Upon comparison with the result of Example 3-3, we see that by taking the traffic $a = 0.1$ into account, the size of switches is slightly larger and the number of crosspoints is twice as many as when the traffic is not considered.

3-5. NONBLOCKING NETWORKS

It is possible to build a multi-stage network in which no internal blocking can occur and which still uses fewer crosspoints than a square matrix. This problem was solved by C. Clos in 1953 with three stages. Figure 3-13 shows a Clos-type three-stage nonblocking network.

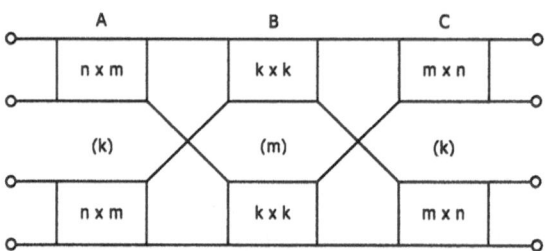

Figure 3-13.   Three-stage nonblocking (Clos) network.

The first stage or A-stage comprises k switches with n inlets each, and the last stage or C-stage comprises k switches with n outlets each. The network has $N = nk$ inlets and N outlets. The second stage or B-stage comprises m switches, each with k inlets and k outlets, so that each B switch has one link to each A and C switch.

For nonblocking in stage A, m must not be less than n. However, this condition is insufficient to ensure the absence of blocking in later stages. Consider the worst possible case for setting up a connection between a pair of inlet and outlet as shown in Figure 3-14. Suppose a free inlet of switch A_1 requires connection to a free outlet of switch C_1. If the remaining $n-1$ inlets of switch A_1 are already connected to outlets of switches other than switch C_1 and the remaining $n - 1$ outlets of switch C_1 are connected to inlets of switches other than switch A_1, then $(n - 1) + (n - 1) = 2n - 2$ switches of stage B are unavailable to the new call. Hence, the minimum number of switches in stage B to ensure nonblocking is $m = 2n - 1$. The required nonblocking network then has k switches in stages A and C, each $n \times (2n - 1)$ and $2n - 1$ switches in stage B, each $k \times k$.

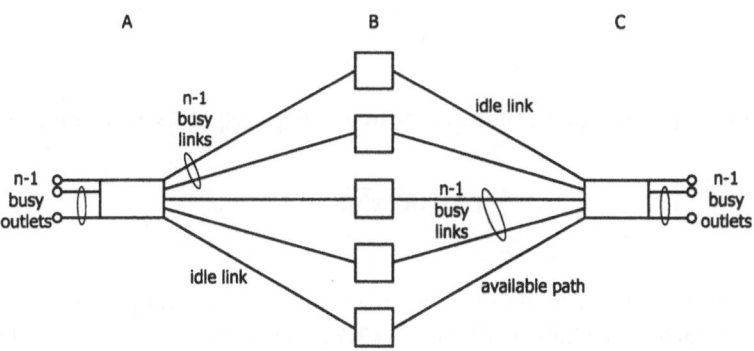

Figure 3-14. Illustration of three-stage nonblocking (Clos) network.

The total number of crosspoints is therefore given by

$$X = k\ n\,(2n - 1) + mk^2 + k\ n\,(2n - 1)$$

$$= (2n - 1)\,(2N + k^2)$$

$$= (2n - 1)\,N\,(2 + N/n^2)$$

where $N = kn$ is the number of inlets or outlets of the network. For a given value of N, the minimum number of crosspoints can be determined by differentiating X with respect to n and letting it be zero, i.e.

$$\frac{dX}{dn} = 0$$

or

$$2n^3 - Nn + N = 0$$

Thus, for large N or $n \gg 1$,

$$N = \frac{2\,n^3}{n - 1} \approx 2\,n^2$$

Taking $n = (N/2)^{1/2}$, the minimum number of crosspoints is approximately

$$X = 4N\,(\sqrt{2N} - 1) \tag{3-11}$$

Note that it is possible to construct a 5-stage nonblocking network by replacing each B switch in Figure 3-13 by a complete 3-stage nonblocking network. In the same way, a 7-stage nonblocking network can be constructed by replacing each C switch on the third stage in the 5-stage network by a 3-stage network.

Example 3-5. Design a three-stage nonblocking network capable of switching any of 1000 inlets to any of 1000 outlets. Suppose that the 1000 inlets and 1000 outlets are both served by 100 switches. Calculate the total number of crosspoints required.

Solution. Since the number of inlets per A-stage switch is $n = \dfrac{N}{k} = \dfrac{1000}{100} = 10$, for the network to be nonblocking, the size of switches on A-stage is 10×19. The B-stage has 19 switches, each of size 100×100. The C-stage is a collection stage consisting of 100 switches, each of size 19×10. The complete nonblocking network is shown in Figure 3-15. The total number of crosspoints is

$$X = (2n - 1)N(2 + \frac{N}{n^2}) = (20 - 1)1000(2 + \frac{1000}{10^2}) = 228 \times 10^3$$

Note that in this simple example, the saving in crosspoints is considerably larger as compared with the 10^6 crosspoints in a simple square switch. However, the saving in crosspoints becomes more prominent when N is larger.

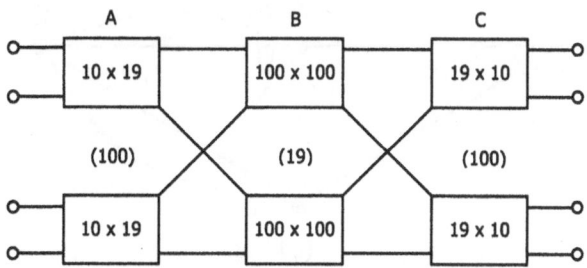

Figure 3-15. Three-stage nonblocking (Clos) network.

3-6. THREE-STAGE NETWORKS WITH RETRIALS

If a call is connected to a trunk group, any idle trunk in the group can serve the purpose. If the first attempted trunk is busy, then retrials may be initiated. This is done by selecting another outlet and reattempting to make a connection. In most cases, the inlet will make use of the same set of $A - B$ links but will attempt to match with a different set of $B - C$ links as illustrated in the network graph and the equivalent channel graph in Figure 3-16(a) and (b), respectively.

For a call from an inlet to x idle outlets to be blocked, it is necessary that either the $A - B$ link is blocked, or the $A - B$ link is not blocked but all the x trials failed. This situation is equivalent to that of having x $B - C$ links, as shown in Figure 3-16(b). The blocking probability is

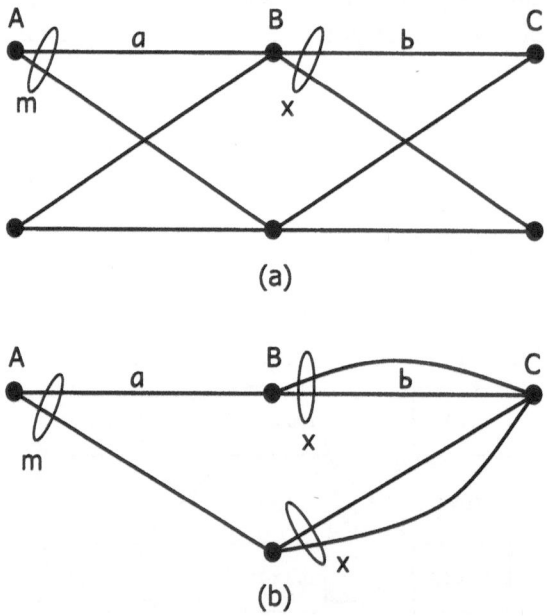

(a)

(b)

Figure 3-16.   (a) Network graph for three-stage network with retrial.
                     (b) Equivalent channel graph.

given by

$$B = [1 - (1 - a)(1 - b^x)]^m \qquad (3\text{-}12)$$

where x is the number of trials, and a and b are the occupancies of the A - B link and $B - C$ link, respectively.

3-7. CONGESTION IN TIME-DIVISION SWITCHING NETWORKS

Time-division switching is a technique of sharing the crosspoints of a switch for shorter periods of time so that individual crosspoints and their associated interstage links are continually reassigned to existing connections. It is equally applicable to either analog or digital signals.

To illustrate the basic principle of time-division switching, we consider the simple analog time-division switching structure shown in Figure 3-17, where a single switching bus is shared by a number of connections. The pulse amplitude modulation samples from receive line interfaces are interleaved to transmit line interfaces. The first cyclic control controls gating of inputs onto the bus, one sample at a time. The second cyclic control operates in synchronism with the first and selects the appropriate output line for each input sample. A complete set of pulses, one from each active input line, is referred to as a frame. The frame rate is equal to the sample rate of each line.

The analog switching matrix in Figure 3-17 is essentially a space-division switching matrix. By continually changing the connections for short periods of time in a cyclic manner, the configuration of the space-division switch is replicated once for each time slot. This mode of operation is referred to as time multiplexed switching.

Usually digital time-division multiplexed signals require switching between time slots, as well as between physical lines. This mode of switching consists of a second dimension of switching and is referred to as time switching. Figure 3-18 shows the basic requirement of a time-division switching network, where channel 3 of the first TDM link is connected to channel 17 of the last TDM link. The return connection is required and realized by transferring information from time slot 17 of the

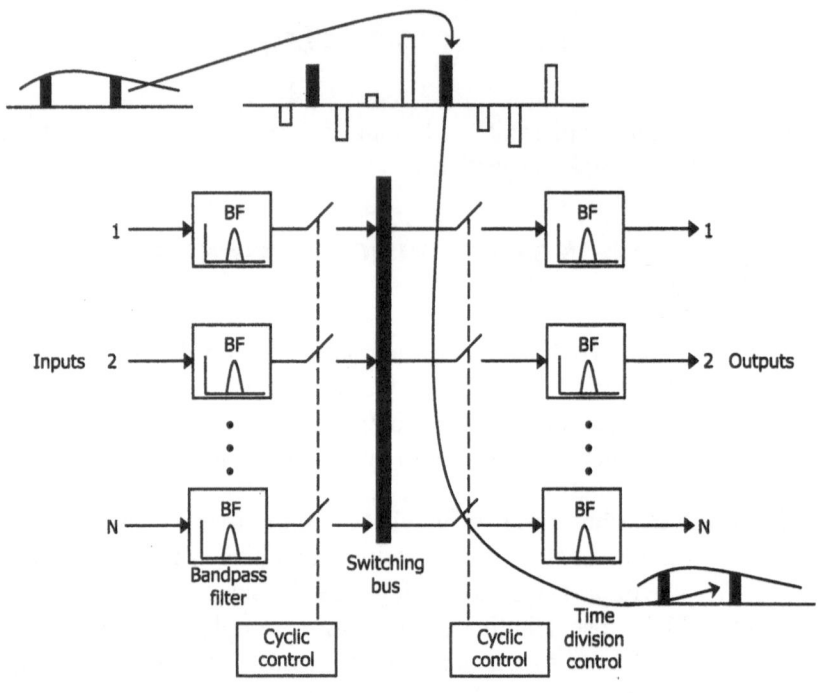

Figure 3-17. Analog time division switching.

last input link to time slot 3 of the first link. Thus each connection involves translations in both time and space.

Time Slot Interchange. The time slot interchange (TSI) unit is the basic building block of many time-division switches. A TSI unit operates on a synchronous TDM stream of time slots, or channels, by interchanging pairs of time slots. This is done by writing data into and reading data out of a digital memory. Figure 3-19(a) shows the basic functional operation of a TSI and Figure 3-19(b) depicts a mechanism for TSI.

The N input lines are synchronously multiplexed to produce a TDM stream with N time slots. To make the interchange of any two slots, the incoming data in a slot must be stored in the memory until they can be sent out on the right channel in the next frame cycle. Hence, the TSI introduces a delay and produces output slots in the desired order. These

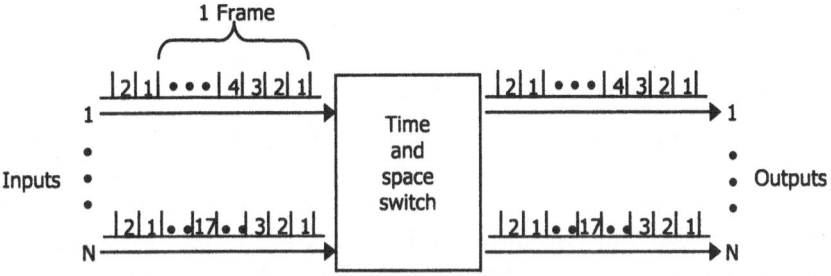

Figure 3-18. Time- and space-division switching [3].

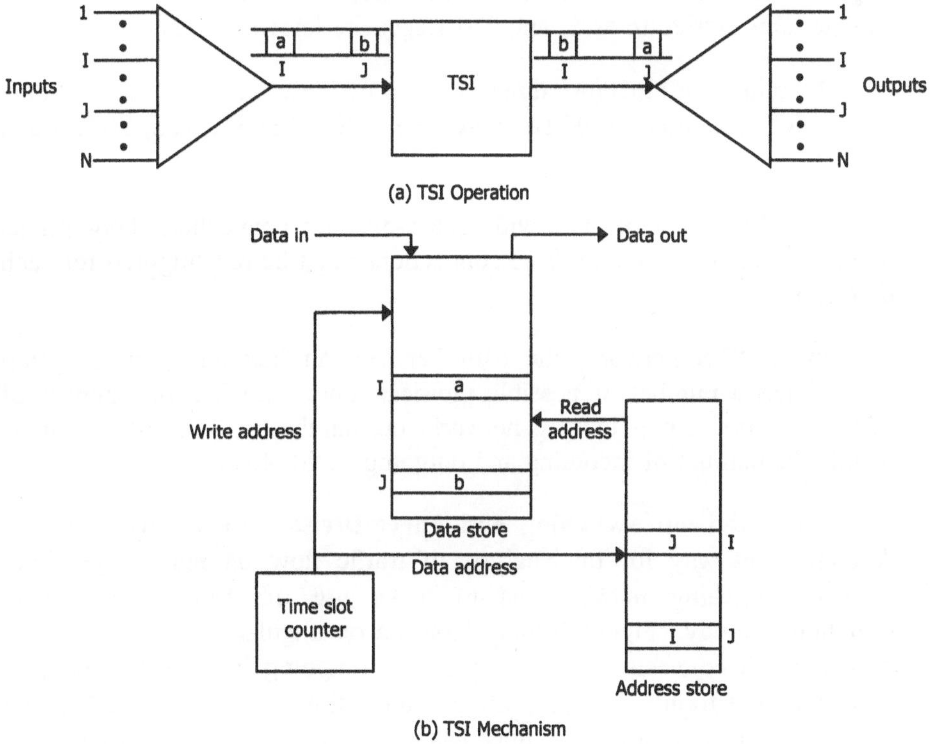

(a) TSI Operation

(b) TSI Mechanism

Figure 3-19. Time slot interchange.

are then demultiplexed and routed to the appropriate output line. Since the frame must provide a time slot for each channel, the size of the TSI unit must be chosen for the capacity of the TDM line.

TSI is a simple, effective way to switch TDM data. However, the size of TSI is limited by the memory access speed. Thus, a TSI unit can support only a limited number of connections.

Time Multiplexed Switching. For a fixed access speed, as the size of the TSI unit grows, the delay at the TSI increases. To overcome these problems, multiple TSI units are used. To interchange time slots between two TDM streams of two different TSI units, some form of space-division multiplexing is needed. This operation is known as time multiplexed switching (TMS).

Multi-stage networks can be built up by connecting TMS and TSI stages: TMS stages, which move time slots from one TDM stream to another, are referred to as S, and TSI stages are known as T.

To minimize blocking, three or more of these stages are used. Some of the more commonly used structures are: TST, TSSST, STS, SSTSS and TSTST.

The TMS unit must provide space-division connections between its input and output lines, and these connections must be reconfigured for each time slot.

In an STS network, the path between an incoming and outgoing channel has a number of possible physical routes equal to the number of TSI units. For a nonblocking network, the number of TSI units must be double the number of incoming and outgoing TDM streams.

Time-Division Switching and Space-Division Switching Analogy. A convenient way for the analysis of traffic flow on multi-stage time-division switching networks is to make use of their space-division switching analogy. Figure 3-20(a) depicts a rectangular $n \times m$ time-division switch which connects n incoming and m outgoing lines, each having k time slots in a frame. In this configuration, a time slot of the TDM stream in any one of the n incoming lines has access to the same time slot in any one of the m outgoing lines. Thus, the $n \times m$ time-division switch is functionally equivalent to k separate $n \times m$ space-division switches as shown in Figure 3-20(b).

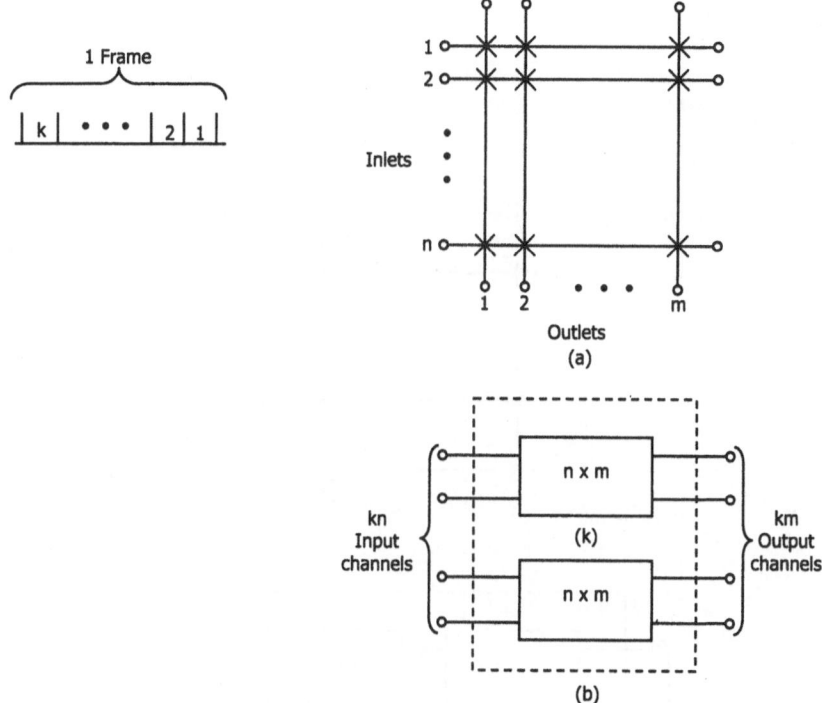

Figure 3-20.   (a) Time-division switch.
(b) Space-division equivalent of (a).

Note that no connection between time slots of different frames is possible with the time-division switch in Figure 3-20(a). However, we can obtain connections between time slots of different frames by connecting the n incoming time-division lines to the m outgoing time-division lines by way of r intermediate time-division lines as shown in Figure 3-21(a). The space-division equivalent of this configuration is shown in Figure 3-21(b).

With this configuration, there is no distribution between the k equivalent space-division switches. In order to transfer data from one time

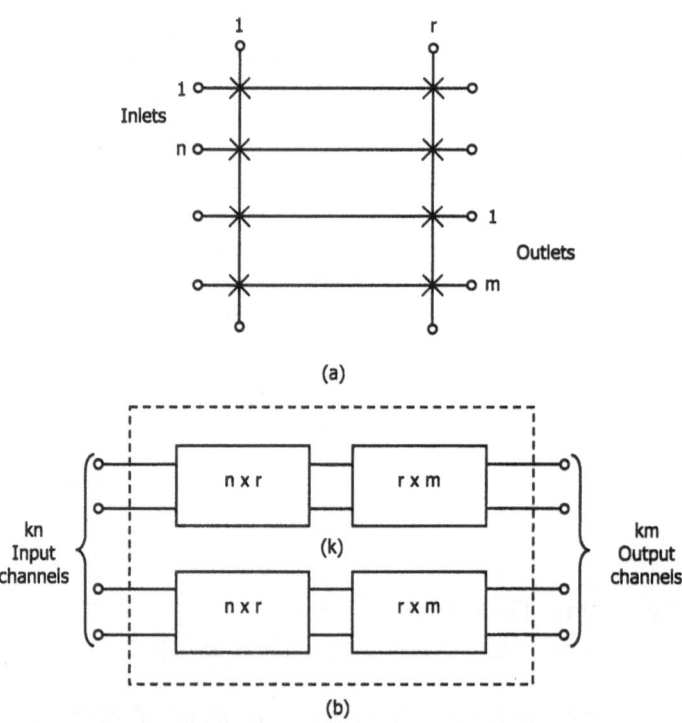

(a)

(b)

Figure 3-21. (a) Time-division switch.
(b) Space-division equivalent of (a).

slot to another time slot, a pulse shifter capable of interchanging an arbitrary time slot at its input to an arbitrary time slot at its output may be employed. Using pulse shifters in the switch in Figure 3-20(a), the preassigned time slot in its first time-division stage may be shifted to any one of the k time slots, so that the network becomes equivalent to a three-stage space-division network as shown in Figure 3-22(a).

Similarly, by employing pulse shifters in the second time-division stage in Figure 3-21(a), we obtain the network shown in Figure 3-23(a).

The networks shown in Figures 3-22 and 3-23 are two basic configurations for time-division switching networks. The first configuration is called the T-S-T type and the second configuration is called the S-T-S type.

Consider the T-S-T (time-space-time) switch in Figure 3-24(a) which is completely analogous to the three-stage space-division switch in Figure 3-12. The corresponding channel graph is shown in Figure 3-24(b).

From the channel graph, we see that there are in total m parallel paths, each having a blocking probability of $1 - (1 - a)^2$, where $a = na_1/m$ and a_1 is the offered traffic per input channel. It follows that the blocking probability is

$$B = \left[1 - (1 - a)^2\right]^m, \quad m < 2n - 1 \tag{3-13}$$

Note that the crosspoint settings of the space switch are changed, each of the m time intervals corresponding to the m time slots of TSIs. The resultant space switch is called a time-multiplexed switch.

If $m \geq 2n - 1$, the switching network becomes nonblocking. However, practical limitations of memory speed limit the size of a time switch, so that some space-division switching is also necessary in large switching networks.

Example 3-6. Calculate the blocking probability of the S-T-S switch in Figure 3-25(a) for $n = 100$, $m = 10$ and the offered traffic per inlet $a_1 = 0.02$ erlangs.

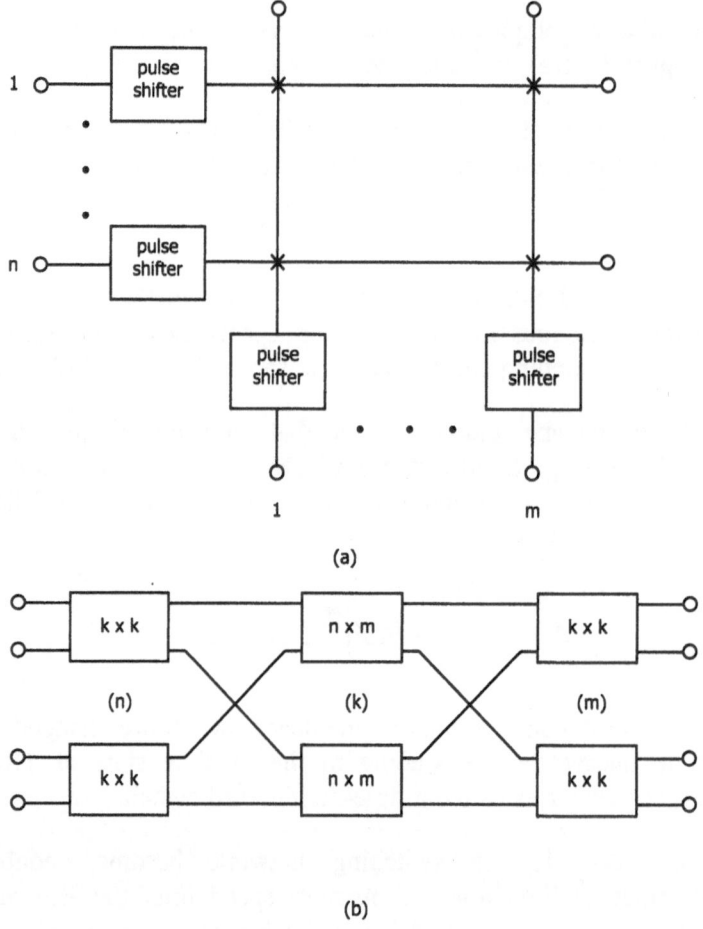

(a)

(b)

Figure 3-22. (a) Time-division switch with pulse shifters in its input and output stages.
(b) Space-division equivalent of (a).

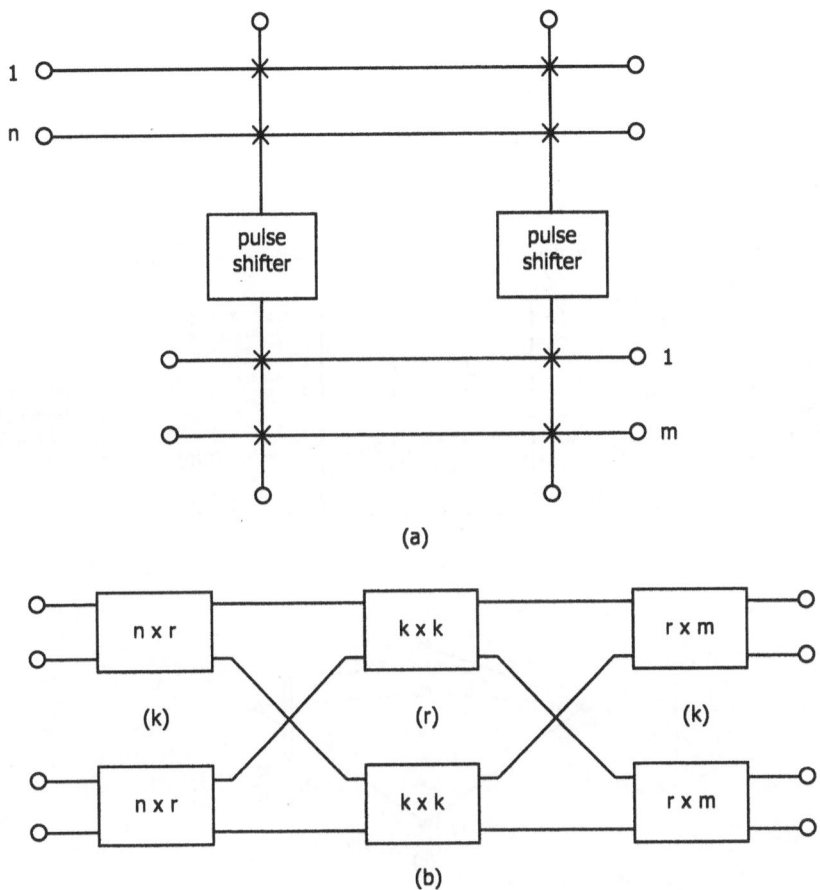

(a)

(b)

Figure 3-23.   (a) Time-division switch with pulse shifter
                              in its second stage.
                         (b) Space-division equivalent of (a).

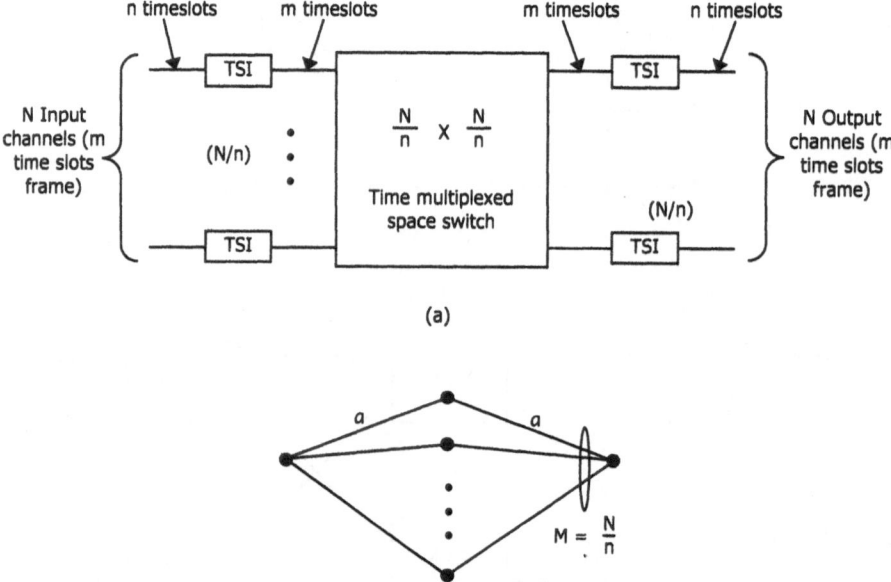

Figure 3-24. (a) T-S-T switching network.
(b) Channel graph.

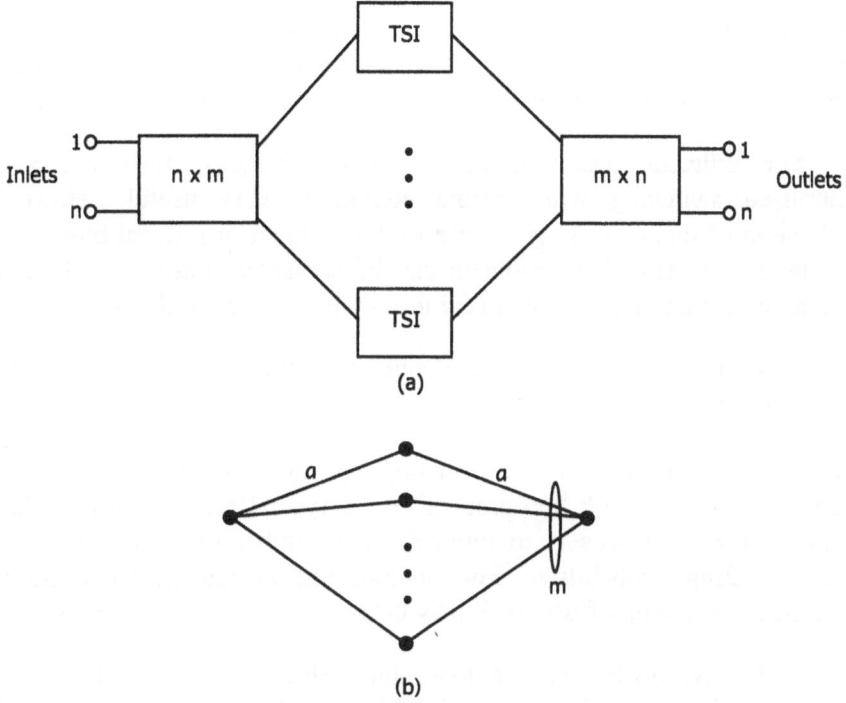

Figure 3-25. (a) S-T-S switch.
(b) Channel graph.

Solution. Since the channel graph for the S-T-S switch in Figure 3-25(b) is exactly the same as that of Figure 3-24(b), formula (3-13) applies. The link occupancy is

$$a = \frac{na_1}{m} = \frac{100 \times 0.02}{10} = 0.2$$

Thus, using (3-13) we obtain the blocking probability

$$B = [1 - (1 - 0.2)^2]^{10} = 3.66 \times 10^{-5}$$

3-8. SUMMARY

This chapter first presents the basic functions required for public telephone networks and then discusses the concepts of circuit switching and store-and-forward switching. Circuit switching is used exclusively in public telephone networks and store-and-forward switching is used in computer communication networks.

For efficient operation and economic reasons, the concept of centralized switching with remote control is very useful. However, application of this concept gives rise to the problem of internal blocking in switching networks. This problem can be solved by trading off between switching cost and transmission cost to minimize the overall cost.

Like signal multiplexing, switching techniques can be divided into space-division, time-division, and frequency-division switchings. In practice, almost all switching networks are multi-stage networks which consist of a concentration, a distribution, and an expansion stage and are subject to blocking with low probability. A typical problem in the design of multi-stage networks is to minimize the number of crosspoints for a given blocking probability. For nonblocking switching networks, the three-stage Clos-type of networks may be used.

Under reasonable assumptions, the calculation of blocking in a switching network can be greatly simplified by the use of network graphs and channel graphs. Several examples are presented for illustration. In this respect the method developed by Lee is found to be very useful.

REFERENCES

[1]   Hills, M.T., Telecommunications Switching Principles, Cambridge, Mass.: MIT Press, 1979.

[2]   Bellamy, J., Digital telephony, 2nd ed., New York, John Wiley & Sons, 1991.

[3]   Stalling, W., Local Networks, 3rd ed., New York, Macmillian Publishing Co., 1990.

[4] Lee, C.Y., "Analysis of Switching Networks", Bell System Tech. J., Vol. 34, 1955, pp. 1287-1315.

[5] Clos, C., "A Study of Nonblocking Switching Networks", Bell System Tech. J., Vol. 32, 1953, pp. 406-424.

[6] Personick, S.D. and Fleckenstein, W.O., "Communications Switching - From Operators to Photonics", Proc. IEEE, Vol. 75, No. 10, October 1987, pp. 1380-1403.

[7] Hebuterne, G., Traffic Flow in Switching Systems, Norwood, MA: Artech House Inc., 1987.

[8] Ahmadi, H. and Denzel, W.E., "A Survey of Modern High-Performance Switching Techniques", IEEE J. Select. Areas Commun., Vol. SAC-7, No. 7, September 1989, pp. 1091-1103.

[9] Broomell, G. and Heath, R.J., "Classification Categories and Historical Development of Circuit Switching Topologies", Computing Surveys, Vol. 15, No. 2, June 1983, pp. 95-133.

PROBLEMS

3-1.   Consider a multiple stage $N \times N$ switching network in which there are K-1 distribution stages and 1 collection stage (the last stage). The network is comprised of $n \times n$ matrices. Determine the optimal size of the matrices such that the total number of crosspoints in the network is minimized. If $N = 1000$, what is the total number of crosspoints?

3-2.   Design a multiple stage $N \times N$ switching network in which there are K - 1 distribution stages and 1 collection stage (the last stage). The offered traffic per inlet is a erlangs and the switch size is $n \times n$. Taking the traffic into account, what is the optimal size of n such that the total number of crosspoints in the network is minimized? If the number of outlets of the network is 1000 and the offered traffic per inlet is 0.05 erlangs, what is the total number of crosspoints in the network?

3-3.   Consider the three-stage switching network as shown in Figure 3-26. Draw the network and channel graphs for the switching network. If $n = 100$, $m = 10$ and the offered traffic per inlet is

0.005 erlangs, calculate the internal blocking in the network.

3-4.   Consider the four-stage network with three distribution stages and one collection stage as shown in Figure 3-27. Draw the network and channel graphs and calculate the internal blocking for $n = 100$, $m = 10$ with the offered traffic per inlet being 0.01 erlangs.

3-5.   Consider the network graph of a switching network as shown in Figure 3-28. If the offered traffic per inlet is 0.02 erlangs and the size of the basic matrices is 10×10, what is the internal blocking?

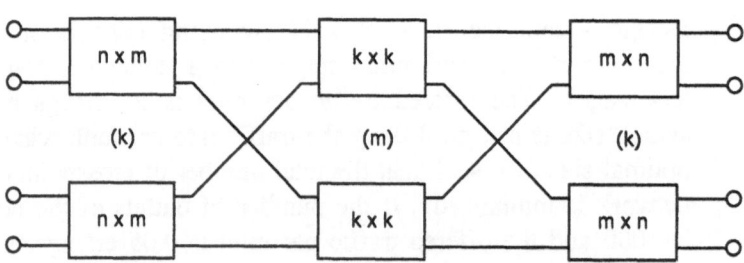

Figure 3-26. N-inlet network with concentration.

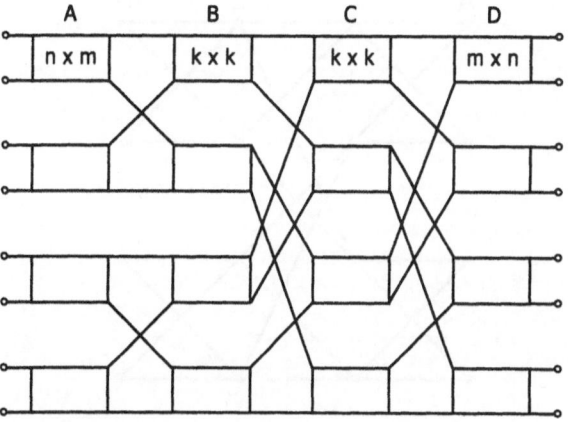

Figure 3-27. Four-stage network.

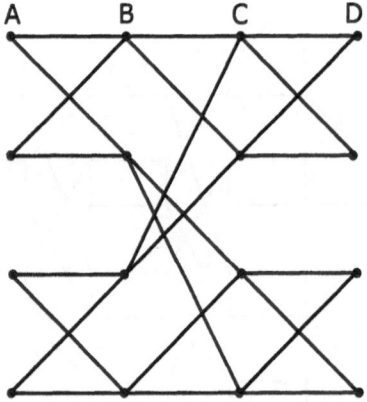

Figure 3-28. Network graph.

CHAPTER 4

MODELING OF TRAFFIC FLOWS, SERVICE TIMES AND SINGLE-SERVER QUEUES

4-1. INTRODUCTION

This chapter discusses the mathematical modeling techniques for input processes or traffic flows and for service times. It also presents the fundamental concepts of single-server queues which have wide applications in teletraffic engineering and in computer communication systems. In the study of teletraffic engineering, it is necessary to not only model the traffic flow and the service time, but also to analyze their statistical characteristics. The model of a physical quantity is a mathematical expression that reasonably and accurately represents the behavior of the quantity under consideration. A physical quantity (or system) may be represented in many different ways and therefore may have many models, depending on the perspective of the investigation.

The traffic flow and the service times of calls or messages in many telecommunication systems, such as telephone systems, data processing systems, computer communication networks and so on, may be described in terms of probability distributions. Such distributions may be obtained by means of probabilistic laws governing a particular circumstance; the methods of the Markov chain and the imbedded Markov chain are appropriate examples.

Mathematical Models. The first step in the analysis of a telecommunication system is to derive its mathematical model. It is important to note that deriving a reasonable mathematical model for the traffic flow, the service time, and the system is the most important part of the performance analysis.

Mathematical models, in general, may assume many different forms. Depending on the particular system and on the particular circumstance, one mathematical model may be more suitable than others. For example, in telephone systems, it is usually assumed that the incoming traffic flow is a Poisson process and that the holding time distribution is exponential. However, when the number of traffic sources is finite, a quasi random input process may be more suitable.

The primary function of a telecommunication system is to provide communication paths between pairs of users upon demand. Since only a small portion of users will be using the facility at any one time, the provision of a permanent communication path between every pair of users would be expensive and unnecessary. Thus, in practice, telecommunication systems are designed in such a way that facilities needed to establish and maintain a communication path between a pair of users are provided in a common pool; these facilities are to be used by a user when required and returned to the pool when no longer needed. However, this arrangement introduces the possibility that the system may be unable to set up a path on demand because of lack of available equipment at that time. Therefore, a question often immediately arises: for a given amount of offered traffic, how much equipment must be provided so that satisfactory service at a reasonable cost can be maintained?

In computer communication systems, small computers are used as communication processors or as message concentrators. They accept messages from data terminals or from computers connected to them, they recode message characters, and they buffer messages while waiting for transmission. A basic question often arises: what size of buffers should be used so that the message loss or blocking probability will be within specified design limits?

In order to investigate this problem, it is necessary to take into account the influence of chance elements on the operation of the telecommunication system. The incoming traffic is not, as a rule, constant, but undergoes occasional fluctuation. Moreover, the time of serving the users is always subject to fluctuation from one case to another. These chance elements may have different characteristics which can be studied using the concepts and methods of the theory of probability.

Characteristics of Queueing Models. Consider the following phenomenon. Customers arrive and request the use of a certain type of equipment (servers). If a server is available, the arriving customer will seize and hold it for some length of time, after which the server will become idle and be available to other customers. If the arriving customer finds no available server, he will wait in a line (queue) or leave immediately. This phenomenon may be regarded as a queueing system shown in Figure 4-1.

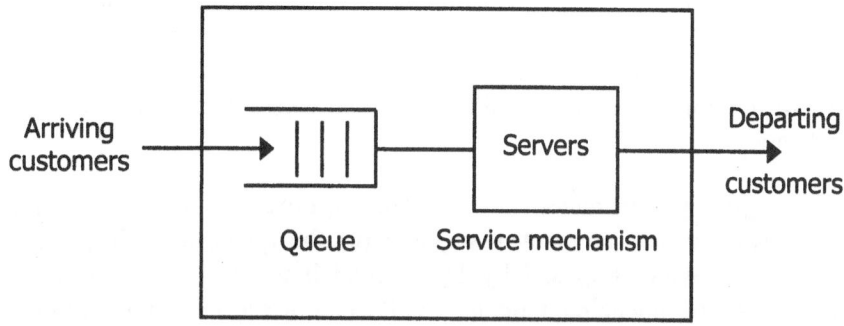

Figure 4-1. A queueing system.

Under certain idealized conditions, many queueing systems may be characterized by random processes such as the input process and the service times of customers. The statistical behavior of a queueing system may be obtained by relating these random processes. The mathematical description of the queueing system characteristics is called a queueing model. The first step in the analysis of a queueing system is to derive its model. The derivation of a reasonable model is the most important part of the entire analysis.

Once a mathematical model of the queueing system is obtained, various analytical and computational tools can be used for analysis and synthesis purposes. In obtaining a model, we must make a compromise between the simplicity of the model and the accuracy of the results of the analysis. Usually, the results obtained from the analysis are valid only to the extent that the model approximates a given real system or phenomenon.

Digital computer simulation provides a method of analysis when mathematical analysis is unmanageable, and it allows the analyst to model a real system in a more detailed and accurate manner than mathematical analysis would permit. However, simulation does not yield the same insight and information that a mathematical model would yield if it could be solved. But in cases where extreme accuracy is not required, it is preferable to obtain only a reasonably simplified model.

The theory of queues is concerned with the analysis of mathematical models representing real queueing processes. The development of queueing theory has its origins in the study of congestion in telephone systems. In this chapter, we shall see that many of the assumptions made in queueing studies are precisely those which seem reasonable and valid in obtaining models of queueing processes associated with the telephone systems. However, these assumptions may not always be applicable to the study of computer communication systems, where we may need some modifications.

The principal characteristics of queueing processes are: (a) the input process, (b) the service mechanism, and (c) the queue discipline. These characteristics were proposed by D.G. Kendall in 1951 and now widely used to describe queueing models. In Kendall's short hand notation, we use A/B/C, where A specifies the arrival (input) process, B specifies the service time, and C is the number of servers. Also, a modified form of Kendall's notation A/B/C/D/E has been used, where D specifies the maximum number of customers who may be present in the system at any one time (including those being served), and E identifies the queue discipline.

Some examples of Kendall's notation are:

M/M/1   represents a queueing model with Poisson (Markov) or random input, exponential (Markov) service times, and one server;

M/G/1   denotes a queueing model with Poisson input, general (arbitrary) service times and one server;

$GI/E_k/s$ denotes general, independently distributed interarrival times, Erlangian service times and s servers.

In theoretical studies of queueing systems, it is always assumed for mathematical convenience that the waiting capacity is infinite. In addition to Kendall's terminology, the queue discipline is usually specified separately. Certain notations are commonly used:

FIFO   represents first into the queue and first out of it into service, or first-come, first-served discipline, or service in order of arrival;

SIRO   means service in random order, or customers are selected randomly from the queue to obtain service;

LIFO   denotes last-come, first-served discipline.

Other types of queue disciplines are associated with priority rules when there are more than one class of customers. Customers belonging to different classes are indicated by a class index. Customers with priority index i are served before those with index $i + 1$, but customers belonging to the same class are served according to a FIFO rule. There are two types of priority schemes: (a) preemptive priority and (b) non-preemptive priority. If a customer of class $i + 1$ is being served, he may be interrupted and displaced back into the queue the moment a customer of higher priority class enters the system. This is called preemptive priority. If, however, he is allowed to complete his service before the priority rules come into operation, this is called non-preemptive priority service.

Note that the FIFO rule is the most natural queue discipline and is the fairest from the point of view of customers, but it may not be the best from the point of view of the servers.

Random Process. A random or stochastic process is a family of random variables $\{X(t)\}$. Here, $X(t)$ is in practice the observation at t, and is called a sample function of the process. A random process may be defined as a random time function which, for a particular value of time, is a random quantity.

4-2. THE POISSON INPUT PROCESS

The incoming traffic flow of a telephone system is constituted by the stream of calls arriving in the central office in an orderly manner. Similarly, the data traffic flow is formed by the stream of messages arriving in a computer network. The time epochs at which individual calls are seen at the central office or at which messages are seen at the data network are called arrival epochs; the intervals between consecutive arrival epochs are called interarrival times. Therefore an input traffic flow may be described by a sequence of time epochs, $T_1, T_2, \cdots$, with T_i being the arrival epoch of the *ith* arrival. The specification of the probabilistic behavior of the input process may be made in several ways. We shall present two models for the input process: the Poisson distribution for the number of arrivals in a fixed time interval, and the exponential interarrival time distribution. In telephone systems, most analysis and design problems are concerned with the busy-hour traffic. During such periods, the input process is considered to be sufficiently stationary.

4-2-1. Distribution for the Number of Arrivals in a Fixed Time Interval

Consider a long record of call arrivals for busy-hour traffic at a telephone exchange. Let the total observation time interval be t, and the total number of calls arriving in t be n_A. The average calling rate is defined by

$$\lambda = \frac{n_A}{t} \tag{4-1}$$

We divide the time interval t into n small and equal subintervals:

$$\Delta t = \frac{t}{n} \tag{4-2}$$

If Δt is sufficiently small such that there is either no or only one call in Δt, then in general, $n > n_A$. For a given interval of length $t = n\ \Delta t$, the number of calls arriving in t can be described by a random variable $N(t)$ with nonnegative integral values 0, 1, 2, ... The important quantity to be studied is the probability that exactly k calls arrive in the interval t, denoted by

$$P_k(t) = P\{N(t) = k\} \quad , \quad k = 0,1,2, \cdots \tag{4-3}$$

Using the statistical (relative frequency) definition of probability, a call is found in an interval Δt with probability

$$P_1(\Delta t) = P\{N(\Delta t) = 1\} = \frac{n_A}{n}$$

$$\tag{4-4}$$

$$= \frac{n_A}{t} \times \frac{t}{n} = \lambda\ \Delta t$$

and a call is not found in Δt with probability

$$P_0(\Delta t) = P\{N(\Delta t) = 0\} = 1 - P\{N(\Delta t) = 1\}$$

$$= 1 - \lambda \Delta t$$

(4-5)

These probabilities are a direct consequence of the way we select Δt; that is, there is either no or only one call arrival in Δt.

To describe the call arrival process mathematically in a more precise way, we make the following assumptions:

1. The arrivals are orderly, stationary, and mutually independent; and

2. The arrival rate λ is constant.

The first assumption implies that a call made by one subscriber has no effect on the calls made by any other subscribers. It also implies that the probability of a subscriber originating a call in a small time interval $(t_0, t_0 + \Delta t)$ is $\lambda \Delta t$, as given by (4-4), and hence, is constant for all t_0. In other words, the probability of a call arrival in the interval $(t_0, t_0 + \Delta t)$ depends on the calling rate λ and on the length of the interval Δt, but does not depend on t_0.

The second assumption of constant calling rate implies that the number of sources generating the input traffic must be infinite. For a finite number of sources, the effect of busy sources is to reduce the average arrival rate because busy sources cannot generate calls. Thus, in practice, if the number of sources (subscribers) is large and their average activity is relatively low, the arrival rate may be assumed to be constant.

Since, in each subinterval Δt, there can be no arrival or only one arrival, the event of a call arrival in Δt can be regarded as a Bernoulli trial which has only two possible outcomes. That is, for each trial, an arrival occurs in Δt with probability

$$p = P\{N(\Delta t) = 1\} = \lambda \Delta t = \lambda \frac{t}{n}$$

(4-6)

and no arrival occurs in Δt with probability

$$q = P\{N(\Delta t) = 0\} = 1 - \lambda\,\Delta t = 1 - \lambda\frac{t}{n} \qquad (4\text{-}7)$$

Thus the random variable $N(\Delta t)$ is a Bernoulli random variable. Clearly, the total number of trials is n, for the number of subintervals Δt contained in the interval t is n. It follows that the probability of exactly k arrivals in the interval t is given by the binomial distribution

$$P_k(t) = P\{N(t) = k\}$$

$$= \binom{n}{k}p^k\,(1-p)^{n-k}\;,\quad k = 0, 1, 2, ..., n$$

where $\binom{n}{k} = \dfrac{n!}{k!(n-k)!}$ is the binomial coefficient. The binomial distribution can be rewritten in the form

$$P_k(t) = \frac{n!}{k!(n-k)!}\left[\frac{\lambda t}{n}\right]^k\left[1-\frac{\lambda t}{n}\right]^{n-k}$$

$$= \frac{n(n-1)\,\cdots\,(n-k+1)\,(\lambda t)^k}{k!\,n^k\,(1-\frac{\lambda t}{n})^k}\,(1-\frac{\lambda t}{n})^n$$

As $n \rightarrow \infty$, this expression becomes

$$P_k(t) = \frac{(\lambda t)^k}{k!}\,e^{-\lambda t}\;,\quad k = 0, 1, 2, \cdots \qquad (4\text{-}8)$$

since the limits of the following factors hold:

$$\lim_{n \to \infty} \frac{n(n-1) \cdots (n-k+1)}{n^k (1 - \frac{\lambda t}{n})^k} = 1$$

and

$$\lim_{n \to \infty} (1 - \frac{\lambda t}{n})^n = e^{-\lambda t}$$

In words, formula (4-8) states that the probability of exactly k arrivals in a fixed time interval t is a Poisson distribution with a constant arrival rate λ.

The actual traffic offered to a central office may vary from time to time, and the calling rate λ is certainly not constant. However, the calling rate λ may be considered constant for fairly short time intervals; say, the busy hour. Furthermore, a long time measurement may be obtained by taking many busy-hour intervals of the same weekday and placing them together end to end. Therefore, the assumption of constant arrival rate is justified.

Regarding the independence assumption, certainly not all calls are independent. However, the proportion of those dependent calls is usually very small and hence can be neglected from the population of independent calls. Thus, the number of call arrivals in a fixed time interval can be practically regarded as Poisson distributed in spite of the conditions (assumptions) of call independence and of constant probability of occurrence not being strictly fulfilled.

Example 4-1. Calculate the mean and variance of the Poisson random variable $N(t)$ in (4-8).

Solution. The mean value of $N(t)$ is the mathematical expectation

$$E[N(t)] = \sum_{k=0}^{\infty} k \, P_k(t) = \sum_{k=0}^{\infty} k \frac{(\lambda t)^k}{k!} e^{-\lambda t}$$

$$= (\lambda t)e^{-\lambda t} \sum_{k=1}^{\infty} \frac{(\lambda t)^{k-1}}{(k-1)!}$$

$$= \lambda t$$

since the last summation equals $e^{\lambda t}$.

The variance of $N(t)$ is given by

$$Var[N(t)] = E[N^2(t)] - (E[N(t)])^2$$

$$= E[N^2(t)] - (\lambda t)^2$$

Note that

$$E[N^2(t)] = \sum_{k=0}^{\infty} k^2 P_k(t) = \sum_{k=1}^{\infty} k^2 P_k(t)$$

$$= \sum_{k=1}^{\infty} k(k-1)P_k(t) + \sum_{k=1}^{\infty} kP_k(t)$$

$$= \sum_{k=2}^{\infty} k(k-1)P_k(t) + \lambda t$$

$$= (\lambda t)^2 e^{-\lambda t} \sum_{k=2}^{\infty} \frac{(\lambda t)^{k-2}}{(k-2)!} + \lambda t$$

$$= (\lambda t)^2 + \lambda t$$

Hence

$$Var\,[N(t)] \;=\; \lambda t$$

We see from this example that the Poisson distribution has equal mean and variance and hence has a unity peakedness factor.

4-2-1-1. Superposition of Independent Poisson Input Traffic Flows. Consider a toll center with a pooled input traffic flow formed by m independent Poisson input traffic flows originating from local centers.

We shall show that if m independent Poisson traffic flows with rates $\lambda_1, \lambda_2, \cdots, \lambda_m$, respectively, are pooled to form a single traffic flow, the pooled traffic flow is also a Poisson flow with the rate $\lambda = \sum_{i=1}^{m} \lambda_i$.

First, consider the case $m = 2$. Let X and Y be two mutually independent Poisson random variables with rates λ_1 and λ_2, respectively, and let $Z = X + Y$. We observe that the event $\{Z = k\}$ is the sum (or union) of the mutually independent events $\{X = j, \ Y = k - j\}$ for $j = 0, 1, \cdots, k$. We have

$$P\{Z = k\} = \sum_{j=0}^{k} P\{X = j, Y = k - j\} = \sum_{j=0}^{k} P\{X = j\}\, P\{Y = k - j\}$$

$$= \sum_{j=0}^{k} \frac{(\lambda_1 t)^j}{j!}\, e^{-\lambda_1 t} \frac{(\lambda_2 t)^{k-j}}{(k-j)!}\, e^{-\lambda_2 t}$$

$$= e^{-(\lambda_1 + \lambda_2)t} \sum_{j=0}^{k} \frac{(\lambda_1 t)^j}{j!} \frac{(\lambda_1 t)^{k-j}}{(k-j)!}$$

$$= \frac{e^{-(\lambda_1 + \lambda_2)t}}{k!} \sum_{j=0}^{k} \frac{k!}{j!(k-j)!}\, (\lambda_1 t)^j (\lambda_2 t)^{k-j}$$

$$= \frac{[(\lambda_1 + \lambda_2)t]^k}{k!} \, e^{-(\lambda_1+\lambda_2)t} \, , \, k = 0, 1, \, \cdots \, .$$

By induction, in the general case, we have

$$P\{Z = k\} = \frac{(\lambda t)^k}{k!} \, e^{-\lambda t} \, , \, k = 0, 1, \, \cdots \, , \tag{4-9}$$

where $\lambda = \sum_{i=1}^{m} \lambda_i$.

Example 4-2. Consider the Poisson distribution of (4-8). Calculate the probability generating function of N(t) and then show that if m independent Poisson flows with rates $\lambda_1, \lambda_2, \, \cdots \, , \lambda_m$, are pooled to form a single input flow, the resultant pooled input flow is also a Poisson flow with rate $\lambda = \sum_{i=1}^{m} \lambda_i$.

Solution. By definition, the probability generating function of the Poisson random variable $N(t)$ in (4-8) is given by

$$G(z) = E[z^{N(t)}] = \sum_{k=0}^{\infty} z^k P_k(t), \, |z| \le 1$$

For the *ith* Poisson flow with rate λ_i, the corresponding probability generating function is then given by

$$G_i(z) = \sum_{k=0}^{\infty} e^{-\lambda_i t} \, \frac{(\lambda_i t z)^k}{k!}$$

$$= e^{-\lambda_i t(1-z)}$$

Since

$$N(t) = \sum_{i=1}^{m} N_i(t)$$

and $N_i(t)$ are independent Poisson random variables, the generating function of $N(t)$ is equal to the product of the individual generating functions $G_i(z)$:

$$G(z) = \prod_{i=1}^{m} G_i(z) = \prod_{i=1}^{m} e^{-\lambda_i t(1-z)}$$

$$= e^{-\sum_{i=1}^{m} \lambda_i t(1-z)} = e^{-\lambda t(1-z)}$$

where $\lambda = \sum_{i=1}^{m} \lambda_i$. We see that the generating function $G(z)$ has the same form as that of $G_i(z)$. Therefore, the resultant pooled flow $N(t)$ is also a Poisson flow with rate λ.

Example 4-3. Calculate the mean and variance of the Poisson random variable $N(t)$ using the generating function $G(z)$ obtained in Example 4-2.

Solution. From Example 4-2, we have

$$G(z) = \sum_{k=0}^{\infty} P_k(t)z^k = e^{-\lambda t(1-z)}$$

Differentiating $G(z)$ once with respect to z, we get

$$G'(z) = \sum_{k=0}^{\infty} kP_k(t)z^{k-1} = \lambda t e^{-\lambda t(1-z)}$$

Thus, by letting $z = 1$, we obtain

$$E[N(t)] = G'(1) = \sum_{k=0}^{\infty} kP_k(t) = \lambda t$$

Similarly, we also have

$$G''(z) = \sum_{k=0}^{\infty} k(k-1) P_k(t) z^{k-2} = (\lambda t)^2 e^{-\lambda t(1-z)}$$

Hence, letting $z = 1$ yields

$$G''(1) = \sum_{k=0}^{\infty} k(k-1) P_k(t) = (\lambda t)^2$$

Since

$$\sum_{k=0}^{\infty} k(k-1) P_k(t) = E[N^2(t)] - E[N(t)] = (\lambda t)^2$$

it follows that

$$Var[N(t)] = E[N^2(t)] - (E[N(t)])^2 = E[N(t)] = \lambda t$$

From this example, we see that a probability generating function can be very useful for calculating the moments of a random variable.

4-2-1-2. Decomposition of a Poisson Flow. Suppose that calls arrive at a central office according to a Poisson process with rate λ. These calls are distributed to two machines for processing. If each call is directed to the *kth* machine with probability λ_k/λ, where $\lambda = \lambda_1 + \lambda_2$, then the original Poisson arrival process is to be decomposed into two processes according to these probabilities. Let these processes be denoted by $N_i(t)$, $i = 1, 2$, and the original Poisson process by $N(t)$. It follows that

$$N(t) = N_1(t) + N_2(t)$$

In addition, we shall show that $N_1(t)$ and $N_2(t)$ are independent Poisson processes with rates λ_1 and λ_2 respectively. First, we compute the joint probability

$$P\{N_1(t) = n_1, N_2(t) = n_2\} = P\{N_1(t) = n_1, N_2(t) = n_2 | N(t) = n\}$$

$$\times P\{N(t) = n\}$$

Since

$$P\{N_1(t) = n_1, N_2(t) = n_2 | N(t) = n\} = \binom{n}{n_1}(\lambda_1/\lambda)^{n_1}(\lambda_2/\lambda)^{n-n_1}$$

where $n = n_1 + n_2$, $\lambda = \lambda_1 + \lambda_2$, and

$$P\{N(t) = n\} = \frac{(\lambda t)^n}{n!}\, e^{-\lambda t}$$

we obtain

$$P\{N_1(t) = n_1, N_2(t) = n_2\} = \frac{n!}{n_1!(n-n_1)!}\,(\lambda_1/\lambda)^{n_1}(\lambda_2/\lambda)^{n-n_1}\frac{(\lambda t)^n}{n!}\, e^{-\lambda t}$$

$$= \frac{(\lambda_1 t)^{n_1}}{n_1!}e^{-\lambda_1 t}\frac{(\lambda_2 t)^{n_2}}{n_2!}e^{-\lambda_2 t}$$

$$= P\{N_1(t) = n_1\}\, P\{N_2(t) = n_2\}$$

This result asserts that the decomposition of a Poisson process into two processes according to probabilities λ_1/λ and λ_2/λ results in two independent Poisson processes with rates λ_1 and λ_2, respectively. Notice that by repeating the same argument, we can show the decomposition of a Poisson process into m independent Poisson processes with rates λ_i, $i = 1, 2, \cdots, m$, where $\lambda = \lambda_1 + \lambda_2 + \cdots + \lambda_m$.

Example 4-4. Consider a Poisson input traffic flow with rate λ. Each arrival of this traffic flow is randomly directed to the *kth* central office with probability λ_k/λ, as shown in Figure 4-2. In total, there are m such offices. Show that the decomposed traffic flows are independent Poisson flows with the corresponding rate λ_i, $i = 1, 2, ..., m$.

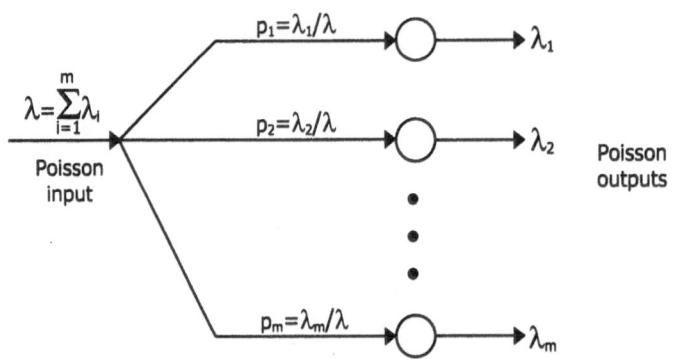

Figure 4-2. Decomposition of a Poisson input traffic flow.

Solution. Let $N_k(t)$ denote the number of calls directed to the *k*th office in a fixed time interval t. The conditional joint probability $P\{N_1(t) = n_1, N_2(t) = n_2, \cdots, N_m(t) = n_m \mid N(t) = n\}$ is the multinomial distribution

$$P\{N_1(t) = n_1, N_2(t) = n_2, \cdots, N_m(t) = n_m \mid N(t) = n\}$$

$$= \frac{n!}{n_1! \, n_2! \, \cdots \, n_m!} \, p_1^{n_1} \, p_2^{n_2} \, \cdots \, p_m^{n_m}$$

where

$$p_k = \lambda_k/\lambda, \quad k = 1, 2, \cdots, m \; ; n_1 + n_2 + \cdots n_m = n \; .$$

Since $N(t)$ is a Poisson traffic flow, we have

$$P\{N(t) = n\} = \frac{(\lambda t)^n}{n!} e^{-\lambda t}$$

where $\lambda = \lambda_1 + \lambda_2 + \cdots + \lambda_m$ for $p_1 + p_2 + \cdots + p_m = 1$.

Multiplying the conditional joint probability by $P\{N(t) = n\}$, we obtain

$$P\{N_1(t) = n_1, N_2(t) = n_2, \cdots, N_m(t) = n_m\}$$

$$= \frac{n!}{n_1! \, n_2! \, \cdots \, n_m!} \, p_1^{n_1} \, p_2^{n_2} \cdots p_m^{n_m} \, \frac{(\lambda t)^n}{n!} e^{-\lambda t}$$

$$= \prod_{k=1}^{m} \frac{(\lambda_k t)^{n_k}}{n_k!} e^{-\lambda_k t}$$

Since the joint probability $P\{N_1(t) = n_1, N_2(t) = n_2, \cdots, N_m(t) = n_m\}$ is a product of m Poisson distributions, the random variables $N_1(t), N_2(t), \cdots, N_m(t)$ are independent Poisson random variables. Therefore, the decomposed traffic flows are independent Poisson traffic flows with the corresponding rate λ_k, $k = 1, 2, ..., m$.

4-2-2. The Interarrival Time Distribution.

The interarrival time distribution provides another useful model for the specification of input traffic process. Let T_i, $i = 0, 1, 2, \cdots$, be the *ith* arrival epoch and let $X_i = T_i - T_{i-1}$, $i = 1, 2, \cdots$, with $T_0 = 0$, be the *ith* interarrival time. Assume that the interarrival times are mutually independent and identically distributed with the common distribution function

$$F(t) = P\{X_i \le t\}$$

Let $F_c(t) = 1 - F(t)$; $F_c(t)$ is the probability that the interarrival time X_i is greater than t. Consider the possible events in the small time interval $(t, t + \Delta t)$. The probability

$$F_c(t + \Delta t) \; = \; P\{X_i > t + \Delta t\}$$

is equal to the conditional probability $P\{X_i > t + \Delta t \mid X_i > t\}$ multiplied by the probability $P\{X_i > t\}$. Hence,

$$P\{X_i > t + \Delta t, X_i > t\} \; = \; P\{X_i > t + \Delta t\}$$
$$= P\{X_i > t + \Delta t \mid X_i > t\} \, F_c(t)$$

Let the arrival rate be λ. Suppose that the probability of one arrival in $(t, t + \Delta t)$ is $\lambda \Delta t + o(\Delta t)$, and that the probability of more than one arrival in $(t, t + \Delta t)$ is $o(\Delta t)$. Then

$$P\{X_i > t + \Delta t \mid X_i > t\} \; = \; 1 - \lambda \Delta t + o(\Delta t)$$

where $o(\Delta t)$ is an infinitesimal quantity of order higher than Δt.

Hence,

$$F_c(t + \Delta t) \; = \; [1 - \lambda \Delta t + o(\Delta t)]F_c(t)$$

or

$$\frac{F_c(t + \Delta t) - F_c(t)}{\Delta t} \; = \; -\lambda F_c(t) + \frac{o(\Delta t)}{\Delta t} F_c(t)$$

Passing to the limit $\Delta t \to 0$ gives the differential equation

$$\frac{d}{dt} F_c(t) = -\lambda F_c(t)$$

whose solution is

$$F_c(t) = F_c(0) \, e^{-\lambda t} = e^{-\lambda t}$$

for $F_c(0) = 1$. Finally, we obtain

$$F(t) = 1 - F_c(t)$$

$$= 1 - e^{-\lambda t}, \qquad t \geq 0 \tag{4-10}$$

This result shows that under the given assumptions, the interarrival time distribution of the input traffic flow with constant arrival rate λ is an exponential distribution.

The exponential distribution has some important properties that play a central role in teletraffic analysis and in queueing theory. The expected value of the interarrival time with distribution (4-10) is equal to

$$E[X_i] = \int_0^\infty t \; dP\{X_i \leq t\}$$

$$= \int_0^\infty t \, \lambda \, e^{-\lambda t} \; dt$$

$$= \frac{1}{\lambda}$$

and the variance of the interarrival time is given by

$$Var[X_i] = E[X_i^2] - (E[X_i])^2$$

$$= \int_0^\infty t^2 \, \lambda \, e^{-\lambda t} \; dt - \frac{1}{\lambda^2}$$

$$= \frac{2}{\lambda^2} - \frac{1}{\lambda^2}$$

$$= \frac{1}{\lambda^2}$$

4-2-2-1. The Markov Property or Memoryless Property. A continuous nonnegative random variable X is said to have the Markov property if for every $t > 0$ and every $x > 0$,

$$P\{X > t + x \mid X > t\} = P\{X > x\} \qquad (4\text{-}11)$$

A random variable with the Markov property (4-11) is often said to have no memory. The Markov property plays an important role in teletraffic theory.

To illustrate the Markov property, consider the exponential distribution (4-10). The conditional probability on the left-hand side of (4-11) yields

$$P\{X > t + x \mid X > t\} = \frac{P\{X > t + x, X > t\}}{P\{X > t\}}$$

$$= \frac{P\{X > t + x\}}{P\{X > t\}} = \frac{e^{-\lambda(t + x)}}{e^{-\lambda t}}$$

$$= e^{-\lambda x} = P\{X > x\}$$

This result shows that the exponential distribution (4-10) has the Markov property. The exponential distribution can be used to describe the holding time distribution of telephone calls.

4-2-2-2. Relationship between the Poisson Process and the Exponential Interarrival Times. Recall that the input traffic flow can be modeled by the Poisson process $N(t)$ or by the exponential interarrival time $X_i = T_i - T_{i-1}$, with $X_1 = T_1$ and $T_0 = 0$. Furthermore, observe that the events $\{X_i > t\}$ and $\{T_i - T_{i-1} > t\}$ are equivalent:

$$\{X_i > t\} \sim \{T_i - T_{i-1} > t\} \sim \{T_i > T_{i-1} + t\}$$

and that the event $\{T_i > T_{i-1} + t \mid T_{i-1} = t_{i-1}\}$ occurs if and only if the value of $N(t)$ does not change during the interval $(t_{i-1}, t_{i-1} + t)$:

$$\{T_i > T_{i-1} + t \mid T_{i-1} = t_{i-1}\} \sim \{N(t_{i-1} + t) - N(t_{i-1}) = 0\}$$

Since equivalent events have equal probabilities, we have

$$P\{X_i > t \mid T_{i-1} = t_{i-1}\} = P\{N(t_{i-1} + t) - N(t_{i-1}) = 0\}$$

$$= P\{N(t) = 0\}$$

for the random process $N(t)$ has stationary and independent increments. The conditional probability on the left-hand side of the above expression is therefore independent of the value of i and T_{i-1}, and hence, we obtain

$$P\{X_i > t\} = P\{N(t) = 0\} = e^{-\lambda t}.$$

As this result is independent of the value of the index i, it follows that if the random process $N(t)$ has stationary increments, then the interarrival times X_i will all have the same exponential distribution. That is,

$$P\{X_i \le t\} = 1 - e^{-\lambda t}. \tag{4-12}$$

Comments. In this section, we have presented two mathematical models for the representation of the input traffic flow. The Poisson distribution (4-8) represents the probability distribution of the number of calls, $N(t)$, in a fixed time interval t. It has the property that the mean and variance are equal. The exponential distribution (4-10) for the call interarrival time distribution, on the other hand, is a continuous time distribution. We shall show in the following section that the exponential distribution can also be used to describe the holding time distribution of calls.

The importance of the exponential distribution is that it has the Markov property. Hence, when the present is known, its future behavior does not depend on the past history. An input process obeying (4-8) or (4-10) is called a Poisson input with rate λ.

4-3. THE SERVICE TIME DISTRIBUTION

In the study of teletraffic engineering, the service time is the duration of time for which a subscriber makes use of the equipment for communication purposes. It is called the holding time of the call. In many cases, such as with trunk lines, the holding time is roughly equal to the duration of the call. In the case of control devices, such as markers, the holding time equals the time necessary to set up a path and is thus much less than the duration of the call.

Let T denote the holding time with mean $1/\mu$. In teletraffic applications, it is usually assumed that T follows an exponential distribution:

$$P\{T \le t\} = 1 - e^{-\mu t}, t \ge 0 \qquad (4\text{-}13)$$

According to the measurement made by the American Telephone and Telegraph Company, the holding time distribution is close to exponential. Here, the holding time of a call is defined as the duration of occupancy of a traffic path by a call. Sometimes the holding time is also referred to as the average duration of occupancy of one or more paths by calls. A typical distribution of the holding time is shown in Figure 4-3.

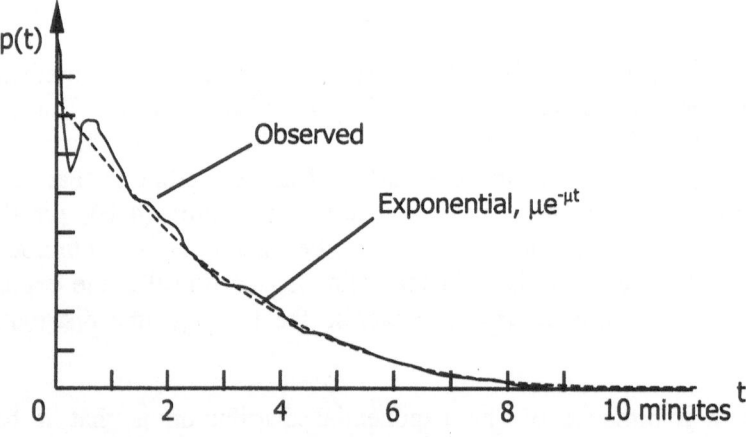

Figure 4-3. Distribution of holding times.

From Figure 4-3, we see that the observed curve follows an exponential distribution very closely except for a dip for t between 20 and 30 seconds. This dip corresponds to a time that is too short for a successful call and too long for a misdialed or unsuccessful call. In practice, relatively few calls have holding times between these two limits. Furthermore, the distribution of holding times in long-distance traffic is also close to exponential. Therefore, for practical applications, in most cases, the holding times are very nearly exponential with a mean of approximately 3 minutes.

The Geometric Distribution. Consider a sequence of Bernoulli trials. Let p be the probability of success at any particular trial. Define the discrete random variable N as the number of trials required up to and including the first success. Thus, N assumes the possible values 1, 2, 3, $\cdots$. Since $N = k$ if and only if the first $k - 1$ trials result in failures while the kth trial results in a success, the probability that the first success occurs at the kth trial is given by

$$P\{N = k\} = q^{k-1}p \ , \ k = 1, 2, 3, \ \cdots \tag{4-14}$$

where $q = 1 - p$. The probability distribution (4-14) is known as a geometric distribution. The geometric distribution can be used to describe the number of packets of a message in data communications.

The random variable N has the probability generating function

$$G(z) = E[z^N] = \sum_{k=1}^{\infty} z^k \, P\{N = k\}$$

$$= \frac{pz}{1 - qz} \ , \ |z| \leq 1$$

as well as the mean

$$E[N] = G'(1) = 1/p$$

and the variance

$$Var\,[N] = q/p^2$$

The geometric distribution is the discrete counterpart of the exponential distribution. It also has the Markov property. To show this property, we consider the probability that the number of trials required for the first success is greater than k

$$P\,\{N > k\} = \sum_{j=k+1}^{\infty} pq^{j-1} = q^k$$

The conditional probability that N will be greater than $j+k$, given that N is greater than j is, by definition,

$$P\,\{N > j + k \mid N > j\} = \frac{P\,\{N > j + k, N > j\}}{P\,\{N > j\}}$$

$$= \frac{P\,\{N > j + k\}}{P\,\{N > j\}}$$

$$= \frac{q^{j+k}}{q^j} = q^k$$

which is independent of j. Therefore, we deduce that

$$P\,\{N > j + k \mid N > j\} = P\,\{N > k\}$$

Thus, the geometric distribution has the Markov property.

4-4. THE RESIDUAL SERVICE TIME DISTRIBUTION

Consider a single-server queue where an arbitrary arriving customer finds a partially-served customer in the system. The remaining portion of service time for the partially-served customer is called the residual service time. Let R denote the residual service time, and let $F(t)$ and $f(t)$ be its distribution function and probability density function, respectively. We shall determine $F(t)$ and $f(t)$ in terms of the original service time distribution function $H(x) = P\{X \le x\}$ and mean service time τ. In particular, we shall obtain an expression for the mean residual service time as the second moment of the original service time divided by twice the original mean service time.

Now let us consider the case of the arbitrary customer arriving in a special interdeparture time interval Y. First, we determine the probability density function $g(y)$ of this interdeparture time. Observe that the service time of the partially-served customer must be within Y and hence, the length of Y must be proportional to the length of the service time $X = y$ as well as to the probability of occurrence of such a service time [which is given by $h(y)dy$]. Thus, using this fact, we can write

$$g(y)dy = Kyh(y)dy$$

where the left-hand side is $P\{y < Y \le y + dy\}$ and the right-hand side expresses the linear weighting with respect to the service time length y and to a constant K, which must be evaluated so as to properly normalize the density function $g(y)$.

Integrating both sides of this expression from 0 to ∞, we find that $K = 1/\tau$. It follows that the probability density function of the special interdeparture time (interval) Y is given by

$$g(y) = \frac{yh(y)}{\tau} \tag{4-15}$$

Since customers can arrive randomly at any point in the special interdeparture time, the occurrence of the residual service time R must be uniformly distributed with Y. Therefore, for a given value of $Y = y$, we have

$$P\{R \le t \mid Y = y\} = \begin{cases} t/y & , \ 0 \le t \le y \\ 0 & , \ otherwise \end{cases}$$

Thus, the joint probability density function of R and Y can be expressed as

$$P\{t < R \le t + dt, y < Y \le y + dy\} = \left[\frac{dt}{y}\right]\left[\frac{yh(y)dy}{\tau}\right] = \frac{h(y)}{\tau}dydt,$$

$$0 \le t \le y$$

Integrating over y from t to ∞, we obtain $f(t)$, which is the unconditional probability density function of R, namely,

$$f(t)dt = \int_t^\infty \frac{h(y)dy}{\tau} dt$$

$$= \frac{1 - H(t)}{\tau} dt$$

or

$$f(t) = \frac{1 - H(t)}{\tau} \qquad (4\text{-}16)$$

It follows that the distribution function of the residual service time R is given by

$$F(t) = \frac{1}{\tau} \int_0^t [1 - H(x)] \, dx$$

The Laplace-Stieltjes transform of $F(t)$ is equal to

$$\hat{F}(s) = \int_0^\infty e^{-st} \, dF(t)$$

$$(4\text{-}17)$$

$$= \frac{1}{\tau s} [1 - \hat{H}(s)]$$

from which we can calculate the average residual service time

$$E[R] = - \frac{d}{ds} \hat{F}(s) \bigg|_{s=0}$$

$$= \frac{1}{\tau s^2} [1 - \hat{H}(s) + s\hat{H}'(s)] \bigg|_{s=0}$$

$$= \frac{1}{2\tau} \hat{H}''(0)$$

$$(4\text{-}18)$$

$$= \frac{1}{2\tau} E[X^2]$$

$$= \frac{1}{2} \frac{\overline{X^2}}{\overline{X}}$$

where $\hat{H}(s)$ is the Laplace-Stieltjes transform of $H(t)$. This result indicates that the average residual service time is equal to the second moment of the original service time divided by twice the original average service time.

4-5. THE BIRTH AND DEATH PROCESS

The birth and death process is a special case of the discrete-state continuous-time Markov process, which is often called a continuous-time Markov chain. The basic feature of the method of Markov chains is the Kolmogorov differential-difference equation, whose solution, for the limiting case, can provide a solution to the state probability distribution for the Erlang systems and the Engset systems.

In applications, it is usually convenient to describe Markov chains in terms of random variables. Let the state of a stochastic process be described by the random variable $N(t)$, $t \geq 0$, taking on nonnegative integral values. We say that the random variable $N(t)$ is a continuous-time Markov chain if for all $s, t \geq 0$ and nonnegative integers $i, j, n(u)$, $0 \leq u < s$,

$$P\{N(t+s) = j \mid N(s) = i, N(u) = n(u), 0 \leq u < s\}$$

$$= P\{N(t+s) = j \mid N(s) = i\}$$

In words, this expression states that a continuous-time Markov chain is a stochastic process with the Markov property that given the present $N(s)$ and the past $N(u), 0 \leq u < s$, the conditional probability of the future $N(t+s)$ depends only on the present and is independent of the past. If, in addition, the conditional probability $P\{N(t+s) = j \mid N(s) = i\}$ is independent of s, then the continuous-time Markov chain is said to have stationary or homogeneous transition probabilities.

Let $N(t)$ be a random variable specifying the size of the population at time t. The process $N(t)$ is said to be a birth and death process if $N(t)$ can take only nonnegative integral values and can have positive and negative jumps. Furthermore, the process is said to be in state k at time t if $N(t) = k$; that is, the size of the population at time t is equal to k.

For a complete description of a birth and death process, we assume that $N(t)$ is in state k at time t and has the following properties:

(a) The probability of transition from state k to state $k + 1$ in the time interval $(t, t + \Delta t)$ is $\lambda_k \Delta t + o(\Delta t)$, where λ_k is called the birth rate in state k.

(b) The probability of transition from state k to state $k - 1$ in the time interval $(t, t + \Delta t)$ is $\mu_k \Delta t + o(\Delta t)$, where μ_k is called the death rate in state k.

(c)   The probability of transition in the time interval $(t, t + \Delta t)$ from state k to a state other than a neighboring state $k + 1$ or $k - 1$ is equal to $o(\Delta t)$.

(d)   The probability of no change of state in the time interval $(t, t + \Delta t)$ is equal to $1 - (\lambda_k + \mu_k)\Delta t + o(\Delta t)$.

According to these properties, we can draw a diagram for the process $N(t)$ as shown in Figure 4-4(a) and a state-transition rate diagram as shown in Figure 4-4(b), where the numbered circles denote the states of the process at t.

Note that in the state-transition rate diagram Figure 4-4(b), the self-loop from state k back to state k is not included because such a diagram displays only the rate of transition to the neighboring states and not the probabilities.

In many cases, the behavior of the process $N(t)$ for large values of t is of primary interest; that is, we are interested in the behavior of the process after it has been in operation for a long period of time. Under this condition, the state probabilities of the process will be independent of the initial conditions of the state, and the process is said to be in equilibrium.

Let $P_k(t)$ denote the probability that the process is in state k at time t. Clearly, by definition,

$$P_k(t) = P\{N(t) = k\}$$

Also, let $P_{ik}(t)$ denote the conditional probability that the process in state i at a certain moment after the passage of t time units changes to state k. These transition probabilities satisfy the following conditions:

$$P_{ik}(t) \geq 0, \ \sum_{k=0}^{\infty} P_{ik}(t) = 1$$

According to the Markov theorem, which states that for any Markov process characterized by the transition probability $P_{ik}(t)$, the limit

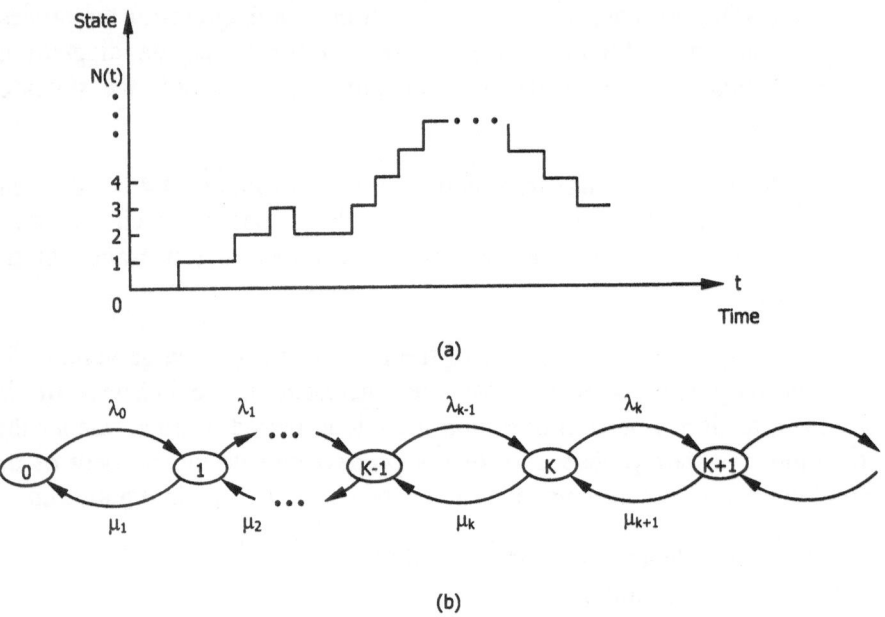

(a)

(b)

Figure 4-4.   (a) Birth and death process $N(t)$, (b) State-transition rate diagram for the birth and death process.

$$\lim_{t \to \infty} P_{ik}(t) = p_k$$

exists and does not depend on i, the probability $P_k(t)$, if at the starting moment 0, the process is in state i, is given by the formula of total probability.

$$P_k(t) = \sum_{i=0}^{\infty} P_i(0) \, P_{ik}(t)$$

As $t \to \infty$, we obtain

$$\lim_{t \to \infty} P_k(t) = \lim_{t \to \infty} \sum_{i=0}^{\infty} P_i(0) \, P_{ik}(t)$$

$$= \sum_{i=0}^{\infty} P_i(0) \, p_k \qquad\qquad (4\text{-}19)$$

$$= p_k$$

since $\displaystyle\sum_{i=0}^{\infty} P_i(0) = 1$.

It follows from properties (a) - (d) of the birth and death process and from the formula of total probability that

$$P_k(t + \Delta t) = \sum_{i=0}^{\infty} P_i(t) \, P_{ik}(\Delta t)$$

$$= P_k(t) \, P_{kk}(\Delta t) + P_{k-1}(t) \, P_{k-1,k}(\Delta t) + P_{k+1}(t) \, P_{k+1,k}(\Delta t) + o(\Delta t)$$

$$= P_k(t)[1 - (\lambda_k + \mu_k)\Delta t + o(\Delta t)] + P_{k-1}(t)[\lambda_{k-1}\Delta t + o(\Delta t)]$$

$$\quad + P_{k+1}(t)[\mu_{k+1}\Delta t + o(\Delta t)] + o(\Delta t)$$

$$= P_k(t)[1 - (\lambda_k + \mu_k)\Delta t] + \lambda_{k-1}P_{k-1}(t)\Delta t + \mu_{k+1}P_{k+1}(t)\Delta t + o(\Delta t)$$

or

$$\frac{P_k(t + \Delta t) - P_k(t)}{\Delta t} = -(\lambda_k + \mu_k)P_k(t) + \lambda_{k-1}P_{k-1}(t) + \mu_{k+1}P_{k+1}(t)$$

$$\quad + \frac{o(\Delta t)}{\Delta t}$$

As $\Delta t \to 0$, we have

$$\frac{d}{dt} P_k(t) = -(\lambda_k + \mu_k) P_k(t) + \lambda_{k-1}P_{k-1}(t) + \mu_{k+1}P_{k+1}(t) ,$$

$$(4\text{-}20)$$

$$k = 0, 1, 2, \cdots$$

with $\lambda_{-1} = \mu_0 = P_{-1}(t) = 0$.

This set of differential-difference equations represents the dynamic behavior of the birth and death process. To solve these equations for $P_k(t)$, the initial conditions $P_k(0)$ are required. However, if the primary interest is the behavior of the process in equilibrium, then only the probabilities p_k are of main concern. In practical applications, this is indeed the case. Therefore, we shall determine only the equilibrium distribution $\{p_k\}$ in the sequel.

Before solving for p_k, it is important to note that, as $t \to \infty$, all derivatives

$$\lim_{t \to \infty} \frac{d}{dt} P_k(t) = 0$$

since, if any $\frac{d}{dt}P_k(t)$ tended to values other than zero, then as $t \to \infty$, $|P_k(t)|$ would increase infinitely.

Consequently, as $t \to \infty$, (4-20) becomes

$$-(\lambda_k + \mu_k) p_k + \lambda_{k-1} p_{k-1} + \mu_{k+1} p_{k+1} = 0 \qquad (4\text{-}21)$$

with $\lambda_{-1} = \mu_0 = p_{-1} = 0$.

This set of linear equations, together with the normalization condition

$$\sum_{k=0}^{\infty} p_k = 1 \qquad (4\text{-}22)$$

uniquely determine the required state probabilities p_k. If we let

$$z_k = \lambda_{k-1} p_{k-1} - \mu_k p_k,$$

then (4-21) can be written in the form

$$z_k - z_{k+1} = 0 \ , \ \ k = 0, 1, 2, \ \cdots$$

Since $z_0 = \lambda_{-1} p_{-1} - \mu_0 p_0 = 0$, then

$$z_k = 0 \ , \ \ k = 0, 1, 2, \ \cdots$$

and hence

$$p_k = \frac{\lambda_{k-1}}{\mu_k} p_{k-1} \tag{4-23}$$

Consequently,

$$p_k = \frac{\lambda_0 \lambda_1 \cdots \lambda_{k-1}}{\mu_1 \mu_2 \cdots \mu_k} p_0 \ , \ \ k = 1, 2, 3, \ \cdots \tag{4-24}$$

The probability p_0 can be determined by the normalization condition (4-22):

$$p_0 = (1 + \sum_{k=1}^{\infty} \frac{\lambda_0 \lambda_1 \cdots \lambda_{k-1}}{\mu_1 \mu_2 \cdots \mu_k})^{-1} \tag{4-25}$$

where the infinite series on the right-hand side is assumed to be convergent.

It is interesting to point out that the set of differential-difference equations can also be obtained by observing the state-transition-rate diagram. Consider a dynamic situation. Note that the difference between the rate at which the process enters state k and the rate at which the process leaves state k must be equal to the rate of change of flow into that state. This notion provides for us a simple means for writing down the equations of motion for $P_k(t)$ of the process. Now we observe that the rate at which probability flows into state k at time t is given by

$$Flow\ rate\ into\ state\ k = \lambda_{k-1}P_{k-1}(t) + \mu_{k+1}P_{k+1}(t)$$

and the flow rate out of that state at time t is given by

$$Flow\ rate\ out\ of\ state\ k = (\lambda_k + \mu_k)\,P_k(t)$$

The difference between these two flow rates is the rate of change of $P_k(t)$ at time t; that is,

$$\frac{d}{dt}P_k(t) = \lambda_{k-1}P_{k-1}(t) + \mu_{k+1}P_{k+1}(t) - (\lambda_k + \mu_k)P_k(t), k = 0, 1, 2, \cdots$$

which is the same as (4-20).

By using the state-transition-rate diagram, we can also write down the equilibrium equation (4-21) by observing that

$$Flow\ rate\ into\ state\ k = \lambda_{k-1}p_{k-1} + \mu_{k+1}p_{k+1}$$

and

$$Flow\ rate\ out\ of\ state\ k = (\lambda_k + \mu_k)p_k$$

In equilibrium, these two rates must be equal:

$$\lambda_{k-1}p_{k-1} + \mu_{k+1}p_{k+1} = (\lambda_k + \mu_k)p_k$$

which is the same as (4-21). Furthermore, the relationship (4-23), which leads to the solution (4-24), can be written down by considering the conservation of flow across a vertical plane separating the adjacent states. Let us imagine a vertical plane separating state $k-1$ and state k as shown in Figure 4-5.

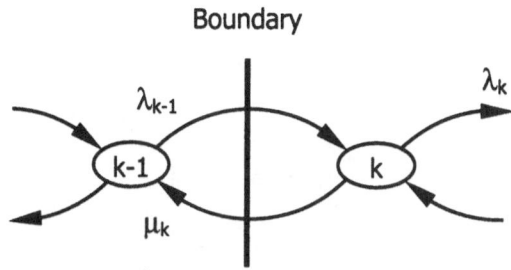

Figure 4-5. Flow rates at a boundary.

Note that

$$Flow\ rate\ into\ boundary\ =\ \lambda_{k-1}p_{k-1}$$

and

$$Flow\ rate\ out\ of\ boundary\ =\ \mu_{k}p_{k}$$

Equating these two flow rates yields the relationship (4-23).

The major significance of the birth and death process is that there are many queueing systems whose behavior can be modeled by this process.

Example 4-5. Consider the birth and death process for which the birth and death rates are, respectively,

$$\lambda_{k}\ =\ \lambda\quad and\quad \mu_{k}\ =\ 0$$

for $k = 0, 1, 2, \cdots$. This process is called a pure birth process. Determine the state probability distribution $P_k(t)$, $k = 0, 1, 2, ...$, where $P_0(0) = 1$, $P_k(0) = 0$, for $k = 1, 2, 3, \ ...$

Solution. Applying the birth and death rates to the differential-difference equations (4-20) gives the differential-difference equation

$$\frac{dP_k(t)}{dt} = -\lambda\, P_k(t) + \lambda P_{k-1}(t)\,, \; k = 0, 1, 2, \cdots \qquad (4\text{-}26)$$

In particular, for $k = 0$, (4-26) reduces to

$$\frac{dP_0(t)}{dt} = -\lambda P_0(t) \qquad (4\text{-}27)$$

for $P_{-1}(t) = 0$ by definition. The solution to this equation is

$$P_0(t) = C\, e^{-\lambda t}$$

Since

$$P_0(0) = C = P\,\{N(0) = 0\} = 1$$

then

$$P_0(t) = e^{-\lambda t} \qquad (4\text{-}28)$$

This result shows that as the length of the time interval $(0, t)$ increases without limit, the probability of no birth in this interval becomes zero.

To calculate $P_k(t)$ for $k \geq 1$, let us introduce the notation

$$Q_k(t) = P_k(t)\, e^{\lambda t}$$

and

$$\frac{d}{dt} Q_k(t) = e^{\lambda t}\, \frac{d}{dt} P_k(t) + \lambda e^{\lambda t}\, P_k(t)$$

We can then write (4-26) in the form

$$\frac{d}{dt} Q_k(t) = \lambda\, Q_{k-1}(t) \qquad\qquad (4\text{-}29)$$

It follows that for $k = 1$,

$$\frac{d}{dt} Q_1(t) = \lambda\, Q_0(t) = \lambda e^{\lambda t}\, P_0(t) = \lambda$$

Hence,

$$Q_1(t) = \lambda\, t + C_1$$

where $C_1 = Q_1(0) = 0$, since $Q_k(0) = P_k(0) = 0$ for $k \geq 1$. Consequently, we obtain

$$Q_1(t) = \lambda\, t$$

For $k = 2$, we obtain from (4-29),

$$\frac{d}{dt} Q_2(t) = \lambda\, Q_1(t) = \lambda^2 t$$

whose solution is

$$Q_2(t) = \frac{1}{2!} (\lambda t)^2 + C_2$$

or

$$Q_2(t) = \frac{1}{2!} (\lambda t)^2$$

for $C_2 = Q_2(0) = 0$.

In general, we have

$$Q_k(t) = \frac{1}{k!} (\lambda t)^k \, , \, k = 0, 1, 2, \cdots$$

Hence,

$$P_k(t) = e^{-\lambda t} \, Q_k(t) = \frac{(\lambda t)^k}{k!} \, e^{-\lambda t}, \, k = 0, 1, 2, \cdots \qquad (4\text{-}30)$$

This result shows that the random variable $N(t)$ for the pure birth process has a Poisson distribution with rate λ.

4-6. LITTLE'S FORMULA FOR MEAN VALUES FOR A GENERAL QUEUE

Consider a general queueing system with a service facility. The interarrival time distribution and the service time distribution are completely arbitrary and assumed to be independent of each other. The order of service (the queue discipline) and the number of servers are also arbitrary.

Let us define some important random processes for the system. We denote

$$\Lambda(t) = \textit{the number of arrivals in } (0, t)$$

$$\nabla(t) = \textit{the number of departures in } (0, t)$$

The number of customers in the system at t is then given by

$$N(t) = \Lambda(t) - \nabla(t)$$

Sample functions for $\Lambda(t)$ and $\nabla(t)$ are shown in Figure 4-6.

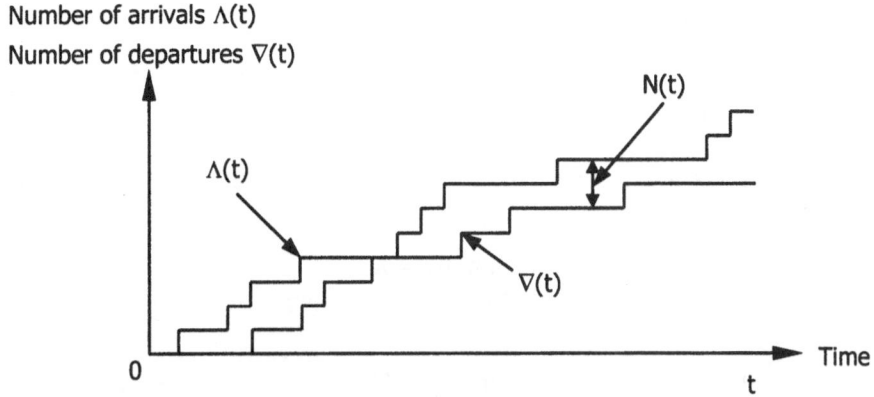

Figure 4-6. Arrival process $\Lambda(t)$ and departure process $\nabla(t)$.

The total area between the two curves of $\Lambda(t)$ and $\nabla(t)$ up to t represents the total time that the customers have spent in the system (measured in units of customer-seconds) during the interval $(0, t)$. Let this cumulative area be $A(t)$, where $A(t) = \int_0^t N(x)dx$. Moreover, let Λ_t be the average arrival rate (customers per second) during the interval $(0, t)$; that is,

$$\Lambda_t = \frac{\Lambda(t)}{t} \tag{4-31}$$

Also, let N_t be the average number of customers in the system during the interval $(0, t)$; that is,

$$N_t = \frac{A(t)}{t} \tag{4-32}$$

The time spent in the system per customer, T_t, in the interval $(0, t)$ is given by

$$T_t = \frac{A(t)}{\Lambda(t)} \tag{4-33}$$

From these last three equations, we have

$$N_t = \Lambda_t \, T_t \tag{4-34}$$

Let us assume that as $t \to \infty$, the following limits exist:

$$\lim_{t \to \infty} \Lambda_t = \lambda$$

$$\lim_{t \to \infty} T_t = T$$

It follows that the following limit also exists

$$\lim_{t \to \infty} N_t = N$$

Taking the limits in (4-34), we have

$$N = \lambda T \tag{4-35}$$

This result is known as Little's formula. It states that the average number of customers in a queueing system is equal to the product of the average arrival rate and the average time a customer spends in the system.

When Little's formula is applied to the queue, the relationship (4-35) becomes

$$N_q = \lambda W \tag{4-36}$$

where N_q represents the average number of customers in the queue and W is the average waiting time.

If Little's formula is applied to the service facility, the relationship (4-35) would be

$$N_s = a' = a = \lambda\tau = \lambda/\mu \tag{4-37}$$

where N_s represents the average number of customers in the service facility (average number of busy servers), $\tau = 1/\mu$ is the average service time, and a' is the carried traffic.

Note that for a single-server queue, the utilization factor or server occupancy ρ, which is defined as the carried traffic per server, and the offered traffic a are equal. Hence, $\rho = a = \lambda\tau$ for a single-server queue.

Since the total time delay in the system is equal to the sum of the waiting time in the queue and the service time, it follows that their mean values satisfy the relation

$$T = W + \tau \tag{4-38}$$

Multiplying this expression by λ and using (4-36) and (4-37) gives

$$N = N_q + N_s \tag{4-39}$$

or

$$N = \lambda W + a \tag{4-40}$$

These formulas are useful for the calculation of the mean queue length N_q and the mean waiting time W when the average number of customers N in the system and the carried traffic a' are known.

Now let us consider a general single-server queueing system (a G/G/1 queue) with average arrival rate λ and average service time $1/\mu$. For the system to be stable, it is necessary that λ/μ be smaller than unity. Suppose that the system is in equilibrium. Let p_0 be the probability that the server is idle at any instant t, or the fraction of the time interval $(0, t)$ during which the server is idle. Then, during the interval $(0, t)$, the server is busy for

$(1 - p_0)t$ seconds. The number of customers served during the interval $(0, t)$ is equal to $(1 - p_0)t\mu$. Thus, for large t, the number of arrivals is equal to the number of customers served during this interval; that is, when the system is in statistical equilibrium:

$$\lambda t = (1 - p_0)t\mu$$

As $t \rightarrow \infty$, we have

$$\lambda = (1 - p_0)\mu$$

or
$$a = \frac{\lambda}{\mu} = 1 - p_0 \qquad\qquad (4\text{-}41)$$

Therefore, for a single-server queueing system, the traffic intensity ρ is equal to the offered traffic a and hence is also equal to $1 - p_0$, which is the fraction of time the server is busy.

Note that formulas (4-35) and (4-36) are especially useful for the calculation of the mean delay time and the mean waiting time in a single-server queueing system, since the quantities N and N_q can be calculated from the state probabilities p_k , $k = 0, 1, \cdots$.

4-7. THE M/M/1 QUEUE

The M/M/1 queue is characterized by a Poisson input process with rate λ, exponential service times with mean $1/\mu$, a single-server, and infinite waiting capacity. It is important to note that state-transitions in the $M/M/1$ queue satisfy all four properties of the birth and death process described in Section 4-5. The behavior of this single-server queue can be described by the birth and death process. We shall determine the equilibrium probabilities p_k, the mean and variance of the number of customers in the system, and the average waiting time.

For the single-server queue in question, we note that the birth rate λ_k corresponds to the arrival rate λ,

$$\lambda_k = \lambda \, , \, k = 0, 1, 2, \, \cdots$$

and the death rate μ_k corresponds to the service rate μ,

$$\mu_k = \begin{cases} 0 & , \quad k = 0 \\ \mu & , \quad k = 1, 2 \quad \cdots \end{cases}$$

Using these birth and death rates in (4-24), we have

$$p_k = \rho^k p_0 \, , \, k = 0, 1, 2, \, \cdots$$

where the utilization factor $\rho = \dfrac{\lambda}{\mu} < 1$ and p_0 is given by (4-25)

$$p_0 = \frac{1}{1 + \displaystyle\sum_{k=1}^{\infty} \rho^k} = 1 - \rho$$

Finally, we have the equilibrium probability distribution

$$p_k = \rho^k (1 - \rho) \, , \, k = 0, 1, \, \cdots \tag{4-42}$$

The waiting probability is defined as the probability that upon arrival, a customer finds the server busy and waits for service in the queue,

$$P_{waiting} = \sum_{k=1}^{\infty} \pi_k = \sum_{k=1}^{\infty} p_k$$

$$= \sum_{k=1}^{\infty} (1 - \rho) \rho^k = \rho \tag{4-43}$$

where $\{\pi_k\}$ is the arriving customer's distribution. Equivalence of π_k and p_k follows from the Poisson arrival assumption and the fact that Poisson arrivals see time averages. These concepts will be elaborated upon in

Section 5-5.

The average number of customers in the system is given by

$$E[N] = \sum_{k=0}^{\infty} k \, p_k = (1 - \rho) \sum_{k=0}^{\infty} k\rho^k$$

$$= \rho(1 - \rho) \sum_{k=0}^{\infty} k\rho^{k-1} = \rho(1 - \rho) \frac{d}{d\rho} \left[\sum_{k=0}^{\infty} \rho^k \right] \qquad (4\text{-}44)$$

$$= \frac{\rho}{1 - \rho}$$

The behavior of $E[N]$ is plotted in Figure 4-7.

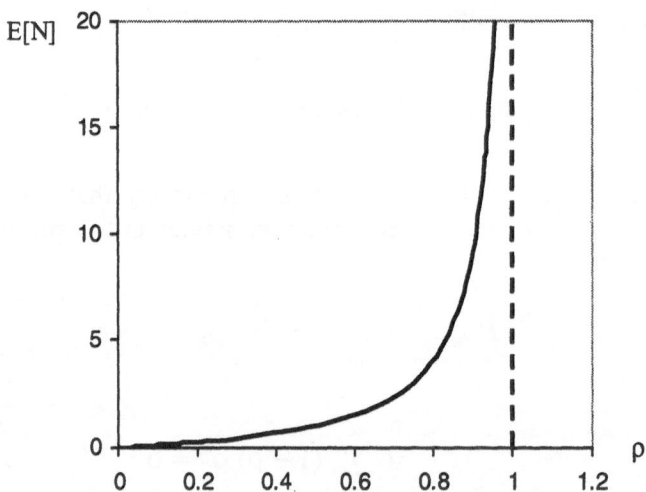

Figure 4-7. Average number of customers $E[N]$ versus offered load ρ.

The variance $Var[N]$ is given by

$$Var[N] = \sum_{k=0}^{\infty} k^2 p_k - (E[N])^2$$

The second moment of N is

$$\sum_{k=0}^{\infty} k^2 p_k = \sum_{k=0}^{\infty} k(k-1)p_k + E[N]$$

$$= \rho^2(1-\rho)\frac{d^2}{d\rho^2} \sum_{k=0}^{\infty} \rho^k + E[N]$$

$$= \frac{2\rho^2}{(1-\rho)^2} + \frac{\rho}{1-\rho}$$

$$= \frac{\rho^2 + \rho}{(1-\rho)^2}$$

Consequently, we find

$$Var[N] = \frac{\rho^2 + \rho}{(1-\rho)^2} - \frac{\rho^2}{(1-\rho)^2}$$

$$\tag{4-45}$$

$$= \frac{\rho}{(1-\rho)^2}$$

By means of Little's formula (4-35) and (4-44), we obtain the average time delay T in the system:

$$T = \frac{E[N]}{\lambda} = \frac{\tau}{1-\rho} \tag{4-46}$$

The dependence of the average time delay T on the utilization factor ρ is shown in Figure 4-8.

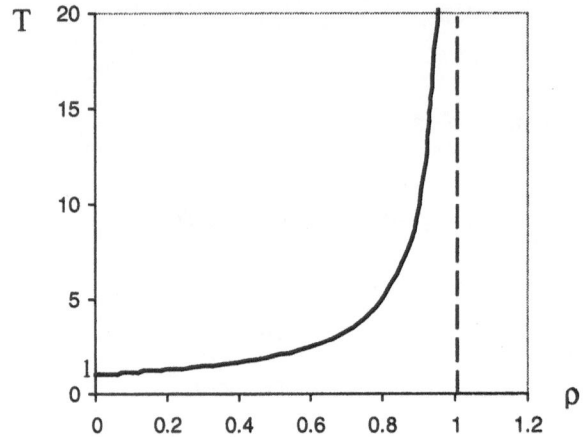

Figure 4-8. Average time delay T versus offered load ρ.

We see from Figure 4-7 and Figure 4-8 that both the average number of customers in the system and the average time delay in the system grow in a unbounded fashion as $\rho \to 1$. This type of behavior is characteristic of almost all queueing systems.

It follows from (4-38) and (4-46) that the mean waiting time W is given by

$$W = T - \tau$$

$$= \frac{\rho\tau}{1 - \rho}$$

(4-47)

Note that when the single-server queue is used to model a data communication channel with a bit rate C in bits per second (bps), average arrival rate λ in packets per second, and average packet length $\bar{L}$ in bits per packet, the average packet service rate becomes $\mu = C/\bar{L}$ in packets per second. Thus, the average service time per packet is given by

$$\tau = \frac{1}{\mu} = \frac{\bar{L}}{C} \quad seconds$$

Example 4-6. A computer supplier proposes a centralized data processing system to a company for entry of orders. The system consists of six visual display terminals which will be used for entry of orders seven hours per day. To investigate whether the system can handle the job, the manager of computing services may apply the queueing theory for analysis.

Suppose that there are four types of orders to be entered as shown in the following table.

Table 4-1. Characteristics of messages.

Message Type				Quantity
A	B	C	D	
160	520	800	520	messages/day
800	2600	4000	2600	messages/week
8	26	40	26	%
100	110	160	20	characters (input)
200	200	210	280	characters (output)
3	3	3	5	disk reads/message
9	9	8	1	disk writes/message
5000	5200	5300	4000	CPU instructions/message
5	5	5	20	Operator think time (sec./message)

The terminal operators are assumed to have an average keying rate of 2 characters/sec. The transmission line has a speed of 4800 bits/sec., as well as a line overhead due to polling and hardware delays of 100 ms/message. The average waiting time for a poll is about 130 ms, and the computer processing time for a message is 1 second.

Messages arriving at the system are assumed to follow a Poisson process with an average interarrival time consisting of the average keying time, the average think time, and the average system response time. The message service times are assumed to be independent, identical and exponentially distributed with 13 control characters (8 bits per character) per message.

Solution. A schematic diagram of the proposed system may be illustrated by Figure 4-9.

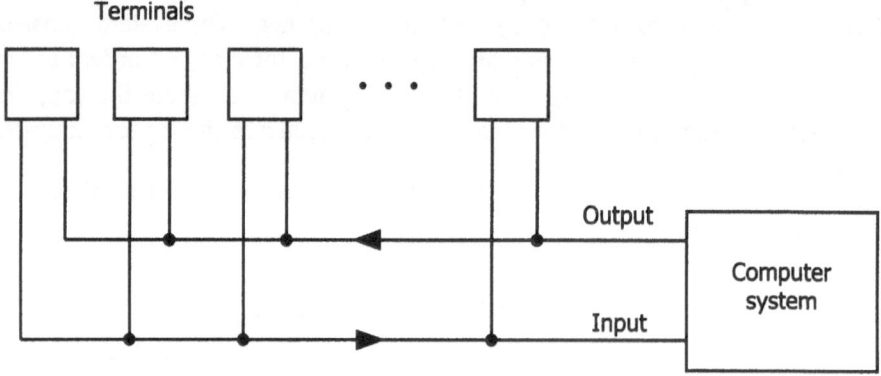

Figure 4-9. System diagram.

In this example, the transmission line and the computer together are regarded as the server. The input traffic is generated by the messages from the keyboard.

In order to model this system by the M/M/1 queue, we have to identify the average input rate, the average service time, and the server. Clearly, the input messages correspond to the customers. To calculate the average input rate, we shall first determine the characteristics of an equivalent message type by combining all four types of messages into one. The result is shown in the following table:

According to the assumptions, we calculate the average service time as follows:

The average time required to transmit an input message is given by

Table 4-2. Equivalent message characteristics.

Messages per day	2000
Characteristics	Mean value/message
characters (input)	106
characters (output)	225
disk reads	3.5
disk writes	6.5
CPU instructions	4912
Operator think time	9 sec.

(Input character + Control character) × *Character length* ×

$$\frac{1}{Transmission\ rate} + Transmission\ line\ overhead$$

$$= (106 + 13) \times \frac{8 \times 1000}{4800} + 100$$

$$= 300\ ms$$

The average time required to transmit an output message is equal to

(Output character + Control character) × *Character length*

$$\times \frac{1}{Transmission\ rate} + Transmission\ line\ overhead$$

$$= (225 + 13) \times \frac{8 \times 1000}{4800} + 100$$

$$= 500\ ms$$

Since the number of input and output messages are the same, the total transmission time per message is $\tau_t = 300 + 500 = 800\ ms$. This transmission time constitutes part of the average message service time. The average system response time consists of the average waiting time for a poll of $130ms$, the average transmission time of 800 ms, and the average computer processing time of 1 $second$. Thus, the average system response time or the average service time τ_s is equal to

$$\tau_s = 0.13 + 0.8 + 1 = 1.93\ sec.$$

and the average message interarrival time per terminal is given by

$$Input\ character \times \frac{1}{Keying\ rate} + Think\ time + System\ response\ time$$

$$= 106 \times \frac{1}{2} + 9 + 1.93$$

$$= 64\ sec.$$

Therefore, the total average arrival rate from all six terminals is

$$\lambda = 6 \times \frac{1}{64} = 0.0938\ message/sec.$$

Suppose that messages are served in the order of arrival. The utilization factor of the system is

$$\rho_s = \lambda\tau_s = 0.0938 \times 1.93 = 0.181$$

The average waiting time in the input queue is given by

$$W = \frac{\rho_s \tau_s}{1 - \rho_s} = \frac{0.181 \times 1.93}{1 - 0.181} = 0.43\ sec.$$

The utilization factor of the transmission line is

$$\rho_t = \lambda \; \tau_t = 0.0938 \times 0.8 = 0.075$$

Since there are 2000 message entries per day, the total number of seconds required is

$$2000 \times 64 = 128,000 \; sec.$$

With 6 terminals available for 7 hours per day, the total number of seconds available is

$$6 \times 7 \times 3600 = 151,200 \; sec.$$

However, if there were 5 terminals, the total time available would only be

$$5 \times 7 \times 3600 = 126,000 \; sec.$$

Thus, 5 terminals would be insufficient to handle the work load. With 6 terminals, the utilization of the terminals is

$$\frac{128,000}{151,200} = 0.85$$

Note that in this example, the interarrival time depends on the service time. The average interarrival time is 64 *sec.*, and the average service time is 1.93 *sec.*, which is also included in the calculated 64 *sec.* We see that in practice, the assumption of independence of interarrival times and service times may not be true in a strictly theoretical sense. However, if the effect is not significantly large, the independence assumption may be considered valid in an approximation sense.

4-8. THE POLLACZEK-KHINCHIN FORMULAS FOR THE M/G/1 QUEUE

In this section, we shall investigate the M/G/1 queue using the method of the imbedded Markov chain, which was introduced by D.G. Kendall in 1951. We shall say that a time epoch is a regeneration point of a random process if the Markov property holds at that point. Thus, at a regeneration point, the future evolution of the random process depends only on the state at that regeneration point. It follows then that every time point of a continuous-time Markov chain is a regeneration point.

Discrete-Time Markov Chain. A stationary discrete-time Markov chain is a random process $\{X_n, n = 0, 1, 2, \cdots\}$ whose values are countable, and for which the transition probabilities p_{ij} satisfy

$$P\{X_{n+1} = j \mid X_n = i, X_{n-1} = i_{n-1}, \cdots, X_1 = i_1, X_0 = i_0\}$$

$$= P\{X_{n+1} = j \mid X_n = i\} = p_{ij}$$

where $p_{ij} \geq 0$ and $\sum_{j=0}^{\infty} p_{ij} = 1$; $i, j = 0, 1, 2, \cdots$.

Imbedded Markov Chain. An imbedded Markov chain is a Markov chain whose states are defined at a discrete set of regeneration points that are imbedded in the nonnegative time points.

Consider a single-server queueing system with a Poisson input of arrival rate λ and a general service time distribution $H(t)$ with mean

$$\tau = \int_0^{\infty} t \, dH(t)$$

We assume that the utilization factor $\rho = \lambda\tau < 1$. This assumption implies that after a sufficiently long time, the system will attain equilibrium. Also, we assume that the waiting capacity is infinite.

By means of the method of the imbedded Markov chain, we shall derive the Pollaczek - Khinchin formula for the mean queue length. Let N_k denote the number of customers in the system (including any customer being served) just after the service completion epoch of the *k*th departing customer, and let X_k denote the number of customers that arrive during the service time of this customer. The idea is to find an equation relating the random variables N_{k+1} and N_k. Depending on the value of N_k , there are two cases, as shown in Figs. 4-10(a) and 4-10(b).

If $N_k > 0$, then the $(k + 1)th$ departure will leave behind those same customers in the queue left by the *k*th customer except for himself, plus all those customers who arrived during the service time of the $(k + 1)th$

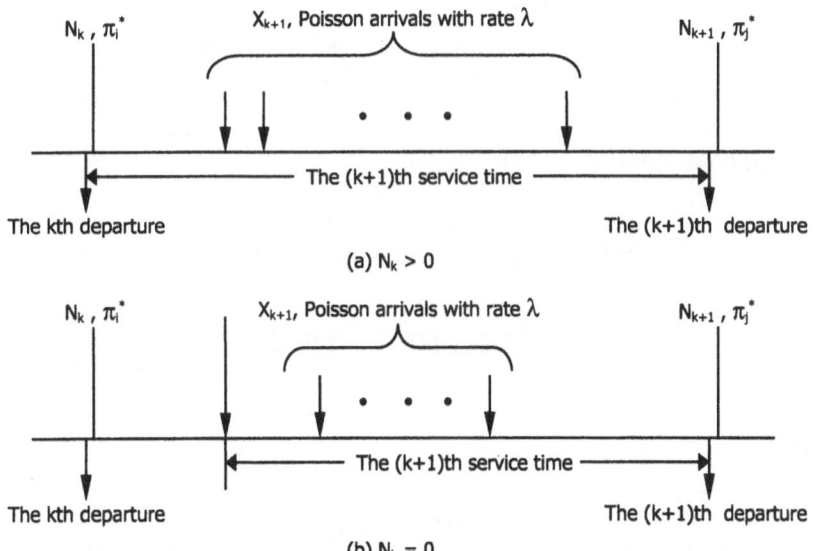

Figure 4-10. Two consecutive regeneration points.

customer. Thus,

$$N_{k+1} = N_k - 1 + X_{k+1} \quad , \quad \text{if} \quad N_k > 0$$

On the other hand, if $N_k = 0$, then the $(k + 1)th$ departure will leave behind just those customers who arrived during his service time. Thus,

$$N_{k+1} = X_{k+1} \quad , \quad \text{if} \quad N_k = 0$$

These two equations can be combined into a single equation:

$$N_{k+1} = N_k - \delta(N_k) + X_{k+1} \tag{4-48}$$

where the random variable $\delta(N_k)$ is defined as

$$\delta(N_k) = \begin{cases} 0 & , \quad \text{if} \quad N_k = 0 \\ \\ 1 & , \quad \text{if} \quad N_k > 0 \end{cases} \tag{4-49}$$

Squaring both sides of (4-48), we get

$$N_{k+1}^2 = N_k^2 + \delta^2(N_k) + X_{k+1}^2 + 2N_k X_{k+1}$$

$$- 2X_{k+1} \, \delta(N_k) - 2N_k \, \delta(N_k) \tag{4-50}$$

Since

$$\delta^2(N_k) = \delta(N_k)$$

and

$$N_k \delta(N_k) = N_k$$

equation (4-50) can be written as

$$N_{k+1}^2 = N_k^2 + \delta(N_k) + X_{k+1}^2 + 2N_k X_{k+1}$$

$$- 2X_{k+1} \delta(N_k) - 2N_k \tag{4-51}$$

Taking expected values of (4-51) and using the fact that N_k and X_{k+1} are independent random variables, we have

$$E[N_{k+1}^2] = E[N_k^2] + E[\delta(N_k)] + E[X_{k+1}^2]$$

$$+ 2E[N_k] E[X_{k+1}] - 2E[X_{k+1}]E[\delta(N_k)] \tag{4-52}$$

$$- 2E[N_k]$$

Similarly, taking expected values through (4-48), we obtain

$$E[N_{k+1}] = E[N_k] - E[\delta(N_k)] + E[X_{k+1}] \tag{4-53}$$

Assuming that an equilibrium distribution exists, then as $k \to \infty$, we have

$$\lim_{k \to \infty} E[N_{k+1}] = \lim_{k \to \infty} E[N_k] = E[N]$$

$$\lim_{k \to \infty} E[N_{k+1}^2] = \lim_{k \to \infty} E[N_k^2] = E[N^2]$$

$$\lim_{k \to \infty} E[X_{k+1}] = \lim_{k \to \infty} E[X_k] = E[X]$$

and

$$\lim_{k \to \infty} E[X_{k+1}^2] = E[X^2]$$

In equilibrium, (4-53) becomes

$$E[\delta(N)] = E[X]$$

and (4-52) reduces to

$$E[N] = \frac{E[X]\{1 - 2E[X]\} + E[X^2]}{2\{1 - E[X]\}} \qquad (4\text{-}54)$$

It remains for us to calculate $E[X]$ and $E[X^2]$. Note that $E[X]$ is simply the average number of arrivals during a service time:

$$E[X] = \int_0^\infty \sum_{k=0}^\infty k \, \frac{(\lambda t)^k}{k!} \, e^{-\lambda t} \, dH(t)$$

$$\qquad (4\text{-}55)$$

$$= \lambda \int_0^\infty t \, dH(t) = \lambda \tau = \rho$$

where τ is the average service time.

Furthermore, the second moment of the number of customers arriving in a service time can be expressed as

$$E[X^2] = \int_0^\infty \sum_{k=0}^\infty k^2 \frac{(\lambda t)^k}{k!} e^{-\lambda t} \, dH(t)$$

$$= \int_0^\infty (\lambda^2 t^2 + \lambda t) \, dH(t) \qquad (4\text{-}56)$$

$$= \lambda^2(\sigma^2 + \tau^2) + \lambda\tau$$

where σ^2 is the variance of the service time and the relation

$$\sigma^2 = \int_0^\infty t^2 dH(t) - \tau^2$$

has been used.

Substitution of (4-55) and (4-56) into (4-54) yields

$$E[N] = \frac{\rho^2[1 + (\sigma^2/\tau^2)]}{2(1 - \rho)} + \rho \qquad (4\text{-}57)$$

Since the average number $E[N]$ of customers in the system is equal to the average number $E[N_q]$ of customers waiting in the queue plus the average number ρ of customers being served,

$$E[N] = E[N_q] + \rho \qquad (4\text{-}58)$$

Comparison of (4-57) and (4-58) gives the Pollaczek - Khinchin formula for the mean queue length:

$$E[N_q] = \frac{\rho^2}{2(1-\rho)} \left[1 + \frac{\sigma^2}{\tau^2} \right] \qquad (4\text{-}59)$$

It follows from Little's formula (4-36) that

$$E[N_q] = \lambda E[W] \qquad (4\text{-}60)$$

Note that in (4-60), instead of using N_q and W as in (4-36), we use the notations $E[N_q]$ and $E[W]$ for the mean queue length and the mean waiting time, respectively. From (4-59) and (4-60), the average waiting time in an $M/G/1$ queue is given by

$$E[W] = \frac{\rho\tau}{2(1-\rho)} \left[1 + \frac{\sigma^2}{\tau^2} \right] \qquad (4\text{-}61)$$

which is the Pollaczek - Khinchin formula for the mean waiting time.

In particular, when the service time is a constant equal to τ, the variance σ^2 of the service time becomes zero and formula (4-61) reduces to

$$E[W] = \frac{\rho\tau}{2(1-\rho)} \qquad (4\text{-}62)$$

A single-server queue with a Poisson input and with constant service time is often referred to as an $M/D/1$ queue. We see that the constant service time gives the smallest mean waiting time in an $M/G/1$ queue.

Table 4-3 shows some important expected values for a single-server queue with Poisson input, FIFO service rule, and various types of service time, where the constant $C_v = \sigma/\tau$ is the coefficient of variation, and where τ and σ are respectively the mean and standard deviation of the service time T_s.

Table 4-3. Expected values for the $M/G/1$ queue.

Expected value	General T_s	Exponential T_s	Constant T_s
	$C_v = \sigma/\tau$	$C_v = 1$	$C_v = 0$
$E[N_q]$	$\dfrac{\rho^2}{2(1-\rho)}(1+C_v^2)$	$\dfrac{\rho^2}{1-\rho}$	$\dfrac{\rho^2}{2(1-\rho)}$
$E[N] = E[N_q] + \rho$	$\dfrac{\rho^2}{2(1-\rho)}(1+C_v^2) + \rho$	$\dfrac{\rho}{1-\rho}$	$\dfrac{\rho^2}{2(1-\rho)} + \rho$
$E[W] = E[N_q]/\lambda$	$\dfrac{\rho\tau}{2(1-\rho)}(1+C_v^2)$	$\dfrac{\rho\tau}{1-\rho}$	$\dfrac{\rho t_s}{2(1-\rho)}$
$E[T] = E[W] + \tau$	$\dfrac{\rho\tau}{2(1-\rho)}(1+C_v^2) + \tau$	$\dfrac{\tau}{1-\rho}$	$\dfrac{(2-\rho)}{2(1-\rho)} t_s$

Probability Generating Function of Number of Customers in System. By squaring the fundamental equation (4-48) and then taking expectations, we obtain (4-57), the Pollaczek-Khinchin formula for mean number of customers in the system. We could find the second moment $E[N^2]$ by first cubing (4-48) and then taking expectations. Similarly, higher moments can be obtained. However, the calculation will be untractable.

We now try to determine the probability distribution for N_{k+1} (actually, we consider the limiting random variable N) as shown in Figure 4-10. As it turns out, we will obtain the probability generating function for N (see reference [4], Vol. 1, p. 191). By definition, the probability generating function for N_k is given by

$$G_k(z) = E[z^{N_k}] = \sum_{j=0}^{\infty} P\{N_k = j\}z^j$$

It follows that the probability generating function for N is given by

$$G(z) = \lim_{k \to \infty} G_k(z) = E[z^N] = \sum_{j=0}^{\infty} \pi_j^* z^j$$

where $\pi_j^* = P\{N = j\}$ and N is the number of customers in the system just after a departure in the steady state.

By using the fundamental equation (4-48), we write

$$G_{k+1}(z) = E[z^{N_{k+1}}] = E[z^{N_k - \delta(N_k) + X_{k+1}}]$$

$$= E[z^{N_k - \delta(N_k)}] E[z^{X_{k+1}}]$$

since X_{k+1} and $N_k - \delta(N_k)$ are independent random variables. Now taking the limit as k goes to infinity yields

$$G(z) = E[z^{N - \delta(N)}] E[z^X]$$

From example 4-7, we have

$$E[z^X] = \hat{H}(\lambda - \lambda z),$$

and hence

$$E[z^{N-\delta(N)}] = P\{N = 0\}z^{0-0} + \sum_{j=1}^{\infty} P\{N = j\}z^{j-1}$$

$$= P\{N = 0\} + \frac{1}{z}\left[\sum_{j=0}^{\infty} P\{N = j\} z^j - P\{N = 0\}z^0\right]$$

$$= P\{N = 0\} + \frac{1}{z}[G(z) - P\{N = 0\}]$$

Thus, we obtain

$$G(z) = \left[P\{N = 0\} + \frac{1}{z}[G(z) - P\{N = 0\}] \right] \hat{H}(\lambda - \lambda z)$$

Using $P\{N = 0\} = 1 - \rho$ and solving for $G(z)$, we find

$$G(z) = (1 - \rho) \frac{(1 - z)\hat{H}(\lambda - \lambda z)}{\hat{H}(\lambda - \lambda z) - z} \qquad (4\text{-}63)$$

This expression is referred to as one form of the Pollaczek-Khinchin transform equation, where $\hat{H}(\lambda - \lambda z)$ is the Laplace-Stieltjes transform of the service time distribution $H(t)$.

In principle, π_j^* can be found by using the inversion formula,

$$\pi_j^* = \frac{1}{j!} G^{(j)}(0) \qquad (4\text{-}64)$$

But, in practice, computing the *jth* derivative is very difficult. However, if the service time has an exponential distribution with mean $1/\mu$, then

$$\hat{H}(\lambda - \lambda z) = \frac{\mu}{\lambda - \lambda z + \mu}$$

and

$$G(z) = \frac{1 - \rho}{1 - \rho z} = \sum_{k=0}^{\infty} (1 - \rho)\rho^k z^k$$

Thus we find $p_k = (1 - \rho)\rho^k$, $k = 0, 1, \cdots$. From the right hand side, we find the state probability distribution for the $M/M/1$ queue as given by (4-42).

Direct Derivation of the Pollaczek-Khinchin Formula for the Mean Waiting Time in the M/G/1 Queue. We have obtained the Pollaczek-Khinchin formula for the mean waiting time by using the method of the imbedded Markov chain. We shall show that the Pollaczek-Khinchin formula can be derived directly through the use of the concept of residual service time discussed in Section 4-4.

Consider the $M/G/1$ queue where customers arrive at the system according to a Poisson process with rate λ. We assume that customers are served in the order of arrival and that the service times Y_k, $k = 1, 2, \cdots$, are mutually independent, identically distributed, and independent of the interarrival times. We also assume that the time average and the statistical average of Y_k are equal:

$$\bar{Y} = E[Y_k] = \tau = \frac{1}{\mu} = Average \ \ service \ \ time$$

$$\overline{Y^2} = E[Y_k^2] = Second \ \ moment \ \ of \ \ service \ \ time$$

$$\bar{R} = E[R] = Average \ \ residual \ \ service \ \ time$$

Depending on the busy-idle condition of the server, the mean waiting time $E[W]$ of a test customer can have two components. Let us first define the following events:

B = on arrival, the test customer finds the server busy with probability ρ and N_q customers waiting;

I = on arrival, the test customer finds the server idle with probability $1 - \rho$.

The mean waiting time of the test customer can then be expressed as

$$E[W] = \rho E[W|B] + (1 - \rho)E[W|I] \qquad (4\text{-}65)$$

Note that for the $M/G/1$ queue under consideration, we have

$$E[W|I] = 0$$

and referring to Figure 4-11,

$$E[W|B] = E[R + Y_1 + Y_2 + \cdots + Y_{N_q}|B]$$

$$= E[R] + \bar{N}_q \, \tau/\rho$$

$$= E[R] + E[W]$$

Substituting this result into (4-65) and solving for $E[W]$, we find the mean waiting time

$$E[W] = \frac{\rho E[R]}{1 - \rho} \qquad (4\text{-}66)$$

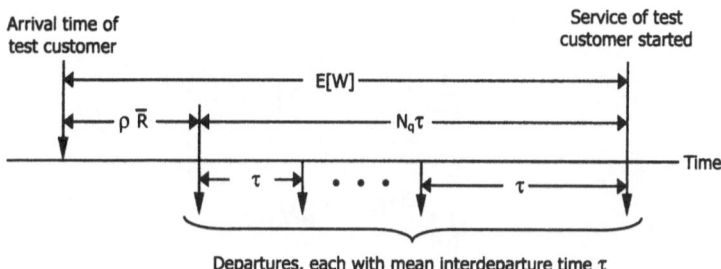

Figure 4-11. Mean waiting time for test customer.

Substitution of the mean residual service time $E[R]$ of (4-18) into (4-66) yields the desired Pollaczek-Khinchin formula (4-61).

Example 4-7. Let the arrival process be Poisson with rate λ, $H(t)$ be the distribution function of service time T_s, and X be the number of arrivals in a service time. Find the probability generating function of X, and calculate the first three moments of X using the probability generating function.

Solution. Since the input process is Poisson with arrival rate λ, X has a Poisson distribution when the service time T_s is t. Thus,

$$P\{X = k \mid T_s = t\} = e^{-\lambda t}\, \frac{(\lambda t)^k}{k!}\ , \ k = 0, 1, 2, \ \cdots$$

Since the service time T_s has the general distribution function $H(t)$, we have

$$P\{X = k\} = \int_0^\infty e^{-\lambda t}\, \frac{(\lambda t)^k}{k!}\, dH(t)$$

The probability generating function of X is given by

$$G_X(z) = E[z^X]$$

$$= \sum_{k=0}^\infty P\{X = k\}\, z^k$$

$$= \sum_{k=0}^\infty \int_0^\infty e^{-\lambda t}\, \frac{(\lambda t z)^k}{k!}\, dH(t)$$

$$= \int_0^\infty e^{-\lambda t(1-z)}\, dH(t)$$

$$= \hat{H}(\lambda - \lambda z)$$

where $\hat{H}(s)$ is the Laplace-Stieltjes transform of the distribution function $H(t)$. Observe that the factorial moment of X is given by

$$E[X(X - 1)\, \cdots\, (X - r + 1)] = \frac{d^r\, G_X(z)}{dz^r}\bigg|_{z=1}$$

In particular, the first three moments of X are given by

$$E[X] = \frac{d}{dz} G_X(z) \mid_{z=1} = \lambda \int_0^\infty t dH(t) = \rho$$

$$E[X^2] = \frac{d^2 G_X(z)\mid_{z=1}}{dz^2} + E[X]$$

$$= \lambda^2 E[T_s^2] + \rho$$

and

$$E[X^3] = \lambda^3 E[T_s^3] + 3\lambda^2 E[T_s^2] + \rho$$

Example 4-8. Information on the expected values of the queue size and the delay time may be insufficient to deduce further details about the behavior of a queueing system. For instance, knowing only the average response time is insufficient to determine whether a certain percentage, say, 95%, of the response time will be less than a certain specified value, say, 3 sec. In this case, the variance and standard deviation may be useful. Show that the variance of N for the M/G/1 queue is given by

$$Var[N] = \frac{\lambda^3 E[T_s^3]}{3(1-\rho)} + \frac{\lambda^4 (E[T_s^2])^2}{4(1-\rho)^2} + \lambda^2 Var[T_s] + E[N]$$

where $\rho = \lambda\tau = \lambda/\mu$ and T_s is the service time.

Solution. It follows from (4-48) that

$$N_{k+1}^3 = [N_k + X_{k+1} - \delta(N_k)]^3$$

$$= N_k^3 + X_{k+1}^3 - \delta(N_k) + 3X_{k+1}N_k^2 + 3N_k X_{k+1}^2 - 3X_{k+1}^2 \delta(N_k)$$

$$+3X_{k+1}\delta(N_k) + 3N_k - 3N_k^2 - 6N_k X_{k+1}$$

Taking expected values on both sides and using the independence of N_k and X_{k+1}, we find

$$E[N_{k+1}^3] = E[N_k^3] + E[X_{k+1}^3] - E[\delta(N_k)] + 3E[X_{k+1}]E[N_k^2]$$

$$+ 3E[N_k]E[X_{k+1}^2] - 3E[X_{k+1}^2]E[\delta(N_k)]$$

$$+ 3E[X_{k+1}]E[\delta(N_k)] + 3E[N_k] - 3E[N_k^2] - 6E[N_k]E[X_{k+1}]$$

which, as $k \to \infty$, reduces to

$$E[N^3] = E[N^3] + E[X^3] - E[\delta(N)] + 3E[X]E[N^2] + 3E[N]E[X^2]$$

$$- 3E[X^2]E[\delta(N)] + 3E[X]E[\delta(N)] + 3E[N] - 3E[N^2]$$

$$- 6E[N]E[X]$$

After substituting the values for $E[\delta(N)]$ by $E[X]$ and using (4-54), (4-55), (4-56), and the results obtained in Example 4-7, we find

$$E[N^2] = \frac{\lambda^3 E[T_s^3]}{3(1-\rho)} + \frac{\lambda^4}{2(1-\rho)^2}(E[T_s^2])^2 + \frac{3\lambda^2}{2(1-\rho)}E[T_s^2] + \rho$$

Finally, using (4-54) and the formula

$$Var[N] = E[N^2] - (E[N])^2$$

we obtain the required result.

Table 4-4 shows some formulas for the variances of N, T, and W for a single-server queue with Poisson input, FIFO queue discipline, and various service time distributions.

Table 4-4. Variances of N, T, and W for the $M/G/1$ queue

Variance	General Service Time T_s	Exponential T_s	Constant T_s
$Var[N]$	$\dfrac{\lambda^3 E[T_s^3]}{3(1-\rho)} + \dfrac{\lambda^4 (E[T_s^2])^2}{4(1-\rho)^2}$ $+ \lambda^2 Var[T_s] + E[N]$	$\dfrac{\rho}{(1-\rho)^2}$	$\dfrac{1}{(1-\rho)^2}(\rho - \dfrac{3\rho^2}{2}$ $+ \dfrac{5\rho^3}{6} - \dfrac{\rho^4}{12})$
$Var[T]$	$Var[T_s] + \dfrac{\lambda E[T_s^3]}{3(1-\rho)} +$ $\dfrac{\lambda^2 (E[T_s^2])^2}{4(1-\rho)^2}$	$\dfrac{\tau^2}{(1-\rho)^2}$	$\dfrac{t_s^2}{(1-\rho)^2}(\dfrac{\rho}{3} - \dfrac{\rho^2}{12})$
$Var[W]$	$\dfrac{\lambda E[T_s^3]}{3(1-\rho)} + \dfrac{\lambda^2 (E[T_s^2])^2}{4(1-\rho)^2}$	$\dfrac{\tau^2(2-\rho)}{(1-\rho)^2}$	$\dfrac{\rho t_s^2}{(1-\rho)^2}\left[\dfrac{4-\rho}{12}\right]$

Example 4-9. In a multi-programmed computer, the available memory is shared by a number of processes. Several programs are held simultaneously in memory. The central processor divides its time among them according to resources such as channels or peripherals. However, a channel control unit of the computer can execute only one channel control program at a time. Since the computer is operating in a multiprogrammed mode, there may be many programs requesting disk accesses via this channel. Suppose that these requests follow a Poisson process with a rate of 10 requests per second, and that the requests which find the channel busy will wait in the queue for service. A sample of 1000 requests collected from observation data of the disk service times are as follows:

(a) Determine the average number of requests in the queue and the average waiting time. (b) Suppose that the input traffic is increased by 10% from 10 to 11 requests per second. How does this increased input traffic affect the channel utilization, the average number of requests in the system, and the average time that a request spends in the system?

Service time, T_s, ms	%
50	10
60	25
70	30
80	20
90	10
100	5

Solution. (a) From the given data, we calculate the average service time:

$$\tau = 50 \times 0.1 + 60 \times 0.25 + 70 \times 0.3 + 80 \times 0.2 + 90 \times 0.1 + 100 \times 0.05$$

$$= 71 \; ms$$

and the variance of T_s:

$$\sigma^2 = E[(T_s - \tau)^2]$$

$$= [(50 - 71)^2 \times 0.1] + (60 - 71)^2 \times 0.25 + (70 - 71)^2 \times 0.3 +$$

$$(80 - 71)^2 \times 0.2 + (90 - 71)^2 \times 0.1 \; + (100 - 71)^2 \times 0.05$$

$$= 169 \; (ms)^2$$

Thus,

$$\sigma = 13 \; ms$$

The coefficient of variation $C_v = \sigma/\tau = 0.18$. Since C_v is not equal to 1, the service time distribution is clearly not exponential. The channel utilization is given by

$$\rho = \lambda\tau = 10 \times 71 \times 10^{-3} = 0.71$$

From Table 4-3, we find the average queue length:

$$E[N_q] = \frac{\rho^2}{2(1 - \rho)}(1 + C_v^2)$$

$$= 0.897 \; requests$$

and the average waiting time:

$$E[W] = \frac{E[N_q]}{\lambda} = 89.7 \; ms$$

(b) For $\lambda = 10 \; requests/sec.$

$$E[T] = E[W] + E[T_s] = 89.7 + 71 = 161 \; ms$$

$$E[N] = E[N_q] + \rho = 0.897 + 0.71 = 1.61 \; requests$$

Since $\rho = \lambda\tau$, ρ will increase by 10 percent if λ is increased by 10 percent. Thus, for $\lambda = 11$, and $\rho = 0.78$, we find

$$E[N_q] = 1.43 \; , \; E[W] = 130 \; ms$$
$$E[T] = 201 \; ms$$
$$E[N] = 2.21 \; requests$$

Therefore, the effect of a 10 percent increase in λ is that $E[T]$ will increase from 161 ms to 201 ms, which is about 25%, and that $E[N]$ will increase from 1.61 to 2.21, which is about 37%. This amplification effect becomes more prominent when the utilization factor ρ approaches 1, for there is a factor $(1 - \rho)$ in the denominator of the expressions for $E[W]$ and $E[N_q]$.

4-9. THE GI/M/1 QUEUE

In Section 4-8, we studied the M/G/1 queue using the method of the imbedded Markov chain, where the service time has an arbitrary distribution. Now we shall study the GI/M/1 queue for which the interarrival times are mutually independent and identically distributed random variables and have an arbitrary distribution, and the service time has an exponential distribution.

Suppose that customers arrive at time epochs $T_1, T_2, \cdots$, corresponding to the arrival times. We assume that the interarrival times $X_{k+1} = T_{k+1} - T_k$, $k = 0, 1, \cdots$; $T_0 = 0$, are mutually independent and identically distributed random variables, with the general distribution function $G(x) = P\{X_{k+1} \le x\}$, $k = 0, 1, \cdots$, and average interarrival time $1/\lambda$. Arriving customers finding the server busy wait as long as necessary for service. The service time has an exponential distribution with mean $1/\mu$.

To analyze the $GI/M/1$ queue, we let N_k be the number of customers in the system just prior to the arrival of the kth customer; thus, N_k denotes the number of customers present at time $T_k - 0$. Figure 4-12 shows two consecutive arrivals:

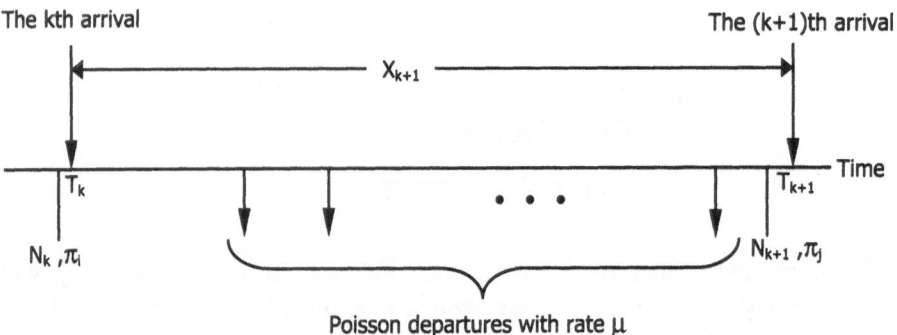

Figure 4-12. Two consecutive arrivals and the departure process.

From the formula of total probability, we write

$$P\{N_{k+1} = j\} = \sum_{i=0}^{\infty} P\{N_{k+1} = j \mid N_k = i\} \, P\{N_k = i\} \, , \; j = 0, 1, \; \cdot(4\text{-}67)$$

Since the conditional probability $P\{N_{k+1} = j \mid N_k = i\}$ depends on the indices i and j, but not on the index k, we write

$$p_{ij} = P\{N_{k+1} = j \mid N_k = i\}$$

According to the theory of Markov chains, if $\rho = \dfrac{\lambda}{\mu} < 1$, then a unique stationary distribution

$$\pi_j = \lim_{k \to \infty} P\{N_k = j\} \, , j = 0, 1, \; \cdots$$

exists. It follows that as $k \to \infty$, (4-67) becomes

$$\pi_j = \sum_{i=0}^{\infty} p_{ij} \, \pi_i \, , j = 0, 1, \; \cdots \tag{4-68}$$

where the stationary probabilities $\{\pi_j\}$ are subject to the normalization condition

$$\sum_{j=0}^{\infty} \pi_j = 1 \tag{4-69}$$

We now proceed to determine the transition probabilities $\{p_{ij}\}$. First, we observe that given that $N_k = i$, the value of N_{k+1} is at most $i + 1$ or the $(k+1)th$ arrival can find at most $i + 1$ customers in the system; hence, we have

$$p_{ij} = 0 \quad , \quad j > i + 1, \; i = 0, 1, \; \cdots \tag{4-70}$$

Next, for $j \le i + 1$, we consider the case where the $(k + 1)th$ arrival finds the server busy; this server must have been continuously busy during the interarrival time $X_{k+1} = T_{k+1} - T_k$. Since the service times are exponentially distributed with mean $1/\mu$ and the server has been continuously busy during the interarrival time $X_{k+1} = x$, the departure process obeys the Poisson distribution with rate μ. It follows that the probability of exactly $i + 1 - j$ departures during a given time interval of length x is

$$P\{N_{k+1} = j \mid N_k = i, X_{k+1} = x\} = \frac{(\mu x)^{i+1-j}}{(i + 1 - j)!} e^{-\mu x}$$

Since X_{k+1} has the distribution function $G(x)$,

$$p_{ij} = \int_0^\infty \frac{(\mu x)^{i+1-j}}{(i + 1 - j)!} e^{-\mu x} \, dG(x), \, i \geq 0, \, 1 \leq j \leq i + 1 \quad (4\text{-}71)$$

Substitution of (4-71) into (4-68) yields

$$\pi_j = \sum_{i=0}^\infty \pi_i \int_0^\infty \frac{(\mu x)^{i+1-j}}{(i + 1 - j)!} e^{-\mu x} \, dG(x)$$

$$\hspace{6cm} (4\text{-}72)$$

$$= \sum_{i=0}^\infty \pi_{i+j-1} \int_0^\infty \frac{(\mu x)^i}{i!} e^{-\mu x} \, dG(x), \, 1 \leq j \leq i + 1$$

To solve (4-72) for π_j, we assume a solution of the form

$$\pi_j = A\sigma^j, \, j = 0, 1, \cdots, 0 < \sigma < 1 \qquad (4\text{-}73)$$

where A and σ are unknown quantities and remain to be determined.

Substitution of (4-73) into (4-72) yields

$$\sigma = \int_0^\infty e^{-(1-\sigma)\mu x} \, dG(x) = \hat{G}((1-\sigma)\mu) \qquad (4\text{-}74)$$

where $\hat{G}((1-\sigma)\mu)$ is the Laplace-Stieltjes transform of the interarrival time distribution function $G(x)$. The quantity σ in (4-73) is a solution of (4-74).

Note that the function $\hat{G} \ ((1-\sigma)\mu)$ has the following properties:

$$\hat{G}((1-\sigma)\mu) \, |_{\sigma=0} = \hat{G}(\mu) = \int_0^\infty e^{-\mu x} \, dG(x) > 0$$

and

$$\hat{G}((1-\sigma)\mu) \, |_{\sigma=1} = \hat{G}(0) = 1$$

Furthermore, for $0 < \sigma < 1$, we have

$$\left. \frac{d\hat{G}}{d\sigma} \right|_{\sigma=1} = \frac{\mu}{\lambda} = \frac{1}{\rho}$$

$$> 1$$

Thus, (4-74) possesses a unique real solution for σ in the range $0 < \sigma < 1$. This is shown in Figure 4-13.

It remains to calculate the constant A. Note that using the normalization condition (4-69) and (4-73) gives

$$\sum_{j=0}^\infty \pi_j = \sum_{j=0}^\infty A \sigma^j = \frac{A}{1-\sigma} = 1$$

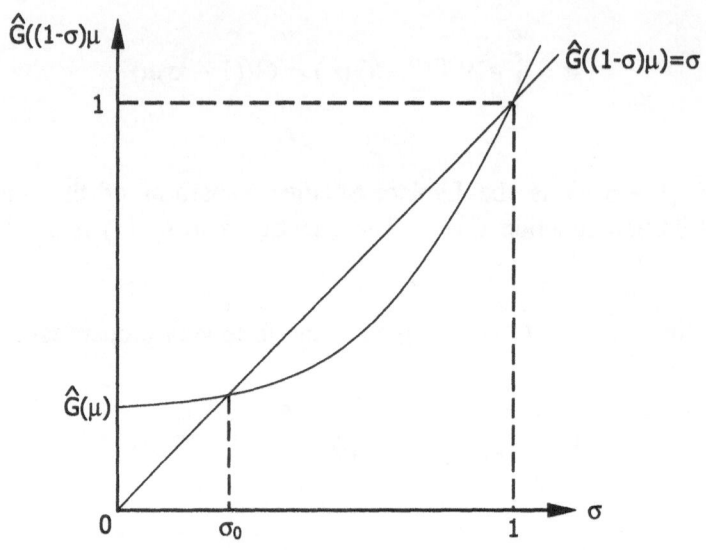

Figure 4-13. Real solution of the equation $\hat{G}((1 - \sigma)\mu) = \sigma$.

Hence,

$$A = 1 - \sigma$$

and

$$\pi_j = (1 - \sigma)\, \sigma^j \, , j = 0, 1, \cdots \qquad (4\text{-}75)$$

Waiting Probability and the Mean Waiting Time for the GI/M/1 Queue. An arriving customer must wait for service if, on arrival, he finds the server busy. Therefore, the waiting probability is given by

$$P\{W > 0\} = P_{waiting} = \sum_{j=1}^{\infty} \pi_j$$

$$= \sum_{j=1}^{\infty} (1 - \sigma)\sigma^j = \sigma \qquad (4\text{-}76)$$

Using exactly the same argument as in the derivation of (4-66), except that the waiting probability is σ instead of ρ and that $E[R]$ is simply the mean service time τ because of the exponential service time, we immediately obtain the mean waiting time for the $GI/M/1$ queue

$$E[W] = \frac{\sigma\tau}{1 - \sigma}. \tag{4-77}$$

4-10. THE M/G/1 QUEUE WITH PRIORITY DISCIPLINE

For priority queueing, customers are classified into different classes, each with a priority index. The usual convention to number the priority classes is that the smaller the priority index, the higher the priority; that is, customers of priority 1 are served before customers of priority $j \geq 2$.

Within each priority class, there are different service disciplines. The most common is head-of-the-line (HOL) service, which implies that customers are served in order of arrival in a given class.

There are two other refinements possible in priority queueing systems:

1. Preemptive

2. Nonpreemptive

For preemptive systems, a customer with higher priority is allowed to disrupt the service of a customer with lower priority. Such a disruption leads to a further classification about whether the disrupted service is to be continued from the point of interruption at a later time or be restarted from the beginning. For nonpreemptive systems, there is no interruption because the higher priority customer must wait until the service in progress is completed.

We shall consider the HOL nonpreemptive priority system with m priority classes, and derive an expression for the average waiting delay W_j encountered by a typical test customer from priority class j, $1 \leq j \leq m$. Figure 4-14 illustrates a nonpreemptive HOL priority queueing system.

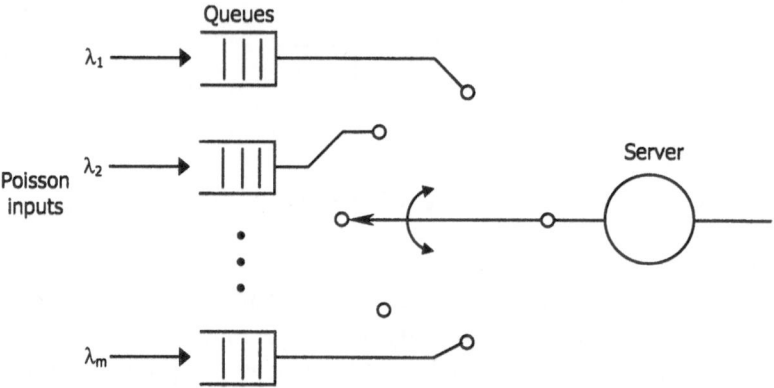

Figure 4-14. Nonpreemptive head-of-the-line (HOL) priority queueing
system

Each priority queue operates on a first-come, first-served basis while
being serviced by the server. After a service is completed, the switch
moves to the highest priority queue which has customers ready for service.

Customers arrive at each priority queue from independent Poisson
processes with arrival rate λ_i, $i = 1, 2, \cdots, m$. Let
$\bar{X}_i$, $i = 1, 2, \cdots, m$, denote the average service time for a customer of
priority i.

Let

$$\rho_i = \text{The offered traffic to the queue of priority } i$$

Then, by definition,

$$\rho_i = \lambda_i \bar{X}_i \qquad\qquad (4\text{-}78)$$

If ρ denotes the total network offered traffic, then under steady-state and

$$\rho = \sum_{i=1}^{m} \rho_i < 1 \qquad (4\text{-}79)$$

The average delay W_j encountered by a test customer of priority j consists of three components:

1. The average delay, if any, due to a customer already in service, W_0;

2. The average delay due to customers of higher or equal priorities that arrive before the test customer and wait in queues, W_A, and

3. The average delay due to customers of higher priorities that arrive after the test customer, W_B.

We shall derive expressions for the mean values of these three components of delay.

By definition, W_0 is precisely the mean value of the residual service time. Let R_i denote the residual service time of a priority i customer. Its mean value $\bar{R}_i$ can then be determined from the service time distribution of the priority i customer.

If X_i is the service time of a priority i customer and has an exponential distribution, then because of the Markov property of the exponential distribution, we have the relationship

$$\bar{R}_i = \bar{X}_i \qquad (4\text{-}80)$$

If X_i is not exponentially distributed, then $\bar{R}_i$ is given by (4-18)

$$\bar{R}_i = \overline{X_i^2} / 2\bar{X}_i \qquad (4\text{-}81)$$

Since there are m priority classes and since the probability of a priority i customer being served at the arrival time of the priority j test customer is ρ_i, the average delay W_0 due to a customer of any priority class found in service is thus given by

$$W_0 = \sum_{i=1}^{m} \frac{\rho_i \overline{X_i^2}}{2\overline{X_i}}$$

$$= \sum_{i=1}^{m} \frac{\lambda_i \overline{X_i^2}}{2}$$

(4-82)

Now consider the average delay W_A due to customers of higher or equal priorities that are in the system before the arrival of the test customer of priority j. Let N_{ij} denote the number of such customers from priority class $i \leq j$.

Since each of the N_{ij} customers requires an average service time $\overline{X_i}$, the total average service time for priority class i is then simply $\overline{N_{ij}}\overline{X_i}$. Note that only those priority classes with an index $i \leq j$ will contribute to W_A. Thus, the total average delay W_A from priority classes with index $i \leq j$ is given by

$$W_A = \sum_{i=1}^{j} \overline{N_{ij}}\overline{X_i}$$

(4-83)

Using Little's formula, we can write

$$\overline{N_{ij}} = \lambda_i W_i \, , \, i = 1, 2, \cdots , j$$

Thus,

$$W_A = \sum_{i=1}^{j} \rho_i W_i$$

(4-84)

To calculate the average delay W_B caused by customers of priority $i < j$ who arrive after the test customer, we let M_{ij} denote the number of such customers. Now only customers with priority $i < j$ will contribute to W_B. Therefore, we obtain

$$W_B = \sum_{i=1}^{j-1} \overline{M}_{ij} \overline{X}_i \tag{4-85}$$

Since $\overline{M}_{ij}$ is caused by the average waiting time W_j, by using Little's formula, we have

$$\overline{M}_{ij} = \lambda_i W_j$$

and

$$W_B = \sum_{i=1}^{j-1} \rho_i W_j \tag{4-86}$$

Hence,

$$W_j = W_0 + W_A + W_B$$
$$= W_0 + \sum_{i=1}^{j} \rho_i W_i + \sum_{i=1}^{j-1} \rho_i W_j$$

Solving this equation for W_j gives

$$W_j = \frac{W_0 + \sum_{i=1}^{j-1} \rho_i W_i}{1 - \sum_{i=1}^{j} \rho_i} \quad , j = 1, 2, \cdots, m \tag{4-87}$$

In particular, when $m = 2$, we have

$$W_1 = \frac{W_0}{1 - \rho_1} \tag{4-88}$$

and

$$W_2 = \frac{W_0 + \rho_1 W_1}{1 - \rho_1 - \rho_2}$$

Substituting W_1 into this expression gives

$$W_2 = \frac{W_0}{(1 - \rho_1 - \rho_2)(1 - \rho_1)} \tag{4-89}$$

For this particular case of only two priority groups, (4-82) yields the average delay W_0:

$$W_0 = \frac{1}{2} (\lambda_1 \overline{X_1^2} + \lambda_2 \overline{X_2^2}) \tag{4-90}$$

Example 4-10. Suppose that a computer system is modeled as a single-server queue with two buffers storing two classes of jobs (messages) while they wait to be served by the CPU which has a processing speed of $10^6 bps$. The computer system is assumed to operate on a HOL nonpreemptive priority discipline so that the jobs in the first buffer have higher priority than those in the second one. Each of the buffers operates on a first-come, first-served basis. Jobs in the second buffer will be served only when the first buffer is empty.

Jobs arrive at each priority buffer from independent Poisson processes with rates $\lambda_1 = 200$ messages/second, and $\lambda_2 = 100$ messages/second. Let the corresponding message length be exponentially distributed with mean $\overline{L}_1 = 640$ bits, and constant message length $\overline{L}_2 = 256$ bits. Calculate the average delay for each class of message.

Solution. Since $\overline{X}_i = \dfrac{\overline{L}_i}{C}$, $i = 1, 2$, and $C = 10^6$ bps, then

$$\bar{X}_1 = 0.64 \; ms$$

$$\bar{X}_2 = 0.256 \; ms$$

$$\rho_1 = \lambda_1 \bar{X}_1 = 0.128$$

$$\rho_2 = \lambda_2 \bar{X}_2 = 0.0256$$

and

$$\rho = \lambda_1 \bar{X}_1 + \lambda_2 \bar{X}_2 = 0.1536$$

Note that for an exponential message length distribution, the second moment $\overline{X^2} = 2 \, \bar{X}^2$. On the other hand, for a constant message length, $\overline{X^2} = \bar{X}^2$. The average residual service time is

$$W_0 = \frac{1}{2} \, (\lambda_1 \overline{X_1^2} + \lambda_2 \overline{X_2^2}) = \frac{1}{2} \, (2\lambda_1 \, \bar{X}_1^2 + \lambda_2 \, \bar{X}_2^2 \,)$$

$$= 0.085 \; ms$$

$$W_1 = \frac{W_0}{1 - \rho_1} = 0.098 \; ms$$

and

$$W_2 = \frac{W_0}{(1 - \rho_1)\,(1 - \rho_1 - \rho_2)} = 0.115 ms$$

4-11. The M/G/1 Queue with Vacations

Generally, a queueing system will have busy periods with at least one customer present and idle periods with no customers present. A busy period is a time interval that begins when an arriving customer finds the system empty and ends when a departing customer leaves the system empty. An idle period is the period between two successive busy periods. Clearly, busy and idle periods occur alternately and form a cycle.

Suppose that at the end of each busy period, the server goes on vacation for a random interval of time with first moment $\bar{V}$ and second moment $\overline{V^2}$. For computer communication networks, vacations correspond to transmissions of various kinds of control and record-keeping packets when there is little traffic.

To derive an expression for the mean waiting time in the $M/G/1$ queue with vacations, the right-hand side of (4-65) must be modified by adding the delay $E[R_V]$. This delay is caused by the mean residual vacation time, which is equal to $\overline{V^2}/2\bar{V}$ with the probability of occurrence $1 - \rho$. With this modification, (4-65) now becomes

$$E[W] = \rho\{\bar{R} + E[W]\} + (1 - \rho)E[R_V] \qquad (4\text{-}91)$$

Solving for $E[W]$ yields the mean waiting time

$$E[W] = \frac{\rho\tau}{2(1 - \rho)}\left[1 + \frac{\sigma^2}{\tau^2}\right] + \frac{\overline{V^2}}{2\bar{V}} \qquad (4\text{-}92)$$

This result is an extension of (4-61) to the $M/G/1$ queue with server vacations. Thus (4-92) can be regarded as the Pollaczek-Khinchin formula for the mean waiting time with server vacations.

4-12. SUMMARY

Performance analysis of telecommunication systems requires a mathematical model for the representation of the systems. This chapter begins with the modeling of the input traffic process for telephone systems, and develops a Poisson distribution with a constant calling rate as the basic model. The Poisson random variable represents the number of calls arriving in the system during a fixed time interval. In addition, an exponential distribution is obtained as a model for the interarrival time. Both models involve the same calling rate as a parameter and can be used to represent the same physical input process.

According to measurements, the holding times have been found to be very close to an exponential distribution with the reciprocal of the average holding time as a parameter. The most important characteristic of the exponential distribution is the Markov property, or the memoryless property. This property plays an important role in the investigation of the performance of telecommunication systems. It facilitates the investigation of the waiting time distribution and the departure process.

The birth and death process can be used as a fundamental model for the study of many telecommunication systems and queueing systems. By choosing the corresponding birth and death rates, the state probability distribution $\{p_i\}$ for the $M/M/1$ queue is determined under steady-state conditions. Using Little's formula, we obtain the mean number of customers in the $M/M/1$ queueing system, as well as the mean time delay in the system.

The method of the imbedded Markov chain is used for the determination of the state probability distribution at the regeneration points for both the $M/G/1$ and $GI/M/1$ queues. We also obtain expressions for the mean waiting time for the $M/G/1$ queue with and without priority discipline and with vacations. The Pollaczek-Khinchin formula is obtained by defining the random variable just after the departure epoch. We determine the probability generating function for the state probability distribution for the $M/G/1$ queue in terms of the Laplace-Stieltjes transform of the service time distribution.

APPENDIX A - Review of Probability Theory

4A-1. Random Experiments

Consider an experiment which may be repeated many times under essentially identical conditions, each individual experiment giving a certain definite result. If the variability of these results are unpredictable, then the experiment is said to be random.

Features of a random experiment:

(a) Each experiment may be repeated indefinitely many times under essentially identical conditions;

(b) Although a particular outcome of the experiment is unpredictable, the set of all possible outcomes of the experiment can be specified; and

(c) The individual outcomes of the experiment seem to occur randomly.

4A-2. The Sample Space and Events

An event is the result of an experiment. Events may be divided into two types: simple events and compound events. Simple events are indecomposable and compound events are aggregates of simple events and hence are decomposable.

In mathematics, the simple events are called sample points. The totality of all sample points (which represents the set of all possible outcomes of a given experiment) is called the sample space of the experiment.

4A-3. Probability and Relative Frequency

If an experiment is repeated n times under the same conditions and the event A occurs n_A times, then the probability of occurrence of the event A is defined by the limit of the relative frequency of occurrence of the event A.

$$P(A) = \lim_{n \to \infty} f_A = \lim_{n \to \infty} \frac{n_A}{n} \qquad (4A\text{-}1)$$

The relative frequency f_A has the following properties:

(a) $0 \leq f_A \leq 1$,

(b) $f_A = 1$ if and only if the event A occurs every time among the n repetitions;

(c) $f_A = 0$ if and only if the event A never occurs among the n repetitions;

(d) If A and B are two mutually exclusive events and if f_{A+B} is the relative frequency associated with the event $A + B$ (A or B), then

$$f_{A+B} = f_A + f_B$$

Since, in any physical experiment, the number n must be finite, $f_A = n_A/n$ cannot be equated to a limit. The existence of the limit is a hypothesis and not an experimental result. In practical applications, when n is very large, we can assume

$$P(A) \doteq f_A = \frac{n_A}{n} \qquad\qquad (4A\text{-}2)$$

Note that the basis of modern probability theory due to A.N. Kolmogorov starts from the assumption that to all possible events in an experiment a (real) numerical value is assigned. Thus the probability of A is a nonnegative real number $P(A)$ assigned to the event A. $P(A)$ satisfies the following postulates:

(a) $0 \leq P(A) \leq 1$;

(b) $P(S) = 1$, where S is the certain event;

(c) If A and B are mutually exclusive events, then $P(A + B) = P(A) + P(B)$.

If A_k, $k = 1, 2, \cdots, n$ are pairwise mutually exclusive events, then

$$P\left[\sum_{k=1}^{n} A_k\right] = \sum_{k=1}^{n} P(A_k)$$

(d)   If A_k, $k = 1, 2, \ldots$, are pairwise mutually exclusive events, then

$$P\left[\sum_{k=1}^{\infty} A_k\right] = \sum_{k=1}^{\infty} P(A_k)$$

This approach to probability is often called the axiomatic approach, where the nonnegative real number $P(A)$ is undefined.

4A-4. The Addition Rule of Probability

If A and B are any two events associated with a random experiment, then

$$P(A + B) = P(A) + P(B) - P(AB) \qquad (4A\text{-}3)$$

This formula is called the addition rule of probability.

In particular, if A and B are mutually exclusive events, that is, they cannot occur together, then $P(AB) = 0$. It follows from (4A-3)

$$P(A + B) = P(A) + P(B) \qquad (4A\text{-}4)$$

4A-5. Conditional Probability and the Multiplication Rule of Probability

The probability of A conditional on B, $P(A|B)$, is the probability that A occurs given that B has occurred. The conditional probability $P(A|B)$ is defined as

$$P(A|B) = P(AB)/P(B) \ , \ \ P(B) \neq 0 \qquad (4A\text{-}5)$$

It follows that $P(AB)$ can be expressed as

$$P(AB) = P(A \mid B)P(B) = P(B \mid A)P(A)$$

This expression is called the multiplication rule of probability. In general, the multiplication rule may be generalized to more than two events as follows.

$$P\left[\sum_{k=1}^{n} A_k\right] = P(A_1)P(A_2 \mid A_1)P(A_3 \mid A_1 A_2) \cdots \qquad (4A\text{-}6)$$

$$P(A_n \mid A_1 A_2 \cdots A_{n-1})$$

4A-6. (Statistically) Independent Events

In A and B are independent events, then

$$P(AB) = P(A)P(B) \qquad (4A\text{-}7)$$

It follows from the definition of conditional probability that if A and B are independent events, then the probability of A does not depend on the occurrence of B. Hence

$$P(A \mid B) = P(A) \quad \text{and} \quad P(B \mid A) = P(B) \qquad (4A\text{-}8)$$

4A-7. The Formula of Total Probability

If $B_1, B_2, \cdots, B_n$ are pairwise mutually exclusive events, which constitute a complete system, that is, the sum or union of all B_i's equals the sample space S,

$$B_i B_j = 0, \; i \neq j; \; i, j = 1, 2, \cdots, n,$$

$$\sum_{i=1}^{n} B_i = S \qquad \text{and} \qquad P(B_i) \neq 0.$$

then
$$P(A) = \sum_{i=1}^{n} P(A \mid B_i)P(B_i) \qquad (4A-9)$$

where A is an arbitrary event with respect to S. This expression is called the formula of total probability.

4A-8. Random Variables

If $X(s)$ is a function of s, where s is an element of the sample space S, then $X(s)$ is called a random variable. In applications, $X(s)$ is commonly called a random variable and is denoted by X. Furthermore, if $s = X(s) = x$, that is, s and x are one-to-one correspondence, then $S = R_X$, where R_X denotes the range space of X. All events associated with a random experiment is denoted by the random variable X in the form $\{X = x\}$ and $P\{X = x\}$ is its corresponding probability. Thus, $P\{X = x\}$ is a function of x and is called the probability function of X.

4A-8-1. Discrete Random Variables

If X takes only a finite or countably infinite number of distinct values x, with corresponding probabilities $p_x = P\{X = x\}$ and $\sum_{x=-\infty}^{\infty} p_x = 1$, then X is said to be a discrete random variable.

4A-8-2. Continuous Random Variables

A random variable is said to be continuous if there exists a function $f(x)$, called the probability density function of X, satisfying the following conditions:

(a) $f(x) \geq 0$ for all x;

(b) $\int_{-\infty}^{\infty} f(x)dx = 1$; and

(c) for any $a < b$, $\int_{a}^{b} f(x)dx = P\{a < X \leq b\} = F(b) - F(a)$

where $F(x) = P\{X \leq x\} = \int_{-\infty}^{x} f(\sigma)d\sigma$, is called the distribution function of X.

4A-9. Mean and Variance

If X is a discrete random variable, its mean $\bar{X}$ or expectation $E[X]$ is defined as

$$\bar{X} = E[X] = \sum_{x=-\infty}^{\infty} x p_x \tag{4A-10}$$

and its variance $Var[X]$ is defined as

$$Var[X] = \sum_{x=-\infty}^{\infty} (x - \bar{X})^2 p_x$$

$$= \sum_{x=-\infty}^{\infty} x^2 p_x - \bar{X}^2 \tag{4A-11}$$

If X is a continuous random variable, its mean $\bar{X}$ or expectation $E[X]$ is defined as

$$\bar{X} = E[X] = \int_{-\infty}^{\infty} x f(x) dx \tag{4A-12}$$

and its variance $Var[X]$ is defined as

$$Var[X] = \int_{-\infty}^{\infty} (x - \bar{X})^2 f(x) dx = \int_{-\infty}^{\infty} x^2 f(x) dx - \bar{X}^2 \tag{4A-13}$$

If X and Y are independent random variables, then

$$E[X \pm Y] = E[X] \pm E[Y] \tag{4A-14}$$

and

$$Var[X \pm Y] = Var[X] + Var[Y] \qquad (4A-15)$$

The standard deviation σ_X is defined as the square root of the variance $Var[X]$, that is.

$$\sigma_X = \sqrt{Var[X]} . \qquad (4A-16)$$

APPENDIX B - Review of Markov Chain Theory

4B-1. Discrete-Time Markov Chains

Consider a random process $\{X_k, k = 0, 1, \cdots \}$ that takes on a finite or countable number of possible values. If $X_k = i$, then the process is said to be in state i at time k. Suppose that whenever the process is in state i, there is a fixed probability p_{ij} that it will next be in state j, that is,

$$P\{X_{k+1} = j | X_k = i, X_{k-1} = i_{k-1}, \cdots, X_1 = i_1, X_0 = i_0\} =$$

$$P\{X_{k+1} = j | X_k = i\} \qquad (4B-1)$$

$$= p_{ij}$$

for all states $i_0, i_1, \cdots, i_{k-1}, i, j$ and all $k \geq 0$. Such a random process is known as a Markov chain. The numbers p_{ij} are called the transition probabilities that must satisfy

$$p_{ij} \geq 0, i, j \geq 0 ; \sum_{j=0}^{\infty} p_{ij} = 1 , i = 0, 1, \cdots$$

The corresponding transition probability matrix denoted by

$$P = \begin{bmatrix} p_{00} & p_{01} & p_{02} & \cdots \\ p_{10} & p_{11} & p_{12} & \cdots \\ p_{i0} & p_{i1} & p_{i2} & \cdots \\ \cdots & \cdots & \cdots & \cdots \end{bmatrix} \qquad (4B-2)$$

is called the one-step transition probability matrix.

The n-step transition probabilities are denoted by

$$p_{ij}^n = P\{X_{n+m} = j \,|\, X_m = i\} \,,\, n \geq 0 \,,\, i, j \geq 0 .$$

Note that $p_{ij}^1 = p_{ij}$. The Chapman-Kolmogorov equations provide a method for computing p_{ij}^n. They are given by

$$p_{ij}^{n+m} = \sum_{k=0}^{\infty} p_{ik}^n p_{kj}^m \,,\, n, m \geq 0 \,,\, i, j \geq 0 \qquad (4B\text{-}3)$$

The n-step transition probability matrix P^n may be obtained by multiplying the matrix P by itself n times. The n-step transition probabilities p_{ij}^n are the elements of the matrix P^n.

Two states i and j are said to communicate if for some integers n and m such that $p_{ij}^n > 0$ and $p_{ji}^m > 0$. If all states communicate, then the Markov chain is said to be irreducible. A state i of a Markov chain is periodic if there exists some integer $m \geq 1$ such that $p_{ii}^m > 0$ and some integer $d > 1$ such that $p_{ii}^n > 0$ only if n is a multiple of d. A state of a Markov chain is said to be aperiodic if none of its states is periodic.

A probability distribution $\{p_k \,,\, k \geq 0\}$ is said to be a stationary distribution for the Markov chain if

$$p_j = \sum_{i=0}^{\infty} p_i p_{ij} \,,\, j = 0, 1, \cdots \qquad (4B\text{-}4)$$

In applications, we will encounter only the irreducible and aperiodic Markov chains. For such a chain, the limit

$$p_j = \lim_{n \to \infty} P\{X_n = j \,|\, X_0 = i\} \,,\, i \geq 0 \qquad (4B\text{-}5)$$

exists and is independent of the initial state $X_0 = i$

Furthermore, the limit

$$p_j = \lim_{k \to \infty} \frac{Number \ of \ visits \ to \ state \ j \ up \ to \ time \ k}{k} \qquad (4B\text{-}6)$$

exists with probability 1. This limit leads to the time average interpretation that p_j is the proportion of time during which the process is in state j. Note that p_j does not depend on the initial state i.

In an irreducible, aperiodic Markov chain, the limit p_j in (4B-5) may have two possible cases:

(a) $p_j = 0$ for all $j \geq 0$, in this case, the chain has no stationary distribution;

(b) $p_j > 0$ for all $j \geq 0$. In this case, $\{p_j\}$ is the unique stationary distribution of the chain.

Case (a) may arise in a queueing system where the arrival rate exceeds the service rate, and then the number of customers in the system increases to ∞, so that the steady-state probability p_j of any finite j is zero. Note that this case can never arise when the number of states is finite.

In case (b), the following equations are often useful for computing p_j for queueing systems. Multiplying the equation $\sum_{i=0}^{\infty} p_{ji} = 1$ by p_j and using (4B-4), we have

$$p_j \sum_{i=0}^{\infty} p_{ji} = \sum_{i=0}^{\infty} p_i p_{ij} \ , j = 0, 1, \ \cdots \qquad (4B\text{-}7)$$

These equations are known as the global balance equations.

4B-2. Continuous-time Markov Chains

A continuous-time Markov chain is a random process $\{X(t), t \geq 0\}$ taking on values from the set of non-negative integers, that for all $t, \tau \geq 0$ and non-negative integers $i, j, x(u), 0 \leq u \leq \tau$,

$$P\{X(t + \tau) = j \,|\, X(\tau) = i, X(u) = x(u), 0 \le u < \tau\}$$

$$= P\{X(t + \tau) = j \,|\, X(\tau) = i\} \tag{4B-8}$$

Thus a continuous-time Markov chain is a random process having the Markov property that given the present $X(\tau)$ and the past $X(u)$, $0 \le u < \tau$, the conditional probability of the future $X(t + \tau)$ depends only on the present and is independent of the past. It has the property that each time it enters state i:

(a) The time it spends in state i is exponentially distributed with mean $1/v_i$; and

(b) When the process leaves state i, it will enter state j with probability p_{ij}.

The rate $q_{ij} = v_i p_{ij}$ at which the process makes a transition to j when in state i, is called the transition rate from state i to state j. For a continuous-time Markov chain, the limit

$$p_j = \lim_{t \to \infty} P\{X(t) = j \,|\, X(0) = i\} \tag{4B-9}$$

exists and is independent of the initial state i. Furthermore, if $T_j(t)$ is the time the process $X(t)$ spent in state j upto time t, then, regardless of the initial state, the limit

$$p_j = \lim_{t \to \infty} \frac{T_j(t)}{t} \tag{4B-10}$$

exists with probability 1.

For continuous-time Markov chains, the Chapman-Kolmogorov equations are

$$P_{ij}(t + \tau) = \sum_{k=0}^{\infty} P_{ik}(t) P_{kj}(\tau) \tag{4B-11}$$

Since the amount of time that the process spent in a state is exponentially distributed, the probability of more than one transition in a small time interval Δt is $o(\Delta t)$. Using this property and the Chapman-Kolmogorov equations, we obtain Kolmogorov's backward equations

$$\dot{P}_{ij}(t) = \sum_{k \neq i} q_{ik} P_{kj}(t) - v_i P_{ij}(t) \qquad (4\text{B-}12)$$

and Kolmogorov's forward equations

$$\dot{P}_{ij}(t) = \sum_{k \neq j} q_{kj} P_{ik}(t) - v_j P_{ij}(t) \qquad (4\text{B-}13)$$

where

$$v_i = \lim_{\Delta t \to 0} \frac{1 - P_{ii}(\Delta t)}{\Delta t}.$$

and

$$q_{ij} = \lim_{\Delta t \to 0} \frac{P_{ij}(\Delta t)}{\Delta t} \;,\; i \neq j$$

Now if we let $t \to \infty$, then assuming that we can interchange limit and summation, we obtain

$$\lim_{t \to \infty} \dot{P}_{ij}(t) = \lim_{t \to \infty} \left[\sum_{k \neq j} q_{kj} P_{ik}(t) - v_j P_{ij}(t) \right]$$

or

$$0 = \sum_{k \neq j} q_{kj} P_k - v_j P_j$$

Thus we have the global balance equations for the continuous-time Markov chain

$$p_j \sum_{i=0}^{\infty} q_{ji} = \sum_{i=0}^{\infty} p_i q_{ij} \ , j = 0, 1, \ \cdots \tag{4B-14}$$

where the relation $v_j = \sum_{i=0}^{\infty} q_{ji}$ has been used.

Note that $p_i q_{ij}$ may be viewed as the long-term frequency of transitions from state i to state j. Thus the global balance equations (4B-7) or (4B-14) express that at equilibrium, the frequency of transitions out of state j (the left-hand side of (4B-7) or (4B-14)) equals the frequency of transitions into state j (the right-hand side of (4B-7) or (4B-14)).

REFERENCES

[1]   Beckmann, P., Elementary Queueing Theory and Telephony, Rev. ed., Batavia, IL.: ABC Tele Training, 1981.

[2]   Khintchine, A.Y., Mathematical Methods in the Theory of Queueing, 2nd ed., New York: Hafner, 1969.

[3]   Cooper, R.B., Introduction to Queueing Theory, 2nd ed., New York: Elsevier Science, 1981.

[4]   Kleinrock, L., Queueing Systems, Vol. 1: Theory, Vol. 2: Computer Applications, New York: John Wiley & Sons, 1975.

[5]   Pritchard, J.A.T., Quantitative Methods in On-Line Systems, New York: Hayden Book, 1976.

[6]   Syski, R., Introduction to Congestion Theory in Telephone Systems, 2nd ed., New York: Elsevier Science, 1986.

[7]   Ross, S.M., (a) Introduction to Probability Models, 4th ed., New York: Academic Press, 1989. (b) Stochastic Processes, New York: John Wiley & Sons, 1983.

[8]   Feller, W., An Introduction to Probability Theory and Its Applications, Vol. 1, 3rd ed., New York: John Wiley & Sons, 1968.

[9]   Cox, D.R. and Smith, W.L., Queues, London: Methuen and Co., 1961.

[10] Boucher, J.R., Traffic System Design Handbook, Piscataway, N.J.: IEEE, 1992.

[11] Bertsekas, D. and Gallager, R., Data Networks, 2nd ed., Englewood Cliffs, N.J.: Prentice Hall, 1992.

[12] Chan, W.C., "Modeling of Data Networks", International J. of Modeling and Simulation, Vol. 13, No. 4, 1993, pp. 121-128.

[13] Kleinrock, L., "On the Modeling and Analysis of Computer Networks", Proc. of IEEE, Vol. 81, No. 8, August, 1993, pp. 1179-1191.

[14] Lavenberg, Steven S. and Sauer, Charles H., "Analytical Results for Queueing Models", Chapter 3, Computer Performance Modeling Handbook, edited by Lavenberg, Steven S., Academic Press, 1983, pp. 55-172.

[15] Akimaru, Haruo and Kawashima, Konosuke, Teletraffic Theory and Applications, New York: Springer-Verlag, 1993.

PROBLEMS

4-1.   A communication network is modeled by three queues connected in series. The channel of each node must be busy if messages are present at its input queue. Under stable operating conditions, the average message arrival rate λ in messages per second, the average queue length N_q in messages, and the channel utilization ρ are given in the figure.

(a)   What is the average time delay for messages entering queue 2 through the network and the output channel?

(b)   What is the average time delay for messages entering queue 1 through the complete network and the output channel?

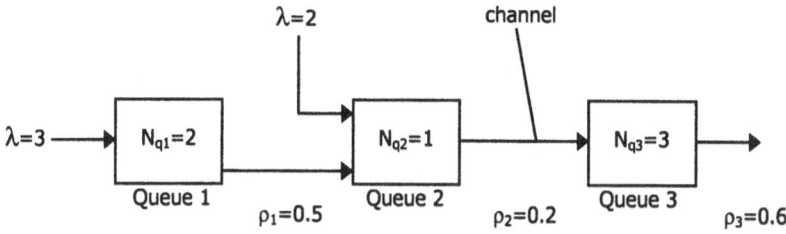

Figure 4-15. Queues in series.

4-2.   Users arriving at a public library check-out counter follow a Poisson process with a rate of 10 users per hour. Each user takes an average of 2 minutes with a standard deviation of 3 minutes for service.

   (a)   What is the average number of users lined up at the counter?

   (b)   What is the total average delay time that a user spends at the counter?

4.3.   Students arrive randomly at a cash-dispensing machine with an average interarrival time of 5 minutes. The length of time that a student spends on the machine is exponentially distributed with an average of 2 minutes.

   (a)   What is the probability that a student arriving at the machine will have to wait?

   (b)   What is the average waiting time for a student?

   (c)   The bank plans to install a second machine when the average waiting time is 5 minutes or longer. At what average arrival rate will this occur? For this arrival rate, what is the total average delay time that students spend at the machine?

   (d)   What is the average queue length for the arrival rate of (c)?

4-4.   A concentrator consists of n input lines and a buffer for storing messages which are transmitted onto an output line with a capacity of 2400 bps. Messages arrive at an input line at an average rate of 0.1 messages per second and have an exponential distribution with an average length of 200 bits.

(a)   If the maximum possible total average delay at the concentrator is 1 second, how many input lines can the concentrator handle?

(b)   If $n = 50$, what is the average number of messages stored in the buffer?

(c)   What is the average waiting time in the buffer?

(d)   What is the output line utilization?

4-5.   A number of terminals are connected to a concentrator, where messages are stored in a buffer and then transmitted in order of arrival to a computer over a 2400 bps synchronous line. All messages from the terminals are exactly 28 bytes long, with each byte having 8 bits, whereas output messages from the concentrator are 30 bytes long (the data of 28 bytes plus 2 bytes for the terminal number). Messages randomly generated at each terminal follow a Poisson process with an average rate of two messages per minute.

(a)   What is the maximum number of terminals that can be connected to the concentrator for a stable operation?

(b)   What is the total average delay time at the concentrator if the number of terminals is 100?

4-6.   Measurements are made for a node of a computer communications network with the following results:

Average number of packets stored in the buffer = 0.1 packets

Average packet length = 500 bits

The output channel has a capacity of 10^6 bps.

(a) If all of the packets are 500 bits long, what is the packet arrival rate?

(b) What is the average transfer delay at the node?

4-7. The capacity of a node in a computer network is 9600 bps. Packets arrive randomly at the node and are all 52 bits long. Under equilibrium conditions, it is found that an average of 4 packets are in the system (in the buffer and the output line).

(a) What is the average input rate of this node?

(b) What is the total average delay time through the system?

(c) What is the average number of packets in the buffer?

(d) What is the average waiting time in the buffer?

(e) What is the channel utilization?

4-8. Consider a single-server queueing system, where messages arrive according to a Poisson process with a rate of λ messages/second. Suppose that the system is idle at $t = 0$. Find the probability that the second arriving message will not have to wait, and also find the average waiting time of this second message W for the following two cases: (a) The service time X is exponentially distributed with a mean of $1/\mu$ seconds, and (b) $X = \bar{X} = \text{constant} = 1/\mu$ seconds.

4-9. Suppose calls arrive at a central office according to a Poisson process with a rate of λ calls per second. Let N be the number of calls that arrive during a time interval of length T seconds, where T is a nonnegative random variable which is independent of the arrival process, and has the distribution function $H(t) = P\{T \le t\}$, mean τ and variance σ^2.

Calculate (a) the first moment of N, (b) the second moment of N, and (c) the variance $Var\,[N]$.

4-10. Telephone calls arriving at a switchboard follow a Poisson process with a rate of 12 calls per minute. (a) What is the probability that the time interval between the next two calls will be less that 7.5 seconds? (b) More than 10 seconds?

4-11.   Consider the birth and death process with the birth and death rates, λ_k and μ_k respectively,

$$\lambda_k = \begin{cases} \lambda \ , & k = 0 \\ 0 \ , & k \neq 0 \end{cases} \quad \text{and} \quad \mu_k = \begin{cases} \mu \ , & k = 1 \\ 0 \ , & k \neq 1 \end{cases}$$

which corresponds to an Erlang loss system with a single server. (a)   Write   the   differential-difference   equations   for $P_k(t) = P\{N(t) = k\}$, $k = 0,1$, and (b) Solve these equations in terms of $P_0(0)$ and $P_1(0)$.

4-12.   Measurements are made in the steady state for a station buffer of a local computer network with channel capacity $C = 10^6$ bps.

The following results are obtained:

Average number of packets stored in the system = 0.1 packets

Average packet waiting time in the buffer = 0.01 milli-seconds

Average packet length = 100 bits and standard deviation = 100 bits.

(*a*)   What is the average packet arrival rate?

(*b*)   What is the offered traffic?

4-13.   The capacity of the output line of a node in a computer network is 10 K bps. Under steady-state conditions, it is known that an average of 10 messages are stored in the buffer of the node. The average message length is 100 bits and the arrival process is Poisson with an average message arrival rate of 20 messages per second.

(a) What is the variance of the message length?

(b) What is the average total time delay through the node?

(c) What is the average number of messages in the system?

4-14.   The link connected to a node in a given network has a bit rate of 10^6 bps. Measurements in the steady state are obtained as follows:

Average packet length = 100 bits;

Average input rate = 10^3 packets/sec.;

Average number of packets in the buffer = 1.5 packets.

(a) What is the traffic intensity?

(b) What is the total average delay time T at the node?

4-15.   Consider a queueing system modeled by a birth and death process with the following birth and death rates:

$$\lambda_k = (k + 2)\lambda, \quad k = 0, 1, 2, \cdots, \text{ and}$$

$$\mu_k = k\mu, \, k = 1, 2, 3, \cdots,$$

where $\lambda/\mu < 1$.

(a)   Determine the equilibrium probability distribution $\{p_k\}$ in terms of λ, μ and k, where $p_k = P\{N = k\}$ and N is the number of people in the system;

(b)   Find the average number of people in the system.

4-16.   A concentrator has m input lines and one output line which has a capacity of 1200 bps. Each input line has a message arrival rate of 0.1 messages/second. The message lengths are identical and exponentially distributed with a mean of 240 bits per message. Suppose the concentrator can be modeled as an $M/M/1$ queue.

(a)   If the total average concentrator delay is less than or equal to 1 second, how many input lines can the concentrator handle?

Let the number of input lines be 30.

(b)   What is the average number of messages stored in the buffer?

(c)   What is the average waiting time?

(d)   What is the offered load?

4-17.   Consider an $M/M/1$ queue in discrete time, where the time axis is divided into subintervals of 1 second in length. We assume that events can occur only at the ends of these discrete time subintervals. The probability of a single arrival is λ, and the probability of no arrival at that discrete time epoch is $1 - \lambda$. Similarly, if the server is busy at an epoch, then the probability that the service will be completed at the next epoch is $1 - \sigma$ and the probability that it will not finish at the next epoch is σ.

(a)   Find the probability that the interarrival time X equals k seconds and the probability that the interdeparture time Y equals k seconds.

(b)   Let $p_k = P\{N = k\}$, where N is the number of customers at a discrete time epoch when the system is in equilibrium. Write down the equilibrium equations that govern the behavior of p_k and solve them.

4-18.   Suppose an $M/M/1$ queue is operated at an offered load of 0.7 erlangs.

(a)   What is the average number of customers in the queue?

(b)   What is the average waiting time if the average time that a customer spends in the system is 10 seconds?

4-19.   Students arrive at a Xerox machine in a Poisson pattern with a rate of 10 students per hour. The length of Xeroxing time is exponentially distributed with a mean of 4 minutes.

(a)   What is the probability that a student arriving at the machine will have to wait?

(b)   What is the average waiting time?

(c)   What is the average number of students (waiting and being served) at the machine?

4-20.   Consider a birth and death process with the following birth and death rates, respectively,

$$\lambda_k = \lambda/(k + 1) \,, \, k = 0, 1, \, \cdots$$

and

$$\mu_k = \begin{cases} 0 \,, & k = 0 \,, \\ \mu \,, & k = 1, 2, \, \cdots \,. \end{cases}$$

Find

(a)   the   steady-state   probability   distribution $\{p_k\}, \, k = 0, 1, \, \cdots$;

(b) the average arrival rate; and

(c) the offered load and the carried load.

4-21.   A concentrator with a 1200 bps output line is connected to a computer.   Messages   from   the   input   terminals   of   the concentrator are all 30 bytes long.   Each terminal generates messages randomly with a rate of 3 messages/minute.

(a)   What is the maximum number of terminals that can be connected to the concentrator?

(b)   If the number of terminals is 80, what is the average message waiting time?

4-22.   Consider an $M/M/1$ queue with the arrival rate of λ customers per second and service rate of μ customers per second.   A customer   in   the   queue   will   leave   without   service   with probability $\alpha \Delta t + o(\Delta t)$ in any time interval of length Δt.

(a) Draw the state-transition-rate diagram.

(b) Express p_k in terms of p_{k-1} by equating the flow rate into a vertical plane separating state $k-1$ and state k to the flow rate out of this plane.

(c) Solve for the p_k for $\alpha = \mu$.

4-23. Consider a cyclic queue in which M customers circulate around and pass through two queues as shown below:

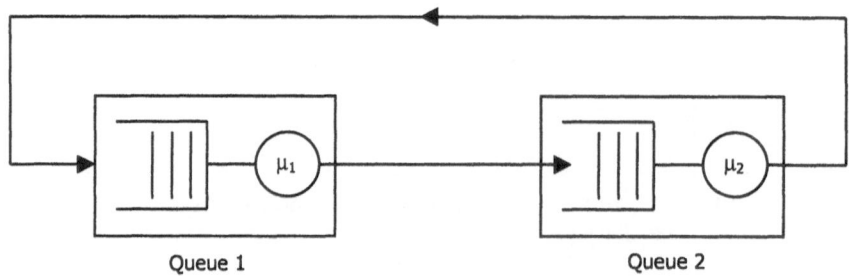

Queue 1   Queue 2

Figure 4-16. A cyclic queue.

Both servers are exponential with rates μ_1 and μ_2, respectively.

Let $p_k = P\{k$ customers in queue 1 and $M-k$ in queue 2$\}$

(a) Draw the state-transition-rate diagram.

(b) Write the relationship between the flow rates into and out of the states.

(c) Find p_k.

4-24. Consider a single-server queue ($M/M/1$) in which there is a maximum possible total of n customers (including the customer in service). Any further arriving customers will be refused and will depart immediately without service. The arrival rate is λ customers per second and the service rate is μ customers per second.

(a)   Find the equilibrium probability p_k.

(b)   Find the average number of customers in the system.

(c)   Calculate the average arrival rate and the average time that a customer spends in the system.

4-25.   Calculate the interdeparture time distribution function, $D(t)$, for an $M/G/1$ queue.

(a)   Find the Laplace-Stieltjes transform $\hat{D}(s)$ of $D(t)$ conditioned firstly on a non-empty queue left behind, and secondly on an empty queue left behind by a departing customer. Combine these results to obtain the Laplace transform of the interdeparture time density, and from this, determine the probability density function.

(b)   For an exponential service time

$$H(t) = \begin{cases} 0 & , \ t < 0 \\ 1 - e^{-\mu t} & , \ t \geq 0 \end{cases}$$

Find the density function $d(t)$ and the distribution function $D(t)$.

(c)   For a constant service time

$$H(t) = \begin{cases} 0 & , \ t < T \\ 1 & , \ t \geq 0 \end{cases}$$

Find the density function $d(t)$ and the distribution function $D(t)$.

4-26.   Consider an $E_2/M/1$ queue for which the interarrival times of customers have the Erlangian distribution

$$G(t) = P\{X \le t\} = 1 - \sum_{k=0}^{r-1} \frac{(r\lambda t)^k}{k!} e^{-r\lambda t}, \, t \ge 0$$

of order r, $r = 1, 2, \cdots$. For E_2, we have $r = 2$.

Calculate the probability σ in (4-74).

4-27. Calculate (a) the waiting-time distribution function in the $GI/M/1$ queue for a first-come, first-served queue discipline, and (b) the mean waiting time.

4-28. Consider a single-server queue with an exponential server of rate μ and hyperexponential interarrival time distribution

$$G(t) = P\{X \le t\} = 1 - \alpha_1 e^{-\lambda_1 t} - \alpha_2 e^{\lambda_2 t}$$

where $\alpha_1 + \alpha_2 = 1$. Suppose that $\lambda_1 = 2$, $\lambda_2 = 1$, $\mu = 2$, and $\alpha_1 = 5/8$. Calculate (a) σ in (4-74), (b) π_k, and (c) the mean waiting time W.

4-29. A communication line capable of transmitting at a rate of 96 Kbps will be used to connect 10 stations, each generating Poisson traffic at a rate of 5 packets per second. Packet lengths are identical and exponentially distributed with a mean of 1000 bits. For each station, calculate (a) the average number of packets in the queue, (b) the average number of packets in the system, and (c) the average transfer delay per packet when the line is allocated to the stations by using (i) 10 equal-capacity time-division multiplexed channels, and (ii) statistical multiplexing.

4-30. A data-storage device consists of a rotating disk which stores blocks of data in sectors. A sensor called a head is located close to the rotating surface and can read or write data blocks at some predetermined sector. All blocks begin on a sector boundary. Suppose that read/write requests arrive at the head queue according to a Poisson process with an average rate of λ requests per second. Each of these requests demands a block of data of fixed length b sectors beginning at a random sector boundary. Calculate the mean waiting time for a disk access, given that the disk rotates at a rate of r revolutions per second and has s sectors. Assume that the system can be modeled by an $M/G/1$ queue.

CHAPTER 5

THE ERLANG LOSS AND DELAY SYSTEMS

5-1. INTRODUCTION

Having studied the birth and death process in detail in the last chapter, this chapter will investigate several fundamental questions concerning the service of incoming calls by a number of lines. These incoming calls follow a Poisson process with a constant arrival rate.

The process of service operates in such a way that each incoming call occupies a free line for a period of time, called the holding time or service time. While a line is busy with a call, it is inaccessible to other calls.

If, upon arrival, an incoming call finds all lines busy, the call is said to be blocked. In problems of telephone system design, there are two ways to handle blocked calls: (a) The blocked call is simply refused and is lost as though it had not occurred at all. This type of system is called a loss system; (b) A blocked call remains in the system and waits for a free line. This type of system is known as a delay system. These two types of systems differ from each other not only in details regarding solutions to basic problems, but also in their structures. The important fact is that the grade of service in these two systems are completely different. For a loss system, the grade of service is the probability of blocking. On the other hand, for a delay system, the grade of service is the probability of waiting, and the central problem involves the investigation of the waiting time as a random variable.

In this chapter, we shall investigate both the Erlang loss and delay systems. We shall also present some important features of the Erlang systems and examine the two systems in separate sections.

Full Accessibility. A system with a collection of lines is said to be a fully-accessible system if all of the lines are equally accessible to all incoming calls. For example, the trunk lines for interoffice calls are fully-accessible lines.

Characteristics of the Erlang Systems. The Erlang systems may be defined by the following specifications:

1.   The arrival process of calls is assumed to be Poisson with a rate of λ calls per hour.

2.   The holding times are assumed to be mutually independent and identically distributed random variables following an exponential distribution with a mean of $1/\mu$ seconds.

3.   The system has s lines serving the calls.

4.   Calls are served in order of arrival.

Let the state of the system at time t be defined as the number of calls in the system at t, and consider the probabilities for the state transition in a small time interval $(t, t+\Delta t)$. It follows from the above specifications that the probability of transition from state k to state $k + 1$ in $(t, t + \Delta t)$ is $\lambda \Delta t + o(\Delta t)$; for $0 < k \le s$, the probability of transition from state k to state $k - 1$ in $(t, t + \Delta t)$ is $k \mu \Delta t + o(\Delta t)$, since there are k busy lines and each line becomes free in the time interval $(t, t + \Delta t)$ with a probability $\mu \Delta t + o(\Delta t)$. For $k \ge s$, this transition probability becomes $s \mu \Delta t + o(\Delta t)$ for delay systems. By comparison of these state transition probabilities and those of the birth and death process, we see that the behaviour of the Erlang system in state k can be described by the birth and death process with the birth rate λ and death rate $k \mu$. The Erlang systems can be modeled as an $M/M/s$ queue.

5-2. THE ERLANG LOSS SYSTEM

Consider the Erlang loss system with s fully-accessible lines and exponential holding times. In queueing theory, servers associated with exponential service time distribution are often called exponential servers. A queueing model for the Erlang loss system is shown in Figure 5-1, where the waiting capacity is zero, i.e. no waiting is allowed.

The Erlang loss system can be modeled by the birth and death process with the birth and death rates as follows:

$$\lambda_k = \begin{cases} \lambda, & k = 0, 1, \cdots, s - 1 \\ 0, & k \ge s \end{cases} \tag{5-1}$$

and

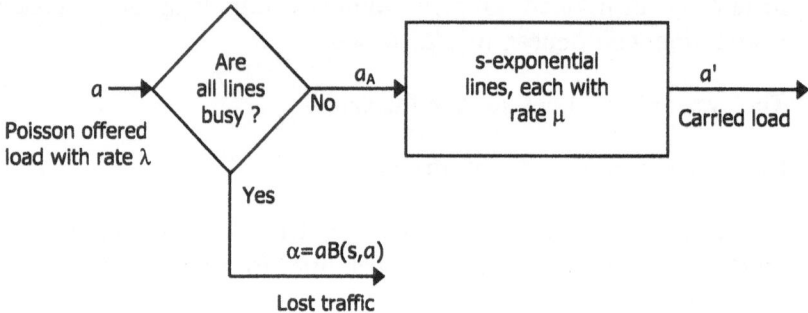

Figure 5-1. Queueing model for the Erlang loss system

$$\mu_k = \begin{cases} k\mu, & k=0,1, \cdots ,s \\ 0, & k>s \end{cases} \tag{5-2}$$

Under equilibrium conditions, the state probability distribution $\{p_k\}$ can be obtained from (4-24) as follows:

Substitution of the birth and death rates (5-1) and (5-2) into (4-24) gives

$$p_k = \frac{1}{k!} \left(\frac{\lambda}{\mu} \right)^k p_0, \quad k = 1, 2, \cdots , s \tag{5-3}$$

By defining the offered traffic

$$a = \lambda/\mu$$

(5-3) can be written as

$$p_k = \frac{a^k}{k!} p_0, \quad k = 0, 1, \cdots , s \tag{5-4}$$

The probability p_0 is determined by the normalization condition

$$\sum_{r=0}^{s} p_r = p_0 \sum_{r=0}^{s} \frac{a^r}{r!} = 1$$

Therefore,

$$p_0 = 1/ \sum_{r=0}^{s} \frac{a^r}{r!} \tag{5-5}$$

Finally, combination of (5-4) and (5-5) yields

$$p_k = \frac{a^k/k!}{\displaystyle\sum_{r=0}^{s} \frac{a^r}{r!}}, \qquad k = 0, 1, 2, \cdots, s \tag{5-6}$$

This probability distribution is called the truncated Poisson distribution or Erlang's loss distribution.

In particular, when k = s, the probability of loss is given by

$$\pi_s = p_s = B(s,a) = \frac{a^s/s!}{\displaystyle\sum_{k=0}^{s} (a^k/k!)}, \quad a = \lambda/\mu \tag{5-7}$$

In the Unites States, (5-7) is called the Erlang loss formula or the Erlang B formula and in Europe, it is called Erlang's first formula and is denoted by $E_{1,s}(a)$.

It is important to point out that the truncated Poisson distribution (5-6) is valid for any holding time distribution with a finite mean of $1/\mu$, even though the Markov property of the exponential holding time distribution was used explicitly in the derivation. This remarkable fact was conjectured by Erlang himself in 1917, and proven by Sevastyanov in 1957. Values for $B(s,a)$ obtained from (5-7) have been plotted against the offered traffic a in erlangs for different values of the number s of lines. Two typical groups of Erlang B curves are given in Figures 5-2 and 5-3.

The Erlang B formula gives the time congestion of the system and relates the probability of blocking to the offered traffic and the number of trunk lines. In design problems, it is usually necessary to find the number of trunk lines needed for a given offered traffic and a specified grade of service.

The ratio $a = \lambda/\mu$ is called the offered load or offered traffic. It is a dimensionless quantity that is numerically equal to the average number of incoming calls that occur during an average holding time. The international unit for the offered traffic is the erlang. One erlang represents the amount of traffic that occupies one piece of traffic-sensitive equipment (a circuit or a trunk) busy for one hour. Thus one erlang is equal to one

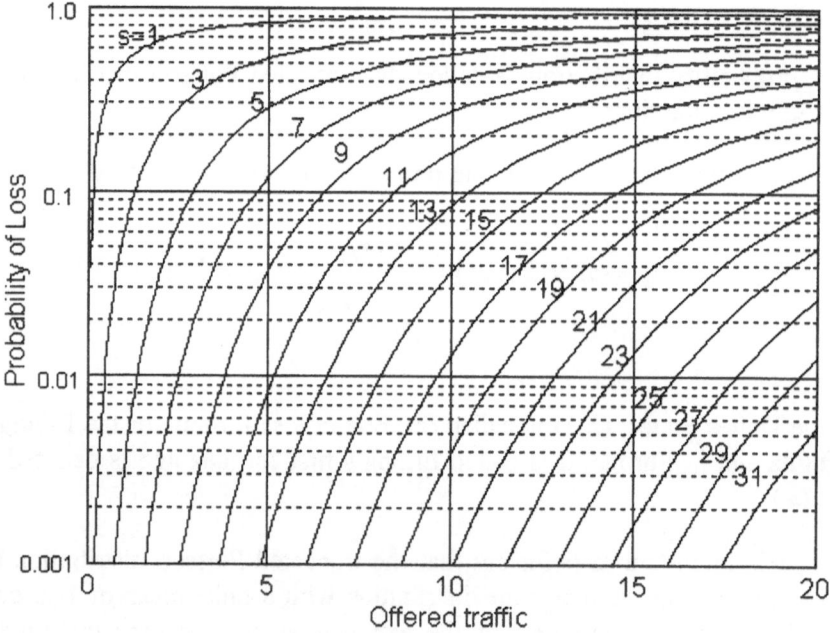

Figure 5-2. Erlang loss formula $B(s,a)$ plotted against offered load a in erlangs for different values of the number s of servers.

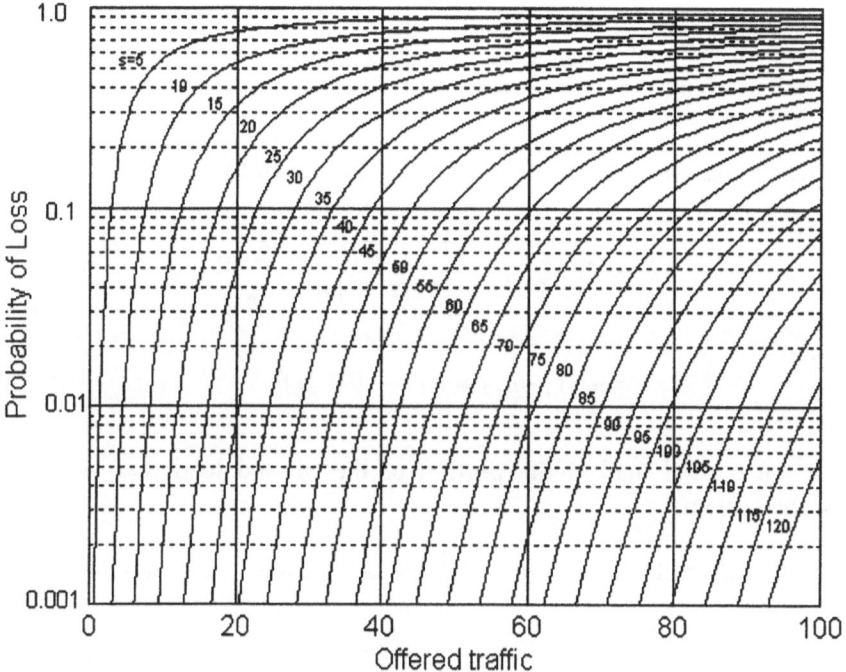

Figure 5-3. Erlang loss formula $B(s,a)$ plotted against offered load a in erlangs for different values of the number s of servers.

call-hour per hour and it is a dimensionless unit.

More generally, the offered load generated by a Poisson input process with a rate of λ calls per hour may be defined as the average number of incoming calls per holding time; that is,

$$a = \int_0^\infty \lambda t \ dH(t) = \lambda \tau \tag{5-8}$$

where λt is the average number of incoming calls in a fixed time interval of t hours and $H(t)$ is the holding time distribution and τ is the average holding time.

The average number of occupied or busy trunks is defined as the carried load:

$$a' = \sum_{k=1}^{s-1} k \, p_k + s \sum_{k=s}^{\infty} p_k \qquad (5\text{-}9)$$

For the Erlang loss system, the carried load is given by

$$a' = \sum_{k=1}^{s} k \, p_k = \sum_{k=1}^{s} \frac{a^k}{(k-1)!} \bigg/ \sum_{r=0}^{s} \frac{a^r}{r!}$$

$$= a(1 - p_s) = a[1 - B(s,a)] \qquad (5\text{-}10)$$

Thus, the carried load is the portion of the offered load that is not lost from the system.

The carried load per line is known as the trunk occupancy:

$$\rho = \frac{a'}{s} \qquad (5\text{-}11)$$

The trunk occupancy ρ is a measure of the degree of utilization of a group of lines and is sometimes called the utilization factor.

Investigation of formulas (5-7), (5-10), and (5-11) indicates that as the offered load and the number of lines are increased such that the probability of blocking remains constant, the line occupancy increases. In other words, large line groups are more efficient than small ones. Unfortunately, the size of a line group is limited by hardware complexity, and high-occupancy line groups are more vulnerable to service degradation during periods of overload than the smaller line groups with the same blocking probability but lower occupancy.

To illustrate this point, let us consider a trunk group with an offered load of 4.5 erlangs and a blocking probability of 0.01. From the Erlang B curves in Figure 5-2, we find that 10 trunks are required. If the offered load is increased to 12 erlangs, then in order to keep the blocking probability at 0.01, 20 trunks are needed. Since

$$B(10,4.5) = B(20,12) = 0.01$$

we calculate the trunk occupancies as follows: Using (5-10) and (5-11), we get

$$\rho_{10} = \frac{4.5}{10}\,(1 - 0.01) = 0.4455$$

and

$$\rho_{20} = \frac{12}{20}\,(1 - 0.01) = 0.594$$

Thus, the group of 20 trunks is more efficient than the group of 10 trunks.

For a trunk group of fixed size, occupancy increases with increasing load, thereby increasing trunk group efficiency; unfortunately, the probability of blocking also increases with increasing load. Hence, efficient use of equipment must be balanced against the provision of acceptable grade of service.

The above discussion does not take into account the effect of traffic overload. In designing a telephone system, it is necessary to ensure that the system will operate satisfactorily under a moderate overload condition. Figure 5-4 shows the effect of overload on the blocking probability for various sized groups of trunk lines. For each sized group, the original traffic carried is assumed to be at a 0.01 grade of service. From Figure 5-4, we see that a 20% overload on a 10-line group causes the grade of service to worsen to 0.024. The same overload on a 20-line group causes the grade of service to worsen to 0.035. The larger, more efficient group is therefore more sensitive to overload. Above a certain traffic level, the number of lines needed is determined by the required overload performance rather than the basic offered traffic. For example, a typical design requirement might be that a normal grade of service of 0.01 should increase to no more than 0.02 for a 10% overload. In this case, the group size should be no more than 20 lines.

	s=5	s=10	s=20	s=30	s=50	s=100
0	1	1	1	1	1	1
5	1.19	1.3	1.46	1.58	1.77	2.09
10	1.41	1.67	2.06	2.36	2.84	3.7
15	1.64	2.09	2.78	3.33	4.21	5.73
20	1.9	2.57	3.64	4.49	5.85	8.07
25	2.18	3.12	4.62	5.83	7.69	10.59

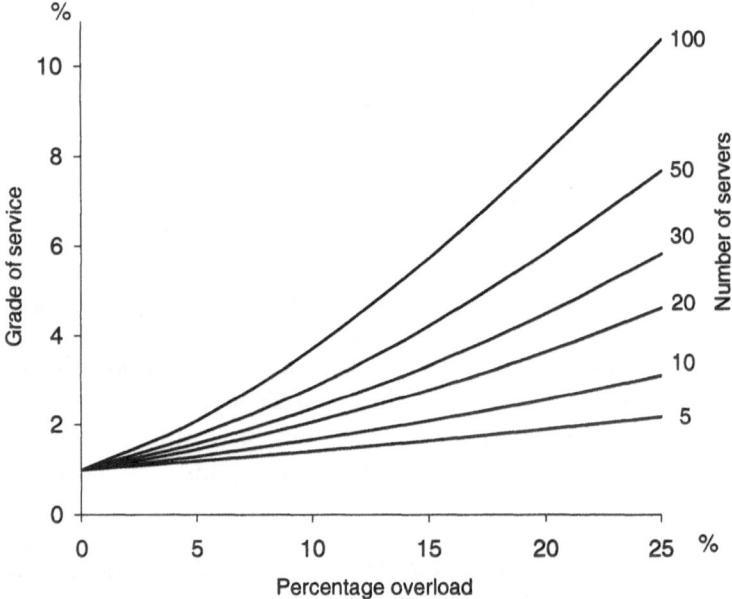

Figure 5-4. Effect of overload on grade of service [6].

In telephone switching networks, the term availability or accessibility is used to define the capacity or number of outlets of a switch to access a given route. For example, in an $m \times n$ switch, the availability is n. The availability is usually designated by the letter k.

The term full availability is applied when any inlet of a switch has access to all of the trunks in a given route. In practice, it is more economical to use smaller switches with restricted availability (restricted blocking). However, it is also desirable to obtain good utilization. This implies the use of a large number of crosspoints in a switching network. Thus, there is a compromise between the utilization and the overload performance.

Example 5-1. Consider the telephone trunking problem. A telephone exchange A is to serve 10,000 subscribers in a nearby exchange B as shown in Figure 5-5. Suppose that during the busy hour, the subscribers generate a Poisson traffic with a rate of 10 calls per minute, which requires trunk lines to exchange B for an average holding time of 3 minutes per call. Determine the number of trunk lines needed for a grade of service of 0.01. Also calculate the lost traffic and the utilization factor.

Solution. Since there are 10,000 telephone subscribers, the input traffic may be assumed to be Poisson. The average calling rate is $\lambda = 600$ calls/hour, and the average holding time is $\tau = 180$ seconds. The offered traffic is $a = \lambda\tau = 30$ erlangs.

Using the Erlang B curves in Figure 5-3, we find that for a grade of service of 0.01, the number of trunk lines needed is 41. Since the carried load is given by

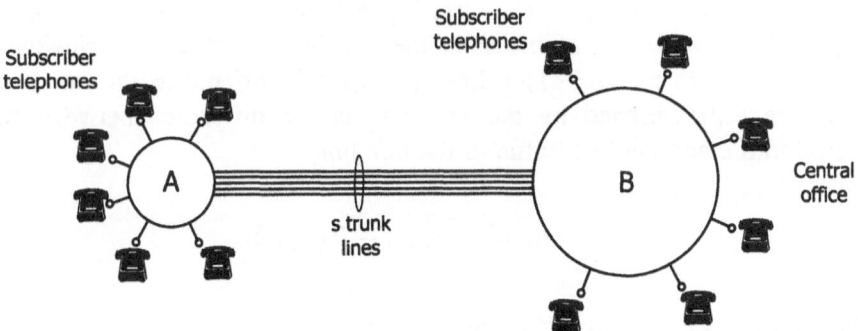

Figure 5-5. Two nearby telephone exchanges with s trunk lines

$$a' = a[1 - B(s,a)]$$

$$= 30[1 - 0.01]$$

$$= 29.7 \ \text{erlangs}$$

the lost traffic is equal to

$$a - a' = aB(s,a) = 30 \times 0.01 = 0.3 \ \textit{erlangs}$$

and the utilization factor is given by

$$\rho = \frac{a'}{s} = \frac{29.7}{41} = 0.724$$

Example 5-2. Consider a switch whose control causes it to search sequentially for a free line. The control always starts searching in the same order and from the same starting position. Suppose that the total traffic offered to the lines is a erlangs. Calculate the traffic carried by the nth line in the system under the condition that the input is Poisson and blocked calls will be cleared.

Solution. This system may be considered as an Erlang loss system with sequential hunting. The traffic lost to the $(n-1)th$ line is $aB(n-1, a)$, and the traffic lost to the nth line is $aB(n,a)$. However, the traffic lost to the $(n-1)th$ line is the traffic offered to the nth line. Thus, the traffic carried by the nth line is the difference between the offered traffic and the lost traffic to the nth line,

$$a_n' = a[B(n-1, a) - B(n, a)] \tag{5-12}$$

where $B(n, a)$ is given by (5-7) with $s = n$.

In particular, if a traffic of 5 erlangs is offered to a number of lines, the traffic carried by the first line is

$$a_1' = 5[B(0, 5) - B(1, 5)]$$

$$= 5\left[1 - \frac{5}{6}\right]$$

$$= 0.833 \text{ erlangs}$$

The traffic carried by the second line is

$$a_2' = 5[B(1, 5) - B(2, 5)]$$

$$= 5\left[\frac{5}{6} - \frac{25}{37}\right]$$

$$= 0.788 \text{ erlangs}$$

and the traffic carried by the third line is

$$a_3' = 5[B(2, 5) - B(3, 5)]$$

$$= 5\left[\frac{25}{37} - \frac{125}{236}\right]$$

$$= 0.730 \text{ erlangs}$$

We see that the traffic carried by latter choice lines are getting smaller, but their presence is essential to provide the required grade of service.

Example 5-3. Four groups of terminals are to be multiplexed and connected to a computer, as shown in Figure 5-6. In Figure 5-6(a), the traffic from each group uses a separate group of circuits. In Figure 5-6(b), all terminals are combined into one group using only one common group of circuits. Determine the total number of circuits required in both cases with a specified grade of service of 0.01. Assume that each terminal offers a traffic of 0.1 erlangs.

Solution. The offered traffic from each group of terminals in Figure 5-6(a) is

$$a = 20 \times 0.1 = 2 \ erlangs$$

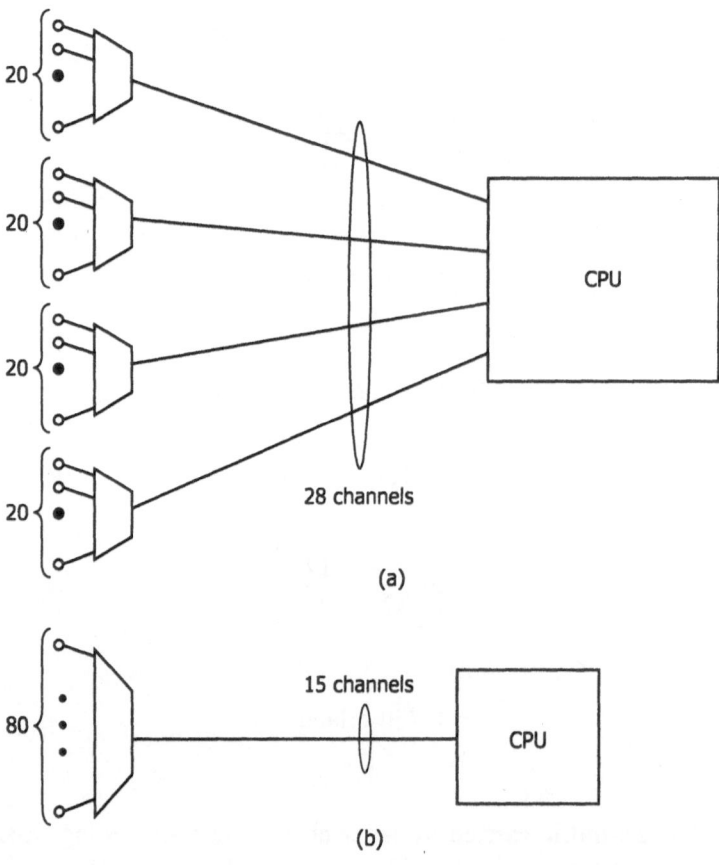

Figure 5-6. Data terminal network (a) Four separate groups; (b) One group.

Since the average number of active terminals is much smaller than the number of terminals in a group, we may assume that the input traffic is approximately Poisson.

Using the Erlang B curves for a grade of service of 0.01 at a loading of 2 erlangs, 7 circuits are required for each group. Thus, the configuration of Figure 5-6(a) requires a total of 28 circuits.

The total offered traffic to the concentrator of Figure 5-6(b) is

$$a = 80 \times 0.1 = 8 \text{ } erlangs$$

From the Erlang B curves, we find that the required number of circuits for a grade of service of 0.01 at a load of 8 erlangs is 15.

This example demonstrates that combining several small traffic groups into one large traffic group can provide significant savings in the total number of circuits required.

5-3. THE ERLANG DELAY SYSTEM

Consider a telephone exchange in which incoming calls follow a Poisson input process with calling rate λ. The calls are served by s fully-accessible lines. The holding times are assumed to be mutually independent, identical, and exponentially distributed with an average holding time of $\frac{1}{\mu}$:

$$H(t) = 1 - e^{-\mu t} \quad , \quad t \geq 0 \qquad (5\text{-}13)$$

An incoming call must wait for service when it finds all s lines occupied. The probability of finding all of the lines occupied is called the probability of waiting. Waiting calls are served in order of arrival. A system satisfying these conditions is called an Erlang delay system.

The objective in the investigation of delay systems is not only concerned with the frequency of delay, but also the distribution function of the waiting time. Thus, the nature of the waiting time W as a random variable plays a decisive role in the study of delay systems.

Suppose that the system is in state k at an arbitrary moment; that is, there are k calls (in service and waiting) in the system. If $k \leq s$, k lines are occupied and no calls are waiting. If $k > s$, all s lines are occupied and there are $k - s$ calls waiting. A queueing model for the Erlang delay system is shown in Figure 5-7, which is an $M/M/s$ queue.

Note that if $k < s$, the Erlang delay system is operated in exactly the same way as the Erlang loss system, since for $k < s$, there are neither lost calls nor waiting calls. However, when $k \geq s$, the operation is completely different. Since the transition is now possible from state s into state $s + 1$ which is impossible for a loss system, by taking the waiting calls into consideration, we can model the Erlang delay system by the birth and death process with the following birth and death rates, respectively,

$$\lambda_k = \lambda, \quad k = 0,1, \cdots \tag{5-14}$$

and

$$\mu_k = \begin{cases} k\mu, & k = 0, 1, \cdots, s - 1 \\ s\mu, & k \geq s \end{cases} \tag{5-15}$$

Under equilibrium conditions, the state probability distribution $\{p_k\}$ can be obtained by substituting these birth and death rates into (4-24) to yield

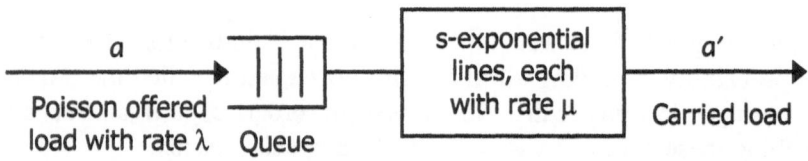

Figure 5-7. Queueing model for the Erlang delay system.

$$p_k = \begin{cases} \dfrac{a^k}{k!} p_0, & 0 \le k < s \\[3mm] \dfrac{a^k}{s!\, s^{k-s}} p_0, & k \ge s \end{cases} \tag{5-16}$$

Using the normalization condition

$$\sum_{k=0}^{\infty} p_k = 1$$

we find

$$\frac{1}{p_0} = \sum_{k=0}^{s-1} \frac{a^k}{k!} + \frac{s^s}{s!} \sum_{k=s}^{\infty} \left(\frac{a}{s}\right)^k$$

$$= \sum_{k=0}^{s-1} \frac{a^k}{k!} + \frac{a^s}{s!} \frac{1}{1 - \dfrac{a}{s}} \tag{5-17}$$

where the condition $a/s < 1$ has been assumed. Furthermore, the probability of waiting (the probability of finding all lines occupied) is equal to

$$P\{W > 0\} = C(s,a)$$

$$= \sum_{k=s}^{\infty} \pi_k \tag{5-18}$$

$$= \sum_{k=s}^{\infty} p_k$$

$$= \frac{s^s}{s!} p_0 \sum_{k=s}^{\infty} \left(\frac{a}{s}\right)^k$$

$$= \frac{a^s}{s!} \frac{p_0}{1 - \dfrac{a}{s}} \, , \, a = \frac{\lambda}{\mu} < s$$

where W is a random variable denoting the waiting time and π_k is the probability that the system is in state k just prior to an arrival. If the input process is a Poisson process, then $\pi_k = p_k$ (see (5-30)). In the United States, (5-18) is called the Erlang delay formula or the Erlang C formula. In Europe, it is known as Erlang's second formula and is denoted by $E_{2,s}(a)$. Two typical groups of Erlang C curves are given in Figures 5-8 and 5-9.

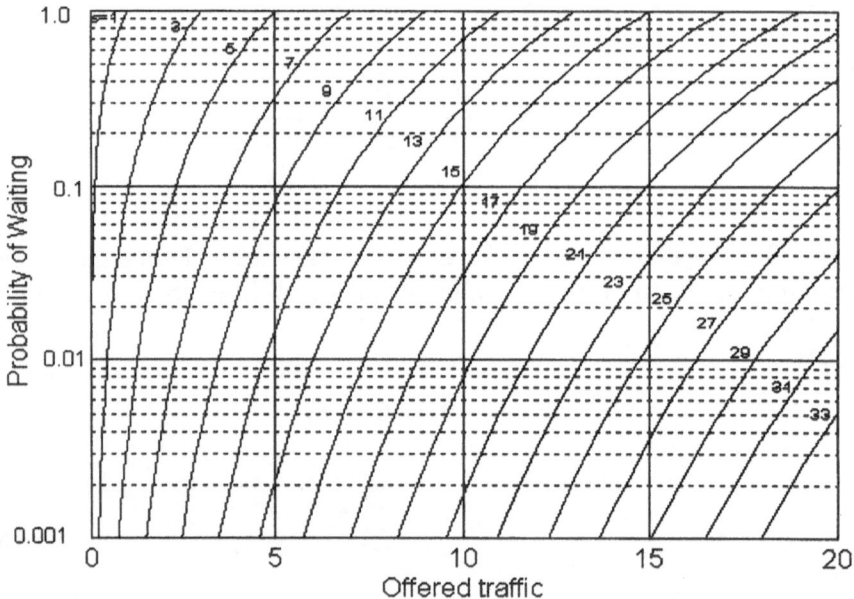

Figure 5-8. Erlang delay formula $C(s, a)$ plotted against offered load a in erlangs for different values of the number s of servers.

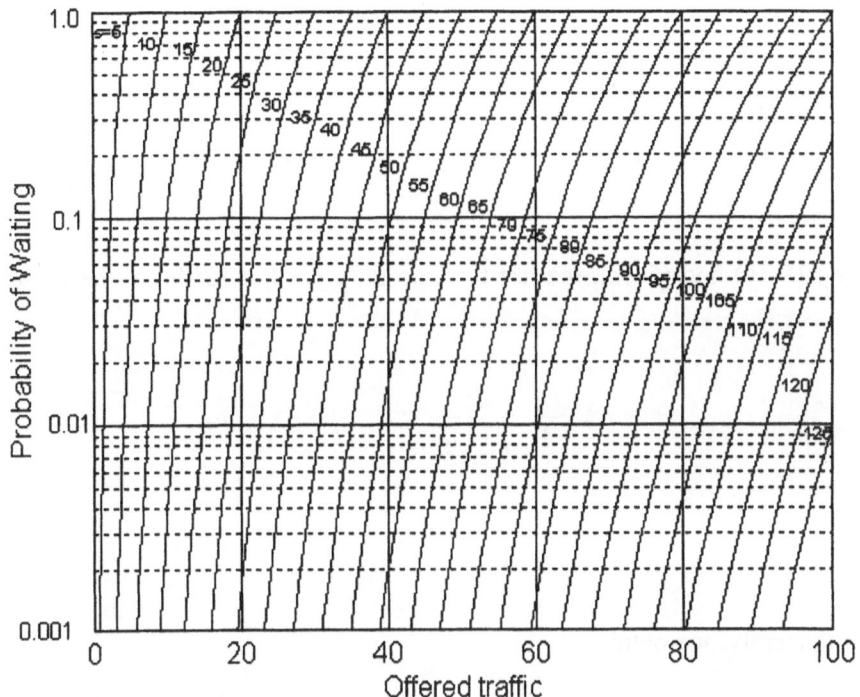

Figure 5-9. Erlang delay formula $C(s, a)$ plotted against offered load a in erlangs for different values of the number s of servers.

Unlike the Erlang loss system, the state probability distribution of the Erlang delay system constitutes a proper distribution only when the offered traffic is less than the number of lines in the system.

From the definition of the carried load, we have

$$a' = \sum_{k=1}^{s-1} k p_k + s \sum_{k=s}^{\infty} p_k$$

$$= \sum_{k=1}^{s-1} \frac{a^k}{(k-1)!} p_0 + s \sum_{k=s}^{\infty} p_k$$

$$= a \sum_{k=0}^{s-2} \frac{a^k}{k!} p_0 + s \sum_{k=s}^{\infty} p_k$$

$$= a \left(1 - p_{s-1} - \sum_{k=s}^{\infty} p_k \right) + s \sum_{k=s}^{\infty} p_k$$

$$= a - a\, p_{s-1} + (s - a) \sum_{k=s}^{\infty} p_k$$

Using (5-16) and (5-18), the last term on the right-hand side can be written as

$$(s - a) \frac{a^s}{s!} \frac{p_0}{1 - \dfrac{a}{s}} = a \cdot \frac{a^{s-1}}{(s-1)!} p_0$$

$$= a\, p_{s-1}$$

Consequently, we obtain

$$a' = a$$

This result shows that for an Erlang delay system, the carried and offered loads are always equal. This is intuitively clear because in a delay system, all calls are served and there is no loss.

The Average Number of Calls in the Erlang Delay System and the Mean Waiting Time. By definition, the average number of calls in a delay system is given by the general expression

$$E[N] = \sum_{k=0}^{\infty} k\, p_k$$

Using (5-16), for the Erlang delay system, we have

$$E[N] = \sum_{k=0}^{s} k \, \frac{a^k}{k!} \, p_0 + \sum_{k=s+1}^{\infty} \frac{k \, a^k}{s! \, s^{k-s}} \, p_0$$

The first summation on the right-hand side is equal to

$$a \, p_0 \sum_{k=1}^{s} \frac{a^{k-1}}{(k-1)!} = a \sum_{k=0}^{s-1} \frac{a^k}{k!} \, p_0 = s \rho [1 - C(s, a)]$$

and the second summation may be written as

$$\frac{s^s}{s!} \, p_0 \rho \left[\sum_{k=s+1}^{\infty} k \, \rho^{k-1} \right] = \rho C(s, a) \left[\frac{s + 1 - s\rho}{1 - \rho} \right]$$

since the quantity in the brackets on the left-hand side is simply equal to

$$\frac{d}{d\rho} \sum_{k=s+1}^{\infty} \rho^k = \frac{d}{d\rho} \left[\frac{\rho^{s+1}}{1 - \rho} \right]$$

$$= \frac{\rho^s (s + 1 - s\rho)}{(1 - \rho)^2}$$

where $\rho = a/s < 1$.

Consequently, we find the average number of calls in the system:

$$E[N] = s\rho [1 - C(s, a)] + \rho C(s, a) \frac{s + 1 - s\rho}{(1 - \rho)}$$

$$\tag{5-19}$$

$$= \frac{\rho}{1 - \rho} \, C(s, a) + s\rho$$

Note that the first term on the right-hand side is the average queue length $E[N_q]$ or the average number of calls waiting in the queue, and the second term is the average number of busy lines. Now, using Little's formula, we find the mean waiting time:

$$E[W] = \frac{E[N_q]}{\lambda} = \frac{C(s,a)}{s\mu(1-\rho)} \qquad (5\text{-}20)$$

Example 5-4. Consider the telephone trunking problem studied in Example 5-1, except that if, on arrival, a call finds all trunk lines busy, this call will wait until a line is freed. In this case, the grade of service is the probability of waiting. Determine the number of lines needed for a grade of service of 0.01, the average number of calls in the system, and the mean waiting time.

Solution. From example 5-1, we have an offered traffic of 30 erlangs. Using the Erlang C curves in Figure 5-9, we find that for a probability of waiting of 0.01, the number of lines needed is 44.

In this example, we see that in order to avoid the loss of 0.3 erlangs as found in Example 5-1, it is necessary to add 3 more lines to the system so that a probability of waiting of 0.01 is maintained.

Since $\rho = \dfrac{a'}{s} = \dfrac{30}{44} = 0.68$, the average number of calls in the system is then given by

$$E[N] = \frac{\rho}{1-\rho} C(s,a) + s\rho$$

$$= \frac{0.68}{1-0.68} \times 0.01 + 30$$

$$= 30.0213 \ calls$$

The mean waiting time is given by (5-20):

$$E[W] = \frac{0.01 \times 180}{44 \times (1-0.68)} = 0.1278 \ seconds \ .$$

5-4. THE COMBINED DELAY AND LOSS SYSTEM

In the Erlang loss system, no waiting is allowed, while in the Erlang delay system, the waiting capacity is assumed to be infinite and hence no loss can occur. In practice, the waiting capacity is always finite. Investigation of this more realistic system is possible by means of the notion of combined delay and loss system. The combined delay and loss system can also be modeled by the birth and death process.

Consider the Erlang delay system of the last section with a finite waiting capacity of $n - s$. Now we have, by taking into consideration the fact that the combined delay and loss system can have only $n + 1$ possible states, the modified birth and death rates

$$\lambda_k = \begin{cases} \lambda, & 0 \leq k < n \\ 0, & k \geq n \end{cases} \tag{5-21}$$

and

$$\mu_k = \begin{cases} k\mu, & 0 \leq k < s \\ s\mu, & s \leq k \leq n \\ 0, & k > n \end{cases} \tag{5-22}$$

Substituting λ_k and μ_k into (4-24), we obtain

$$p_k = \begin{cases} \dfrac{a^k}{k!} p_0, & 0 \leq k < s \\ \dfrac{a^k}{s! \, s^{k-s}} p_0, & s \leq k \leq n \end{cases} \tag{5-23}$$

where

$$\frac{1}{p_0} = \sum_{k=0}^{s-1} \frac{a^k}{k!} + \frac{a^s}{s!} \frac{s}{s-a} \left[1 - (\frac{a}{s})^{n-s+1} \right] \qquad (5\text{-}24)$$

The probability of waiting is given by

$$P\{W > 0\} = C_n(s,a) = \sum_{k=s}^{n} p_k$$

$$= \frac{a^s}{s!} \frac{s}{s-a} \left[1 - (\frac{a}{s})^{n-s+1} \right] p_0 \qquad (5\text{-}25)$$

The probability of loss is equal to

$$p_n = B_n(s,a) = \frac{a^n}{s! \, s^{n-s}} p_0 \qquad (5\text{-}26)$$

Note that when $n = s$, (5-26) reduces to the Erlang B formula (5-7). Therefore,

$$B_s(s,a) = B(s,a)$$

Similarly, when $n \to \infty$, the probability of waiting in (5-25) becomes the Erlang delay formula (5-18):

$$C_\infty(s,a) = C(s, a)$$

Furthermore, the probability p_0 in (5-24) reduces to the p_0 in (5-17). Therefore, the Erlang loss system and the Erlang delay system may be regarded as special cases of the combined delay and loss system.

5-5. THE OUTSIDE OBSERVER'S DISTRIBUTION AND THE ARRIVING CUSTOMER'S DISTRIBUTION

In the investigation of the Erlang systems, the state probability distribution $\{p_k\}$ was derived from equations which relate the proportions of time that the system spends in various states to each other. Therefore, for systems in equilibrium, p_k may be interpreted either as the proportion of time that the system spends in state k or as the probability that the system is in state k at an arbitrary moment. The distribution $\{p_k\}$ of time that the system spends in each state is called the outside observer's distribution. Now let $\{\pi_k\}$ denote the state probability distribution that an arriving call would encounter; that is, the state probability distribution with respect to only those arrival instants of the call. In queueing theory, the distribution $\{\pi_k\}$ is called the arriving customer's distribution. It is important to realize that, in general, the outside observer's distribution $\{p_k\}$ and the arriving customer's distribution $\{\pi_k\}$ are different. However, for systems with Poisson input, these two distributions are identical. This fact is of central importance in the calculation of the waiting time distribution in the Erlang delay system.

Consider a record of the states of a telephone exchange shown in Figure 5-10. Suppose that n calls occur in the observation interval $(0, t)$. The proportion of time that the system spends in state k is given by

$$p_k = \sum_{i=1}^{m_k} t_{k_i}/t = t_k/t \ , \ k = 0, 1, 2, \cdots \tag{5-27}$$

where $t_k = \sum_{i=1}^{m_k} t_{k_i}$ is the total time that the system spends in state k during the interval $(0,t)$ and where m_k is the number of subintervals during which the system is in state k. In a telephone system with s lines, the quantity $\sum_{k=s}^{\infty} p_k = \sum_{k=s}^{\infty} t_k/t$ is known as the time congestion. The observation interval is usually taken as the sum of the BH (busy-hour) traffic on the 10 busiest days of the year.

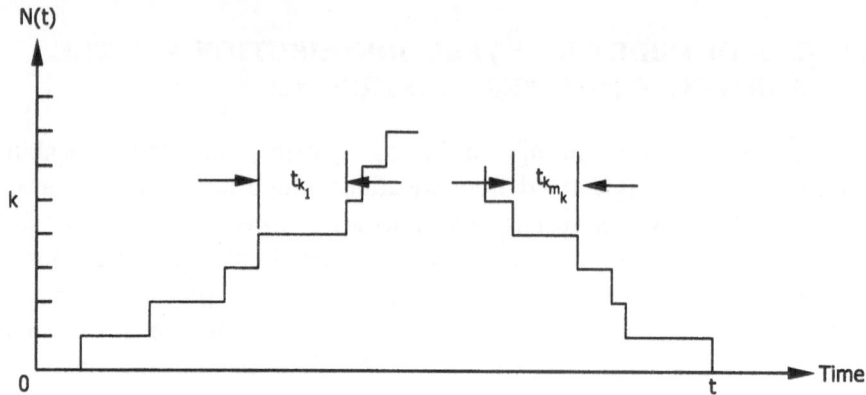

Figure 5-10. States of a telephone exchange

Another important quantity of practical significance is the proportion of calls that, on arrival, find the system in state k:

$$\pi_k = \frac{n_k}{n}, \, k = 0, 1, 2, \cdots \tag{5-28}$$

where n_k is the total number of calls that, upon arrival, find the system in state k during the observation interval (0, t). The quantity $\sum_{k=s}^{\infty} \pi_k = \sum_{k=s}^{\infty} \frac{n_k}{n}$ is known as the call congestion. In general, time congestion and call congestion are not equal; they become equal when the input process is Poisson.

To investigate the relationship between the outside observer's distribution $\{P_j(t)\}$ and the arriving customer's distribution $\{\Pi_j(t)\}$, we let $N(t)$ denote the number of calls in the system at time t, and let $A(t, t + \Delta t)$ be the event that a call arrives in the time interval $(t, t + \Delta t)$. Let

$$P_j(t) \; = \; P\{N(t) = j\}$$

$$= \; \textit{the probability that the outside observer finds}$$

$$\textit{the system in state } j \textit{ at time } t,$$

and

$$\Pi_j(t) \; = \; \textit{the probability that the system is in state } j \textit{ at time } t$$

$$\textit{just prior to an arrival epoch.}$$

By definition, we write

$$\Pi_j(t) = \lim_{\Delta t \to 0} P\{N(t) = j \mid A(t, t + \Delta t)\}$$

Using Bayes' rule of probability theory, we can write

$$P\{N(t) = j \mid A(t\ t, + \Delta t)\} = \frac{P\{A(t, t + \Delta t) \mid N(t) = j\}P_j(t)}{\displaystyle\sum_{k=0}^{\infty} P\{A(t, t + \Delta t) \mid N(t) = k\} \, P_k(t)}$$

It follows that

$$\Pi_j(t) = \lim_{\Delta t \to 0} \frac{P\{A(t, t + \Delta t) \mid N(t) = j\} \, P_j(t)}{\displaystyle\sum_{k=0}^{\infty} P\{A(t, t + \Delta t) \mid N(t) = k\}P_k(t)}$$

If the arrival process described by $A(t, t + \Delta t)$ is a birth and death process, then

$$P\{A(t, t + \Delta t) \mid N(t) = j\} = \lambda_j \, \Delta t + o(\Delta t)$$

and

$$\Pi_j(t) = \lim_{\Delta t \to 0} \frac{[\lambda_j \Delta t + o(\Delta t)] \, P_j(t)}{\sum_{k=0}^{\infty} [\lambda_k \Delta t + o(\Delta t)] \, P_k(t)}$$

$$= \frac{\lambda_j P_j(t)}{\sum_{k=0}^{\infty} \lambda_k P_k(t)}$$

(5-29)

This result shows that for a birth and death process,

$$\Pi_j(t) \neq P_j(t)$$

If the arrival process is a Poisson process, then

$$\lambda_j = \lambda \qquad \text{for } \textit{all} \ j$$

In this case, we obtain from (5-29)

$$\Pi_j(t) = P_j(t)$$

As $t \to \infty$ or when the system is in equilibrium, we obtain

$$\lim_{t \to \infty} \Pi_j(t) = \pi_j$$

$$\lim_{t \to \infty} P_j(t) = p_j$$

and hence,

$$\pi_j = p_j \tag{5-30}$$

This result implies that for a queueing system with a Poisson input, the arriving customer's distribution and the outside observer's distribution are equal, and relation (5-30) is called Poisson Arrivals See Time Averages (PASTA).

5-6. THE WAITING TIME DISTRIBUTION FUNCTION FOR THE ERLANG DELAY SYSTEM WITH SERVICE IN ORDER OF ARRIVAL

Consider the probability $P\{W > t\}$ that a test call entering the system at a random moment has a waiting time greater than t. Let $P_k \{W > t\}$ be the conditional probability of the event $\{W > t\}$ on the assumption that, on arrival, the test call finds the system in state k. Using the formula of total probability, we write

$$P\{W > t\} = \sum_{k=0}^{\infty} \pi_k P_k \{W > t\}$$

where π_k is the probability that the system is in state k just prior to the arrival epoch of the test call. Since $P_k \{W > t\} = 0$ for $k < s$ and $t \geq 0$, we have

$$P\{W > t\} = \sum_{k=s}^{\infty} \pi_k P_k \{W > t\} \tag{5-31}$$

Since the input process is Poisson, then $\pi_k = p_k$, where p_k is given by (5-16). It remains to determine the conditional probabilities $P_k \{W > t\}$ for all $k \geq s$.

Note that if $k \geq s$, there must be $k - s$ calls waiting for service in the queue. If the queue discipline is first-come, first-served or service in order of arrival, then a call finding $k - s$ calls waiting in the queue will obtain service after the $(k - s + 1)$th freeing of line. Thus, the conditional probability $P_k \{W > t\}$ is equal to the probability that during the time interval t after the arrival of the test call, there will be at most $k - s$

freeings of line. It follows that

$$P_k\{W > t\} = \sum_{r=0}^{k-s} f_r(t) \ , \ k \geq s \qquad (5\text{-}32)$$

where $f_r(t)$ denotes the probability of exactly r freeings of line in the time interval t, which remains to be determined. Since the holding time has the exponential distribution

$$H(t) = 1 - e^{-\mu t}, t \geq 0$$

the probability that no freeings occur during the time interval t from the arrival moment of the test call is equal to

$$f_0(t) = (e^{-\mu t})^s = e^{-s\mu t}$$

This implies that the number of freeings of line during the time interval t is a Poisson process with freeing rate $s\mu$. Thus, the interfreeing time follows an exponential distribution with mean $1/s\mu$. Therefore, the probability of exactly r freeings during the time interval t is given by a Poisson process with rate $s\mu$:

$$f_r(t) = e^{-s\mu t} \frac{(s\mu t)^r}{r!}$$

This result indicates that when all s lines are busy, the process of freeing busy lines follows a Poisson process with rate $s\mu$.

Substituting $f_r(t)$ into (5-32) yields

$$P_k\{W > t\} = e^{-s\mu t} \sum_{r=0}^{k-s} \frac{(s\mu t)^r}{r!} \ , \ k \geq s \qquad (5\text{-}33)$$

Since the input process is Poisson, for $k \geq s$, we obtain from (5-16) and (5-30):

$$\pi_k = p_k = \frac{a^k}{s! \, s^{k-s}} \, p_0 \, , \, k \geq s$$

Substituting π_k and $P_k\{W > t\}$ into (5-31) gives

$$P\{W > t\} = \frac{e^{-s\mu t}}{s! \, s^{-s}} \, p_0 \sum_{k=s}^{\infty} \left(\frac{a}{s}\right)^k \sum_{r=0}^{k-s} \frac{(s\mu t)^r}{r!}$$

$$= \frac{e^{-s\mu t}}{s! \, s^{-s}} \, p_0 \sum_{r=0}^{\infty} \sum_{k=s+r}^{\infty} \left(\frac{a}{s}\right)^k \frac{(s\mu t)^r}{r!}$$

$$= \frac{e^{-s\mu t}}{s! \, s^{-s}} \, p_0 \sum_{r=0}^{\infty} \sum_{m=0}^{\infty} \left(\frac{a}{s}\right)^{m+s+r} \frac{(\mu s t)^r}{r!} \qquad (5\text{-}34)$$

$$= \frac{e^{-s\mu t}}{s!} \, a^s \, p_0 \sum_{r=0}^{\infty} \frac{(a\mu t)^r}{r!} \sum_{m=0}^{\infty} \left(\frac{a}{s}\right)^m$$

$$= \frac{e^{-s\mu t}}{s!} \, p_0 \, e^{a\mu t} \, \frac{1}{1 - \dfrac{a}{s}}$$

$$= \frac{a^s \, s}{s! \, (s-a)} \, p_0 \, e^{-(s\mu-\lambda)t}$$

$$= C(s, a) \, e^{-(s\mu-\lambda)t} \, , \, a < s$$

We see that, under the stated conditions, the waiting time has an exponential distribution with a parameter of $s\mu - \lambda$. Note that, as expected

$$P\{W > 0\} = C(s, a)$$

In addition, the waiting time distribution in (5-34) is valid for the Erlang delay system with service in order of arrival discipline and exponential service time distribution.

The mean waiting time is given by

$$E[W] = \int_0^\infty t \; dP\{W \le t\}$$

$$= \int_0^\infty t \; C(s,a) \; (s\mu - \lambda)e^{-(s\mu-\lambda)t} \; dt \qquad (5\text{-}35)$$

$$= \frac{C(s,a)}{s\mu - \lambda}$$

which is the same as (5-20). The second moment of the waiting time is given by

$$E[W^2] = \int_0^\infty t^2 \; dP\{W \le t\}$$

$$= \int_0^\infty t^2 \; C(s,a) \; (\mu s - \lambda) \; e^{-(s\mu-\lambda)t} \; dt$$

$$= \frac{2 \; C(s,a)}{(s\mu - \lambda)^2}$$

Thus, the variance of the waiting time is given by

$$Var[W] = E[W^2] - (E[W])^2$$

$$= \frac{C(s,a)}{(s\mu - \lambda)^2} \; [2 - C(s,a)] \qquad (5\text{-}36)$$

Example 5-5. Consider an airline seat reservation office in which there are five assistants. Each has his own terminal connected to an on-line booking system. Assume that arriving customers follow a Poisson process with a rate of 40 customers during the busy hour of the day, and that the five assistants give similar exponential services with an average of 5 minutes. (a) How long a customer can expect to wait in a queue? (b) How long is the queue expected to be? (c) What is the effect of a 10% increase in the customer arrival rate on the mean waiting time and the mean queue length? (d) What is the variance of the new waiting time?

Solution. (a) The mean arrival rate $\lambda = \dfrac{2}{3}$ customers/min. The mean service time $\tau = 5$ min.

The offered load is $a = \lambda\tau = \dfrac{2}{3} \times 5 = 3.33$ erlangs.

Then the utilization factor $\rho = \dfrac{\lambda\tau}{s} = \dfrac{2}{3} \times \dfrac{5}{5} = \dfrac{2}{3}$. Using the Erlang delay curves, we find

$$C(s,a) = C(5, 3.33) = 0.32$$

Hence,

$$E[W] = \frac{C(s,a)}{s\mu - \lambda} = \frac{0.32}{5 \times \dfrac{1}{5} - \dfrac{2}{3}} = 0.96 \ min.$$

(b) Using Little's formula, the mean queue length is given by

$$E[N_q] = \lambda E[W] = \frac{2}{3} \times 0.96 = 0.64$$

(c) For a 10% increase in λ, the new input rate becomes

$$\lambda = \frac{2.2}{3} = 0.73 \ customers/min.$$

and the corresponding utilization factor now becomes

$$\rho = \frac{\lambda\tau}{s} = \frac{2.2}{3} \times \frac{5}{5} = 0.73$$

Now the new offered load is

$$a = \lambda\tau = \frac{2.2}{3} \times 5 = 3.67 \; \textit{erlangs}$$

Then, the new mean waiting time is

$$E[W] = \frac{0.44}{5 \times \dfrac{1}{5} - 0.73} = 1.63 \; \textit{min.}$$

which increases by $(1.63 - 0.96)/0.96 = 70\%$, and the new mean queue length is

$$E[N_q] = \lambda E[W] = 0.73 \times 1.63 = 1.19$$

which increases by $(1.19 - 0.64)/0.64 = 86\%$.

(d) Using (5-36), we find

$$Var[W] = \frac{0.44}{(5 \times \dfrac{1}{5} - \dfrac{2.2}{3})^2} [2 - 0.44] = 9.6525 \; \text{min}^2.$$

5-7. THE WAITING TIME DISTRIBUTION FOR THE COMBINED DELAY AND LOSS SYSTEM WITH SERVICE IN ORDER OF ARRIVAL

If the Erlang delay system has a finite $n - s$ waiting capacity, arriving calls which find the system in state n will be lost. In this case, the system becomes a combined delay and loss system. Waiting is possible only when $s \leq k < n$. If $k < s$ or $k \geq n$, and $t \geq 0$, then

$P_k\{W > t\} = 0$. Thus, using the formula of total probability, the waiting time distribution $P\{W > t\}$ with service in order of arrival can be written as

$$P\{W > t\} = \sum_{k=0}^{\infty} \pi_k\, P_k\{W > t\}$$

$$= \sum_{k=s}^{n-1} \pi_k\, P_k\{W > t\}$$

$$= \sum_{k=s}^{n-1} p_k\, P_k\{W > t\}$$

Using (5-23) and the conditional probability in (5-33), we find

$$P\{W > t\} = \sum_{k=s}^{n-1} \frac{a^k}{s!\, s^{k-s}}\, p_0 \sum_{r=0}^{k-s} e^{-s\mu t}\, \frac{(s\mu t)^r}{r!}$$

$$= e^{-s\mu t}\, \frac{s^s}{s!}\, p_0 \sum_{r=0}^{n-s-1} \frac{(s\mu t)^r}{r!} \sum_{k=r+s}^{n-1} \left(\frac{a}{s}\right)^k \qquad (5\text{-}37)$$

$$= e^{s\mu t}\, \frac{a^s}{s!}\, p_0 \sum_{r=0}^{n-s-1} \frac{(\lambda t)^r}{r!} \sum_{j=0}^{n-s-r-1} \left(\frac{a}{s}\right)^j$$

$$= e^{-s\mu t}\, \frac{a^s}{s!}\, p_0 \sum_{r=0}^{n-s-1} \frac{(\lambda t)^r}{r!}\, \frac{1-(a/s)^{n-s-r}}{1 - (a/s)}$$

$$= e^{-s\mu t}\, \frac{a^s\, p_0}{s!(1 - a/s)} \left[S_{n-s-1}(\lambda t) - \left(\frac{a}{s}\right)^{n-s} S_{n-s-1}(s\mu t) \right]$$

where

$$S_k(x) = \sum_{r=0}^{k} \frac{x^r}{r!}$$

and p_0 is given by (5-24).

5-8. OVERFLOW TRAFFIC

For the Erlang loss system, it was assumed that blocked calls (those calls which find all circuits busy) would be cleared from the system and would not return. In practical applications, blocked calls are routed to another similar group of circuits. For example, calls in an alternate routing network can take alternate routes when all of the circuits in the direct routes are busy. The traffic overflowed from the primary or direct routes (first-choice routes) is called the overflow traffic, and will be handled by an alternate trunk group.

To study the overflow traffic, we shall determine the mean and variance of the overflow distribution. First, we shall consider a queueing system composed of an s-server primary group and an infinite-server overflow group. The queueing model is shown in Figure 5-11.

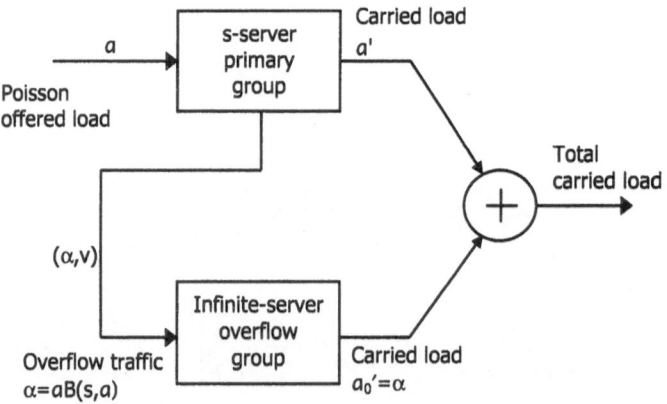

Figure 5-11. Overflow traffic from an Erlang loss system

We assume that customers request service first from the s-server primary group, and that customers who find all servers busy in the primary group, overflow to and are handled by the infinite-server overflow group. The input traffic is Poisson with rate λ, and the service times are independent, identical and exponentially distributed with mean $1/\mu$.

Let M and N denote the number of customers that are being served simultaneously in the primary and overflow groups, respectively. Furthermore, we let

$$p(j, k) = P\{M = j \,, N = k\}$$

be the statistical-equilibrium state probability of the system.

Consider the possible changes of state during a small time interval $(t, t + \Delta t)$ and display the possible transitions from state (j, k) to its neighboring states by using a two-dimensional state-transition-rate diagram as shown in Figures 5-12(a) and (b).

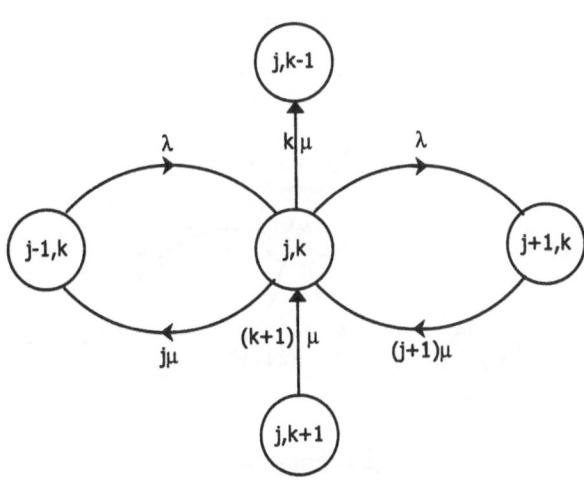

(a) For $0 \le j \le s - 1$

Using Figures 5-12(a) and (b), we write the following probability flow rates. For $0 \le j \le s - 1$, we have

Flow rate into state $(j, k) = \lambda p(j - 1, k) + (j + 1)\mu p(j + 1, k)$

$$+ (k + 1)\mu p(j, k + 1)$$

Flow rate out of state $(j, k) = [\lambda + (j + k)\mu]p(j, k)$

For $j = s$, we write

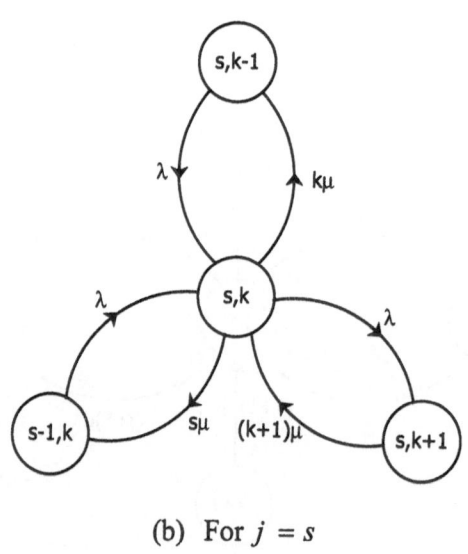

(b) For $j = s$

Figure 5-12. Two-dimensional state-transition-rate diagram

Flow rate into state (s, k) = λp(s − 1, k) + λp(s, k − 1) + (k + 1)μp(s, k + 1)

Flow rate out of state $(s, k) = [\lambda + (s + k)\mu]p(s, k)$

Equating the flow rates into and out of the corresponding states yield, respectively,

$$\lambda p(j - 1, k) + (j + 1)\mu p(j + 1, k) + (k + 1)\mu p(j, k + 1)$$

$$= [\lambda + (j + k)\mu]p(j, k) \quad, \quad 0 \le j \le s - 1, \quad (5\text{-}38)$$

$$k = 0, 1, \cdots,$$

and

$$\lambda p(s - 1, k) + \lambda p(s, k - 1) + (k + 1)\mu p(s, k + 1)$$

$$(5\text{-}39)$$

$$= [\lambda + (s + k)\mu]p(s, k) \quad, \quad k = 0, 1, \cdots,$$

or

$$ap(j - 1, k) + (j + 1)p(j + 1, k) + (k + 1)p(j, k + 1)$$

$$(5\text{-}40)$$

$$= (a + j + k)p(j, k), \, 0 \le j \le s - 1 \quad, \quad k = 0, 1, \cdots,$$

and

$$ap(s - 1, k) + ap(s, k - 1) + (k + 1)p(s, k + 1)$$

$$(5\text{-}41)$$

$$= (a + s + k)p(s, k) \quad, \quad k = 0, 1, \cdots.$$

where $p(s, -1) = 0$.

This set of equations (5-40) and (5-41) is subject to the normalization condition

$$\sum_{j=0}^{s} \sum_{k=0}^{\infty} p(j, k) = 1 \tag{5-42}$$

To solve the equations (5-40), (5-41), and (5-42), we consider the augmented system of equations obtained from (5-40) by extending s to infinity:

$$aq(j-1, k) + (j+1)q(j+1, k) + (k+1)q(j, k+1)$$

$$\tag{5-43}$$

$$= (a + j + k)q(j, k), j = 0, 1, \cdots ; k = 0, 1, \cdots .$$

where $q(-1, k) = 0$.

We shall solve the set of equations in (5-43) in such a way that

$$q(j, k) = p(j, k), j = 0, 1, \cdots s \quad ; \quad k = 0, 1, \cdots . \tag{5-44}$$

Define the two-dimensional probability-generating function

$$Q(x, y) = \sum_{j=0}^{\infty} \sum_{k=0}^{\infty} q(j, k)x^j y^k \tag{5-45}$$

Multiplying (5-43) by $x^j y^k$ and summing for j and k from 0 to ∞ yields the partial differential equation

$$(1-x)\frac{\partial Q(x, y)}{\partial x} + (1-y)\frac{\partial Q(x, y)}{\partial y} = a(1-x)Q(x, y) \tag{5-46}$$

whose general solution is

$$Q(x, y) = e^{-a(1-x)}F(\frac{1-y}{1-x}) \tag{5-47}$$

Using this generating function, R.B. Cooper (see reference [1], p. 139), obtained the following formulas for the mean $E[N] = \alpha$ and the variance $\text{Var}[N] = v$:

$$\alpha = aB(s, a) \qquad (5\text{-}49)$$

and

$$v = \alpha(1 - \alpha + \frac{a}{s + 1 + \alpha - a}) \qquad (5\text{-}50)$$

where $a = \lambda/\mu$ is the offered load in erlangs and $B(s, a)$ is the probability of loss given by the Erlang loss formula (5-7). Expressions (5-49) and (5-50) are called the Wilkinson formulas.

5-9. THE EQUIVALENT RANDOM METHOD

A telephone network where calls to a destination can take alternate routes is called an alternate routing network. A simple alternate routing network for direct distance dialing calls is shown in Figure 5-13. In this figure, switching nodes are denoted by lettered circles; first-choice routes and intermediate routes are represented by dashed lines which are high-usage routes, and final routes are represented by solid lines. The direction of overflow traffic is shown by a curved arrow.

The traffic originating from node A and destined for node D is assumed to be Poisson with a rate of λ calls per second. The holding times of calls are independent, identical and exponentially distributed random variables with a mean of $1/\mu$ seconds. Thus, the primary trunk group of the first choice routes operates like an Erlang loss system. The traffic which overflows the primary trunk group will be handled by the intermediate routes. Similarly, the traffic which overflows the intermediate trunk group will be handled by the final trunk group. Thus, both the intermediate trunk group and the final trunk group may carry the first-attempt Poisson traffic and act as alternate routes. The overflow traffic is not Poisson and will be approximated by its mean and variance.

To investigate the blocking in the alternate routing network, we shall calculate the proportion of calls which overflow the intermediate overflow group, given that these calls overflow the first-choice primary group.

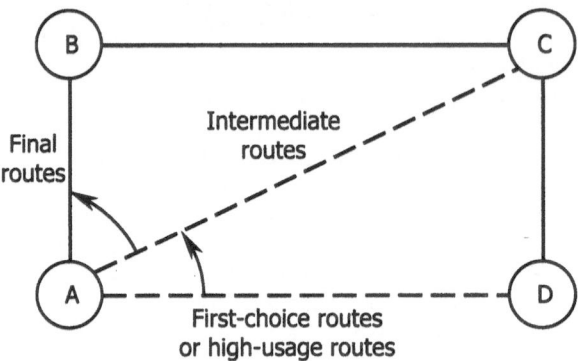

Figure 5-13. Alternate routing network

We shall describe an approximation technique, called the equivalent random method, which was developed by R.I. Wilkinson. We shall also determine the number of servers and the particular server arrangements such that a sufficient number of servers will ensure that a prespecified probability of loss (overflow from both primary and overflow servers) is not exceeded.

Consider the alternate routing network whose model is shown in Figure 5-14. The network has n primary groups; each primary group consists of s_i trunks and receives a direct Poisson offered load of a_i erlangs. All traffic overflowing the n primary groups is handled by a common overflow group of c trunks. The holding times on all groups are assumed to be independent, identical and exponentially distributed.

For the purpose of analysis, we represent approximately the load (traffic) that overflows the kth primary group by two parameters, the mean α_k and variance v_k, which are calculated by (5-49) and (5-50), respectively:

$$\alpha_k = a_k B(s_k, a_k) \, , \, k = 1, 2, \, \cdots , n, \qquad (5\text{-}51)$$

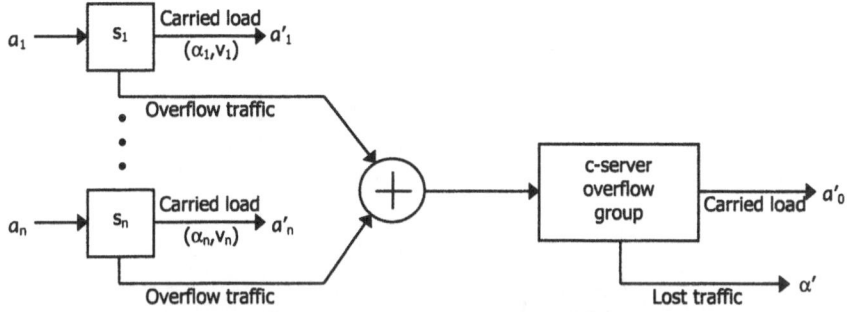

Figure 5-14. Lost traffic from an alternate routing network

$$v_k = \alpha_k(1 - \alpha_k + \frac{a_k}{s_k + 1 + \alpha_k - a_k}), \quad k = 1, 2, \cdots, n, \quad (5\text{-}52)$$

This implies that the load overflowing a primary group is represented approximately by the mean and variance of the overflow distribution that would result if the overflow traffic were handled by an infinite-server overflow group.

We shall calculate the probability that a call arriving at the *kth* primary group and finding all s_k lines busy will also find all c lines busy in the overflow group. We denote by α the mean value of the load that overflows the s-server primary group.

We replace the *n* primary groups and their respective equivalent Poisson offered loads with a single equivalent random primary group and a single equivalent random load so that we have the equivalent random model for Figure 5-14, as shown in Figure 5-15.

Since the *n* primary groups in Figure 5-14 are mutually independent, we choose both the number of *s* servers in the equivalent random primary group and the equivalent random (offered) load *a* in Figure 5-15 such that the total overflow load offered to the c-server overflow group is the same as that in Figure 5-14. Thus, we let

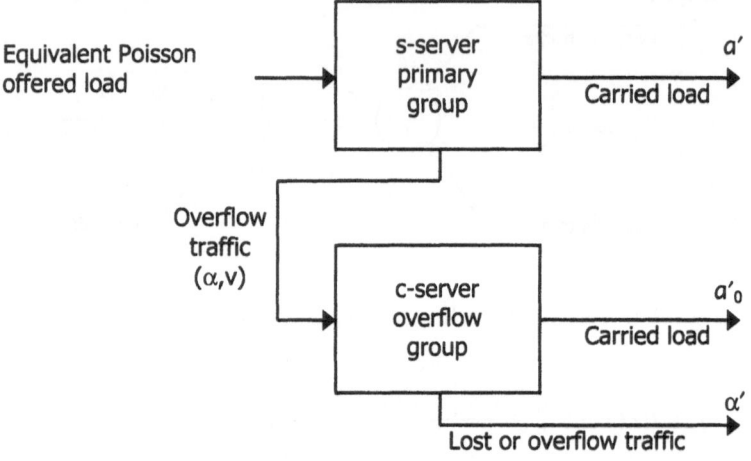

Figure 5-15. Equivalent random load and equivalent random server group

$$\alpha = \alpha_1 + \alpha_2 + \cdots + \alpha_n \qquad (5\text{-}53)$$

$$v = v_1 + v_2 + \cdots + v_n \qquad (5\text{-}54)$$

With the values of α and v calculated by (5-53) and (5-54), respectively, we describe Rapp's approximate formulas for determining the values of a and s from (5-49) and (5-50) as follows. In terms of the peakedness factor $z = v/\alpha$,

$$a = v + 3z(z - 1) \qquad (5\text{-}55)$$

and

$$s = \frac{a(\alpha + z)}{\alpha + z - 1} - \alpha - 1 \qquad (5\text{-}56)$$

These estimates of a and s are generally on the high side of the exact values. By rounding s down to its integral part $\lceil s \rceil$, we obtain the corresponding value of a by solving (5-56):

$$a = \frac{(\lceil s \rceil + \alpha + 1)(\alpha + z - 1)}{\alpha + z} \tag{5-57}$$

It follows that the load overflowing the c-server overflow group is approximately equal to

$$\alpha' = aB(s + c, a) \tag{5-58}$$

In the equivalent random system of Figure 5-15, the proportion of calls overflowing both the equivalent random primary group and the overflow group is given by

$$\gamma_0 = \frac{\alpha'}{\alpha} = \frac{aB(s + c, a)}{aB(s, a)} \tag{5-59}$$

which is the probability of blocking on the overflow group suffered by the overflow traffic from the n primary groups. Similarly, the proportion of traffic that is blocked on both the primary and overflow groups is approximated by

$$\gamma = \frac{aB(s + c, a)}{a_1 + a_2 + \cdots + a_n} \tag{5-60}$$

The equivalent random method has proven to be useful in teletraffic engineering. Given a blocking probability, the values of a and s, we can determine the number of trunks c in an alternate routing network using the Erlang B curves.

The equivalent random method is to replace the network in Figure 5-14 with the network in Figure 5-15, where the values of a and s are chosen approximately such that the mean α and variance v of the overflow traffic remain the same in both networks.

Example 5-6. Consider the outgoing trunk groups on a private switching system. Suppose that blocked calls on these outgoing trunk groups are neither queued nor rerouted so that the callers merely hang up and try again. Trunk groups operated in this way are known as stand-alone trunk groups. They carry first-route traffic only. Thus, the offered traffic to the stand-alone trunk groups is assumed to be random or Poisson. Figure 5-16 shows the traffic flow on a stand-alone trunk group.

Assume that the busy-hour offered traffic to the outgoing trunk group is 5 erlangs.

(a)   What is the apparent offered load if the grade of service or blocking factor β is 0.04?

(b)   How many stand-alone trunks should be installed?

Solution. From Figure 5-16, we write

$$a_A = a + 0.7\beta a_A$$

Thus,

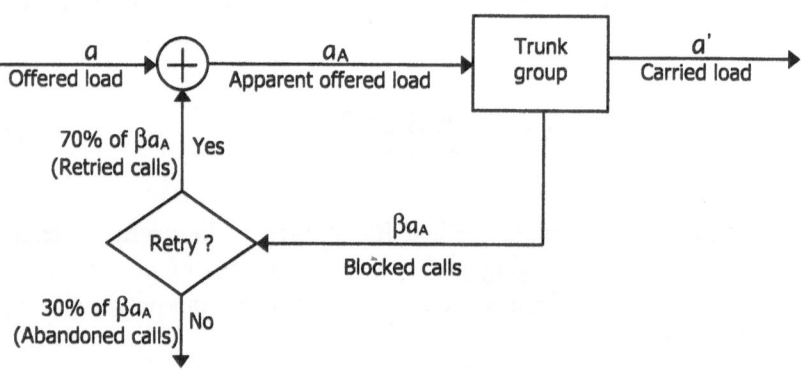

Figure 5-16. Traffic flow on a stand-alone trunk group.

$$a_A = a \beta_A \tag{5-61}$$

where

$$\beta_A = \frac{1}{1 - 0.7\beta} \tag{5-62}$$

is called the apparent offered load adjustment factor. Furthermore, we also write

$$a_A = a' + \beta a_A$$

Hence, using (5-61) and solving for a gives

$$a = a' \beta_0 \tag{5-63}$$

where

$$\beta_0 = \frac{1 - 0.7\beta}{1 - \beta} \tag{5-64}$$

is known as the offered-load adjustment factor.

(a) Since $a = 5$ erlangs and $\beta = 0.04$, using (5-61) and (5-62), we get

$$a_A = 5 \times \frac{1}{1 - 0.7 \times 0.04} = 5.144 \ erlangs$$

(b) Using the Erlang loss curves in Figure 5-2, we find that the number of required stand-alone trunks is equal to 9.

Example 5-7. Poisson traffic of a erlangs is offered to a primary group (or routes) of s trunks. Assume that blocked calls are routed to a secondary route.

(a)   Show that the variance of the carried traffic of the primary route is given by

$$v' = a'\{1 - a[B(s - 1, a) - B(s, a)]\} \qquad (5\text{-}65)$$

(b)   In particular, let $a = 15$ erlangs and the number of primary trunks be 20. Calculate the mean and variance of the carried load. What is the peakedness factor of the carried load?

(c)   Calculate the mean and variance of the overflow traffic that is offered to the secondary route. What is the peakedness factor of the overflow traffic?

Solution. (a) Using (5-4), (5-9) and (5-10), we write

$$v' = \sum_{k=0}^{s} k^2 p_k - a'^2 = \sum_{k=0}^{s} k(k - 1)p_k + a' - a'^2$$

$$= a^2(1 - p_{s-1} - p_s) + a(1 - p_s) - a^2(1 - p_s)^2$$

$$= a(1 - p_s - ap_{s-1} + ap_s - ap_s^2)$$

Writing p_{s-1} in the form

$$p_{s-1} = (1 - p_s)B(s - 1, a)$$

and factoring out $(1 - p_s)$ yield the required expression (5-65).

(b)   Using (5-10), (5-65) and the Erlang loss curves in Figure 5-3, we get

$$a' = 15(1 - 0.046) = 14.31 \ erlangs$$

$$v' = 14.31\{1 - 15[0.064 - 0.046]\}$$

$$= 10.4463$$

The peakedness factor of the carried load is given by

$$z = \frac{v'}{a'} = \frac{10.4463}{14.31} = 0.73$$

Note that $z < 1$. Thus, the carried traffic is smooth.

(c) Using (5-49) and (5-50), we obtain

$$\alpha = 15 \times 0.046 = 0.69 \ erlangs$$

$$v = 0.69\left[1 - 0.69 + \frac{15}{20 + 1 + 0.69 - 15}\right]$$

$$= 1.5471$$

The peakedness factor of the overflow traffic is given by

$$z = \frac{v}{\alpha} = \frac{1.5471}{0.69} = 2.24$$

Since the overflow traffic has a peakedness factor greater than unity, it is rough.

Example 5-8. Consider the alternate routing network shown in Figure 5-17. Suppose that the mean and variance of the overflow traffic are 3.88 erlangs and 6.29, respectively. Determine the equivalent random traffic and the number of trunks in the primary route so that the blocking probability of both the primary and secondary routes will be 0.01. What is the lost traffic of the network?

Solution. Since $\alpha = 3.88$ erlangs and $v = 6.29$, the peakedness factor is

$$z = \frac{6.29}{3.88} = 1.62$$

Using Rapp's approximate formulas (5-55) and (5-56), we get

$$a = 6.29 + 3 \times 1.62(1.62 - 1) = 9.3 \; erlangs$$

$$s = \frac{9.3(3.88 + 1.62)}{3.88 + 1.62 - 1} - 3.88 - 1 = 6.4867$$

By rounding s to 6, we obtain from (5-57) the equivalent Poisson (random) traffic

$$a = \frac{\lceil 6.4867 \rceil + 3.88 + 1)(3.88 + 1.62 - 1)}{3.88 + 1.62} = 8.9 \; erlangs$$

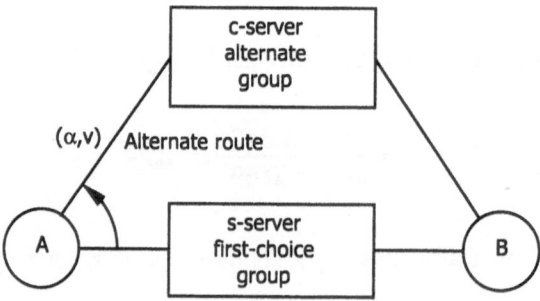

Figure 5-17. Alternate routing network.

Using the Erlang loss curves in Figure 5-2 with the equivalent random traffic of 8.9 erlangs and blocking probability of 0.01, we find

$$s + c = 16 \quad \text{or} \quad c = 10$$

since $s = 6$.

The lost traffic of the network is

$$\alpha' = 8.9 \times 0.01 = 0.089 \; erlangs$$

Example 5-9. To illustrate the application of the equivalent random method, we consider the alternate routing network shown in Figure 5-18, where the offered load and the economic traffic are given as follows:

Offered load	Economic traffic
$a_{AB} = 7.2 \; erlangs$	$a'_{EAB} = 0.409 \; erlangs$
$a_{AC} = 10 \; erlangs$	$a'_{EAC} = 0.425 \; erlangs$

The economic traffic is known as the economic CCS (or ECCS) in the U.S.A. This is the minimum traffic that the last trunk in a trunk group should carry. If the last trunk in a trunk group carries less traffic than the economic traffic, it would be cheaper to reroute the last trunk's traffic to an alternate route.

(a) What is the number of trunks in the two direct routes such that the last trunk in the trunk group will carry a traffic greater than the corresponding economic traffic?

(b) Calculate the mean and variance of the overflow traffic from the direct routes.

(c) What is the peakedness factor of the overflow traffic?

(d) Calculate the equivalent random traffic and the number of trunks required.

(e) Calculate the number of trunks in the alternate trunk group, given that the blocking probability in the final group is 0.01.

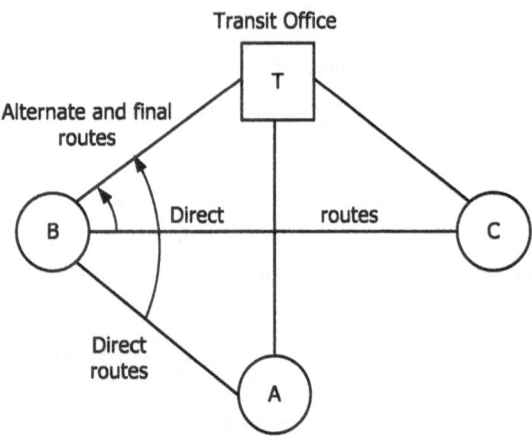

Figure 5-18. Alternate routing network.

Solution. (a) For the direct trunk group between nodes A and B, we find by use of (5-12) and Figure 5-2 that

$$a_9' = 7.2(0.188 - 0.130) = 0.4176 \ erlangs$$

which is greater than a_{EAB}' or $0.409 \ erlangs$. Note that $a_{10}' = 7.2(0.130 - 0.086) = 0.3168 \ erlangs$ which is less than a_{EAB}'. Thus, the number of trunks in this trunk group is 9. Similarly, for the direct trunk group between nodes A and C, we have

$$a_{12}' = 10(0.163 - 0.120) = 0.43 \ erlangs$$

which is greater than a_{EAC}', and hence, the number of trunks in this trunk group is 12.

(b) Using (5-51) and (5-52), we obtain

$$\alpha_{AB} = 7.2 \times 0.13 = 0.936 \; erlangs$$

$$v_{AB} = 0.936 \left[1 - 0.936 + \frac{7.2}{9 + 1 + 0.936 - 7.2} \right] = 1.8638$$

and

$$\alpha_{AC} = 10 \times 0.12 = 1.2 \; erlangs$$

$$v_{AC} = 1.2(1 - 1.2 + \frac{10}{12 + 1.2 - 10}) = 2.6171$$

Thus,

$$\alpha = 0.936 + 1.2 = 2.136 \; erlangs$$

$$v = 1.8638 + 2.6171 = 4.4809$$

(c) The peakedness factor of the overflow traffic is

$$z = \frac{4.4809}{2.136} = 2.0978$$

(d) Using (5-55) and (5-56), we find

$$a = 4.4809 + 3 \times 2.0978(2.0978 - 1) = 11.3898 \; erlangs$$

$$s = \frac{11.3898(2.136 + 2.0978)}{2.136 + 2.0978 - 1} - 2.136 - 1 = 11.7759$$

By rounding this value of s to 11, we obtain the equivalent random traffic using (5-57):

$$a = \frac{(11 + 2.136 + 1)(2.136 + 2.098 - 1)}{2.136 + 2.0978} = 10.7972 \ erlangs$$

and the required number of equivalent trunks is 11.

(e)   For a blocking probability of 0.01 and an equivalent random offered traffic of 10.7972 erlangs, we find from the Erlang B curves

$$s + c = 19$$

or

$$c = 19 - 11 = 8$$

5-10. SUMMARY

The Erlang loss system and the Erlang delay system are investigated using the birth and death process as a model. By specifying the corresponding birth and death rates, the Erlang loss distribution and the Erlang loss formula are obtained. The carried load in the Erlang loss system is then calculated in terms of the offered load and the loss probability. Similarly, the Erlang delay formula for the calculation of waiting probability is derived. In addition, the average number of calls and the mean waiting time in the Erlang delay system are determined.

It is important to note that the Erlang loss formula or the Erlang B formula is valid for arbitrary holding time distributions. However, the Erlang delay formula or the Erlang C formula is true only when the trunk occupancy is less than unity so that the system can reach the steady state or equilibrium conditions. In practical applications, curves for the Erlang B and Erlang C formulas are available and convenient to use.

When calculating the waiting time distribution, the queue discipline must be specified. The Erlang delay system assumes service in order of arrival or FIFO queue discipline. When there are calls waiting for service in the queue, all servers or trunk lines are busy. This leads to the fact that under the condition of all servers being busy, the departing process is Poisson with rate $s \mu$. As a result, the calculation of the waiting time distribution is greatly simplified.

For the study of alternate routing networks, the equivalent random method is useful when the overflow traffic from the first choice routes is approximated by its first two moments or the mean and variance. An example is given to illustrate the application.

REFERENCES

[1] Cooper, R.B., Introduction to Queueing Theory, 2nd ed., New York: Elsevier Science, 1981.

[2] Khintchine, A.Y., Mathematical Methods in the Theory of Queueing, 2nd ed., New York: Hafner, 1969.

[3] Syski, R., Introduction to Congestion Theory in Telephone Systems, 2nd ed., New York: Elsevier Science, 1986.

[4] Fry, T.C., Probability and Its Engineering Uses, 2nd ed., New York: Van Nostrand, 1965.

[5] Bucher, J.R., Traffic System Design Handbook, Piscataway, N.J.: IEEE, 1992.

[6] Hills, M.T., Telecommunications Switching Principles, Cambridge, Mass., MIT Press, 1979.

[7] Martine, R.R., Basic Traffic Analysis, Englewood Cliffs, N.J.: Prentice Hall, 1994.

[8] Wilkinson, R.I., "Theories for Toll Traffic Engineering in the U.S.A.", Bell System Tech. J., Vol. 35, No. 2, 1956, pp. 421-514.

PROBLEMS

5-1.   An average of 150 calls per hour calling rate with an average duration of five minutes per call is offered to a trunk group. The trunks must provide a 0.005 grade of service.

(a) How many stand-alone trunks are required?

(b) How many trunks are needed to ensure a 0.01 grade of service?

(c) What is the grade of service for 20 trunks?

5-2.   The following measurements were made on an outgoing trunk group during the busy hour:

Number of originating calls = 294.

Number of overflow calls = 6.

Number of times that all trunks are busy is 12, with an average of 8 seconds per period.

(a) What is the time congestion?

(b) What is the call congestion?

5-3.   Consider that a connection between the central computer of the head office of a bank and a data terminal of its branch office is established at 1:00 A.M. Assuming that the average calling rate for connection is 1 call per hour and the connection was maintained continuously. If data are transferred at a rate of 9600 bits per second, determine the amount of traffic in erlangs transferred over the established connection between 1:00 A.M. and 1:45 A.M.

5-4.   A group of 10,000 subscribers originate an average rate of 500 calls per hour with a mean holding time of 200 seconds.

(a) What is the offered traffic in CCS?

(b) What is the offered traffic per subscriber in CCS?

(c) How many erlangs of traffic originated by a subscriber?

(d)   If the system is operated on a blocked call cleared basis with a group of 38 trunks, what is the blocking probability?

(e)   What is the carried load per trunk?

5-5.   Show that $B(s,a) = \dfrac{aB(s-1, a)}{s + aB(s-1, a)}$.

5-6.   Show that $C(s,a) = \dfrac{sB(s, a)}{s - a[1 - B(s, a)]}$, $s > a$.

5-7.   Consider an Erlang delay system with s exponential lines of mean service time $\tau = 1/\mu$ seconds and Poisson input of rate λ calls per second. A test call entering the system at $t = 0$ finds all s lines busy with n waiting calls. Waiting calls are served in order of arrival. Service times are assumed to be mutually independent, identical, and exponentially distributed random variables.

(a)   Find the mean waiting time for the test call in the queue.

(b)   Suppose that no new calls arrive at the system after $t = 0$. What is the expected length of time from the arrival of the test call at $t = 0$ until the system finishes serving all the calls?

(c)   Let X be the order of completion of serving the test call; that is, $X = k$ if the test call is the kth call to have been completely served after $t = 0$. Find $P\{X = k\}$.

(d)   What is the probability that the test call has been completely served before the call immediately ahead of it in the queue?

5-8.   A concentrator consists of a number of input lines, each having Poisson input traffic with a constant rate and 20 exponential trunks (output lines) with an average holding time of 3 minutes. Suppose that the total offered traffic is 14 erlangs.

(a)   If blocked calls are cleared or lost, what is the blocking (or loss) probability? How many trunks should be added to ensure a grade of service of 0.01?

(b)   If blocked calls are delayed, what is the call waiting probability? What is the mean waiting time? In order to ensure a waiting probability of 0.02, how many trunks should be added? What is the new mean waiting time?

5-9.   An apparent offered traffic of 45 erlangs is offered to a stand-alone trunk group. Suppose that 70% of blocked calls will retry.

(a) What is the original offered traffic if the blocking factor is 0.02?

(b) What is the abandoned traffic?

(c) What is the carried load?

5-10.   Poisson traffic of 8 erlangs is offered to a stand-alone trunk group. Suppose that 70% of blocked calls are re-attempted.

(a)   Calculate the apparent load for a blocking factor of 0.05.

(b)   Calculate the number of required trunks for the above specified grade of service.

5-11.   Students arrive at a computer terminal at an average rate of 10 per hour. The average length of time using the terminal is 6 minutes. There is one extra chair which is used as a waiting position. Students who find both the terminal and the extra chair occupied will go away.

(a)   Assuming Poisson arrivals and exponential service times, find the steady-state probabilities $p_k = P\{N = k\}$ for the system, where N denotes the number of students in the system.

(b)   Find the average number of students served per hour.

(c)   If there are two terminals (and no waiting positions), then calculate the average number of students served per hour.

5-12.   Calls arrive according to a Poisson process with an average rate of λ calls per hour at a group of s circuits. An arriving call will be held in the system for exactly a random amount of time with an exponential distribution and mean of $1/\mu$ hours, after which the call will depart regardless of whether or not it is being served or is waiting in the queue.

(a)   Find   the   steady-state   probability   distribution $p_k = P\{N = k\}$, where N denotes the number of calls in the system at any time.

(b)   Let q be the steady-state probability that an arriving call does   not   receive   service.   Show   that $q = P(s, a) - \dfrac{s}{a} P(s + 1, a)$, where $a = \dfrac{\lambda}{\mu} < 1$ is the offered load and

$$P(s, a) = \sum_{k=s}^{\infty} \frac{a^k}{k!} e^{-a}$$

(c)   Also show that $q = 1 - \dfrac{a'}{a}$, where a' is the carried load.

5-13.   Calls arrive at a travel agency according to a Poisson process with a mean arrival rate of 6 calls per hour. There are 5 assistants at the agency, each having identical exponential service time distribution with a mean of 15 minutes per call. Services are in order of arrivals.

(a) What is the waiting probability of a call?

(b) What is the probability that a call will wait for more than 2 minutes?

(c) Find the mean waiting time and its standard deviation.

5-14.   A single-server queueing system with an infinite number of waiting positions is modeled as a birth and death process with

$$\lambda_j = \frac{\lambda}{j + 1}, j = 0, 1, 2, \cdots$$

and

$$\mu_j = \mu, j = 1, 2, 3, \cdots$$

where births correspond to arrivals and deaths correspond to service completions.

(a) Find the steady-state probability $p_k = P\{N = k\}$, where N denotes the number of customers in the system.

(b) Show that the arriving customer's distribution is

$$\pi_j = (1 - e^{-\lambda\tau})^{-1} p_{j+1}, j = 0, 1, \cdots$$

where $\tau = 1/\mu$

(c) Calculate the carried load a' and the offered load a.

5-15 An alternate routing network consists of two groups (primary groups) of first-choice routes as shown in Figure 5-19.

The offered Poisson traffic is as follows:

$$a_{AB} = 8.8 \; erlangs \;\; , \;\; s_{AB} = 12 \; trunks$$
$$a_{AC} = 6.6 \; erlangs \;\; , \;\; s_{AC} = 10 \; trunks$$

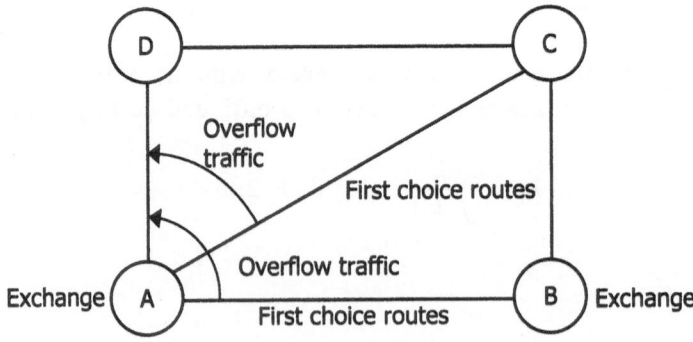

Figure 5-19. An alternate routing network.

The blocking probability for the two primary groups and the alternate routes is 0.01.

(a) Calculate the mean and variance of the overflow traffic.

(b) What is the peakedness factor of the overflow traffic?

(c) Calculate the equivalent random traffic and the number of trunks.

(d) What is the number of trunks required to handle the overflow traffic?

5-16. Poisson traffic of 10 erlangs is offered to a primary group of 15 trunks. Suppose that blocked calls are routed to a secondary alternate group.

(a) Calculate the mean, variance, and peakedness factor of the carried load.

(b) Calculate the mean, variance and peakedness factor of the overflow traffic.

5-17. Consider the $GI/M/1$ queue of Section 4-9.

(a) Using the argument of Section 5-5, show that the arriving customer's distribution and the outsider observer's distribution are related by

$$p_k = \rho \, \pi_{k-1} \, , \, k = 1, 2, \cdots$$

(b) Show that the outside observer's distribution is given by

$$p_k = \begin{cases} 1 - \rho \, , \, k = 0 \\ (1 - \sigma)\rho\sigma^{k-1}, \, k = 1, 2, \cdots \end{cases}$$

where $\rho = 1 - p_o$.

CHAPTER 6

THE ENGSET LOSS AND DELAY SYSTEMS

6-1. INTRODUCTION

If the number of sources which generate the input traffic is finite, the input traffic flow can no longer be modeled by the Poisson process because the arrival rate, in this case, is not constant and depends on the number of idle sources. However, the behaviour of the system can still be described by the birth and death process which was studied in Section 4-5.

This chapter investigates the Engset loss and delay systems using the birth and death process as the basic model. By choosing appropriate birth and death rates, it is possible to obtain the steady-state probability distribution for the Engset loss and delay systems.

Quasi-random Input. If an input traffic flow is generated by a finite number of n sources, with each source independently generating calls at a rate of γ calls per hour when idle and rate zero otherwise, the input process is called a quasi-random input.

6-2. THE ENGSET LOSS SYSTEM

Consider a telephone system with a quasi-random input that is generated by n identical, independent sources and served by s lines, where $s < n$. The holding time distribution of the calls is exponential with a mean of $\dfrac{1}{\mu}$ seconds. An arriving call that finds the system in state s will be cleared from the system. Thus, the system operates on a blocked-call cleared basis. A queueing system with a quasi-random input and exponential servers, which operates on a blocked-call cleared basis, is called an Engset loss system. A queueing model for the Engset loss system is shown in Figure 6-1.

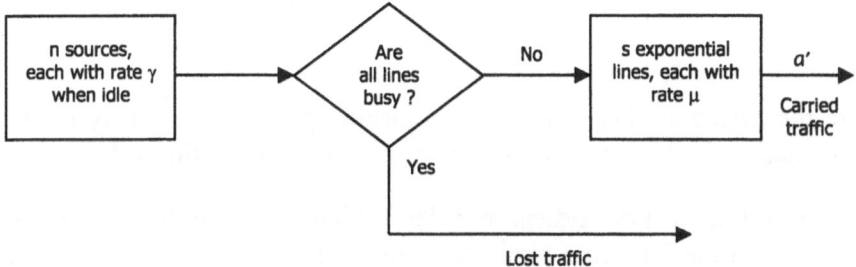

Figure 6-1. Queueing model for the Engset loss system

According to the assumption of quasi-random input with n sources, only an idle source can generate calls at rate γ. The probability of occurrence of a call in a small time interval $(t, t + \Delta t)$ when the system is in state k at time t, is assumed to be

$$(n - k)\gamma\Delta t + o\ (\Delta t),\ \ 0 \le k \le s - 1$$

and zero for $k = s$. Thus, the transition probability for the system to change from state k to state $k+1$ in the interval $(t, t + \Delta t)$ is given by

$$P_{k,k+1}(\Delta t)\ \ = \begin{cases} (n - k)\gamma\Delta t + o\ (\Delta t) & ,\ k \le s - 1 \\ 0 & ,\ k = s \end{cases}$$

Based on the assumption of exponential holding time distribution with service rate μ, the transition probability for the system to change from state k to state $k-1$ in the interval $(t, t + \Delta t)$ is given by

$$P_{k,k-1}(\Delta t)\ =\ k\mu\Delta t + o(\Delta t)\ \ ,\ \ 0 \le k \le s$$

Furthermore, the transition probability that the system does not change state in the interval $(t, t + \Delta t)$ is equal to

$$P_{kk}(\Delta t) = \begin{cases} 1 - (n - k)\gamma\Delta t - k\mu\Delta t + o\ (\Delta t) \ , \ 0 \le k < s \\ \\ 1 - s\mu\Delta t + o\ (\Delta t) \qquad\qquad , \ k = s \end{cases}$$

Like the Erlang systems, these state transition probabilities satisfy the four properties of the birth and death process discussed in Section 4-5.

The Engset loss system can be modeled by the birth and death process with the following birth and death rates when the system is in state k:

$$\lambda_k = \begin{cases} (n - k)\gamma \ , \ k = 0, 1, \ \cdots, s - 1 \\ 0 \qquad\quad , \ k \ge s \end{cases} \tag{6-1}$$

and

$$\mu_k = k\ \mu, k = 0, 1, \ \cdots, s \tag{6-2}$$

Under steady-state conditions, the state probabilities $p_k\ [n]$ of the Engset loss system can be obtained by substituting the birth and death rates of (6-1) and (6-2) into (4-24) to yield

$$p_k\ [n] = \begin{bmatrix} n \\ k \end{bmatrix} \hat{a}^k\ p_0[n] \ , \ k = 1, 2, \ \cdots, s \tag{6-3}$$

where $\hat{a} = \dfrac{\gamma}{\mu}$ is called the offered load per idle source. Using the normalization condition, we find

$$\frac{1}{p_0\ [n]} = \sum_{k=0}^{s} \begin{bmatrix} n \\ k \end{bmatrix} \hat{a}^k$$

Hence,

$$p_k[n] = \frac{\binom{n}{k}\hat{a}^k}{\sum\limits_{r=0}^{s}\binom{n}{r}\hat{a}^r} \quad , \quad k = 0, 1, \cdots, s \qquad (6\text{-}4)$$

and $p_k[n] = 0$ for $k > s$. If we introduce the new notation

$$p = \frac{\hat{a}}{1 + \hat{a}}$$

then

$$\hat{a} = \frac{p}{1 - p}$$

Substituting $\hat{a}$ into (6-4) and multiplying the numerator and denominator by $(1 - p)^n$ yields the truncated binomial distribution,

$$p_k[n] = \frac{\binom{n}{k} p^k (1-p)^{n-k}}{\sum\limits_{j=0}^{s}\binom{n}{j} p^j (1-p)^{n-j}} \quad , \quad k = 0, 1, \ldots, s \qquad (6\text{-}5)$$

Note that $p_1[1] = p$. In other words, p can be interpreted as the probability that an arbitrary source will be busy if there is no interaction among the sources.

6-3. THE ARRIVING CUSTOMER'S DISTRIBUTION FOR THE ENGSET LOSS SYSTEM

Suppose that the system is in equilibrium. Let t_k be the portion of time in which the system is in state k, and let t be the observation time.

The average number of incoming calls during the interval t_k is then equal to

$$(n-k)\gamma \, t_k = (n-k)\gamma \, t \, \frac{t_k}{t} = (n-k)\gamma \, t \, p_k[n]$$

The proportion of incoming calls during the interval t that find the system in state k is then equal to the probability

$$\frac{(n-k)\gamma \, t \, p_k[n]}{\displaystyle\sum_{j=0}^{s} (n-j)\gamma \, t \, p_j[n]} = \frac{(n-k) \, p_k[n]}{\displaystyle\sum_{j=0}^{s} (n-j) \, p_j[n]} \quad , \quad k = 0, 1, \cdots, s$$

which is just the probability $\pi_k[n]$ that an incoming call finds the system in state k:

$$\pi_k[n] = \frac{(n - k) \, p_k[n]}{\displaystyle\sum_{j=0}^{s} (n - j) \, p_j[n]} \quad , \quad k = 0, 1, \cdots, s$$

Substitution of $p_k[n]$ from (6-4) into this expression gives

$$\pi_k[n] = \frac{(n-k) \dbinom{n}{k} \hat{a}^k}{\displaystyle\sum_{j=0}^{s} (n-j) \dbinom{n}{j} \hat{a}^j} = \frac{\dbinom{n-1}{k} \hat{a}^k}{\displaystyle\sum_{j=0}^{s} \dbinom{n-1}{j} \hat{a}^j} \quad ,k = 0, 1, \cdots, s \quad (6\text{-}6)$$

In particular, the probability of loss is given by

$$\pi_s[n] = \frac{\begin{bmatrix} n-1 \\ s \end{bmatrix} \hat{a}^s}{\sum_{j=0}^{s} \begin{bmatrix} n-1 \\ j \end{bmatrix} \hat{a}^j} \quad , \quad \hat{a} = \frac{\gamma}{\mu} \tag{6-7}$$

Formula (6-7) is often called the Engset loss formula. It is interesting to note that the right-hand side of (6-6) is equal to $p_k[n-1]$. Therefore, we see that in the Engset loss system, the arriving customer's distribution with n sources is equal to the outside observer's distribution with $n-1$ sources; that is,

$$\pi_k[n] = p_k[n-1] \quad , \quad k = 0, 1, \cdots, s \tag{6-8}$$

This result exemplifies that in general, the arriving customer's distribution is not equal to the outside observer's distribution. In a finite-source system, the proportion of call arrivals finding all lines busy is smaller than $p_s[n]$ because fewer arrivals occur during periods when all lines are busy. Thus, in a finite-source system, call congestion is always smaller than time congestion.

It is important to point out that the Engset loss formula (6-7) is also valid for arbitrary interarrival time distribution and arbitrary holding time distribution. Values for $\pi_s[n]$ obtained from (6-7) have been plotted against the offered load $\hat{a}$ per idle source for different values of the number of lines s (see, for example, T.C. Fry). Two typical groups of Engset loss curves are given in Figure 6-2 and Figure 6-3 for probability of loss equal to 0.001 and 0.01, respectively.

6-4. THE OFFERED LOAD AND CARRIED LOAD IN THE ENGSET LOSS SYSTEM

Consider the n-source Engset loss system with $s < n$ lines. Suppose that the system is in equilibrium or the steady state. Let N denote the average number of calls in the system. By definition, the carried load is equal to N and can be expressed as

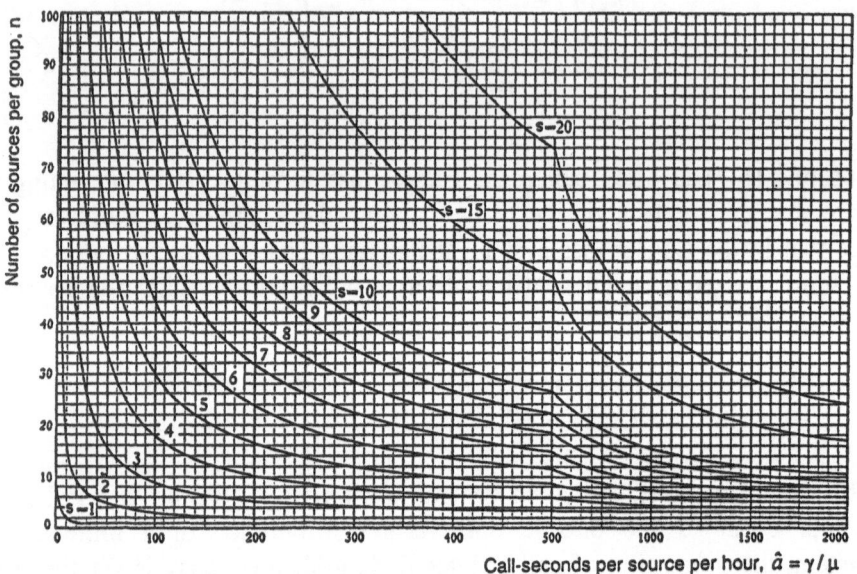

Figure 6-2. Working chart for the Engset formula. Probability of loss=0.001

$$a' = N = \sum_{k=0}^{s} k\, p_k[n] = a(1 - \pi_s[n])$$
(6-9)

Since the average number of idle sources is equal to $n - N$, then the offered load, by definition, is equal to

$$a = (n - N)\,\frac{\gamma}{\mu} = \hat{a}(n - N)$$
(6-10)

Using (6-9) for N, (6-10) can be written as

$$a = \frac{n\hat{a}}{1 + \hat{a}(1 - \pi_s[n])}$$
(6-11)

We see that the offered load generated by a finite number of sources

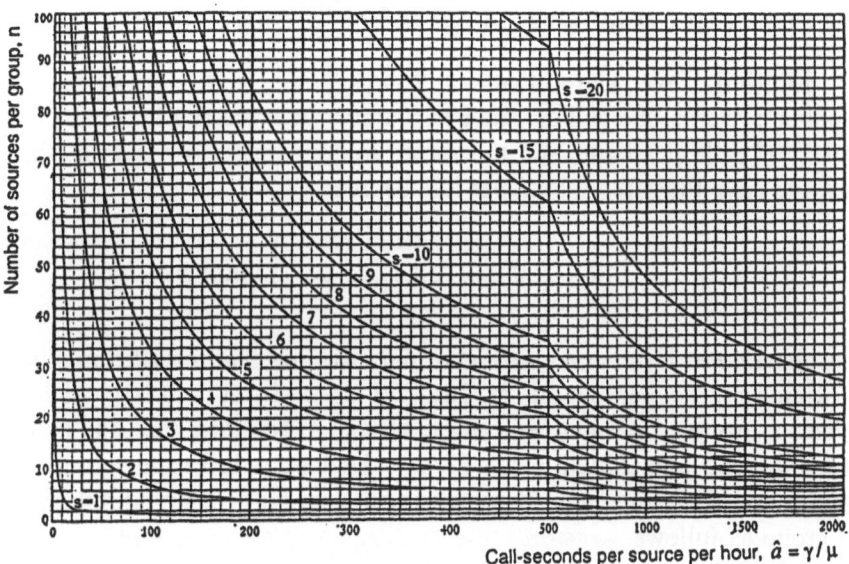

Figure 6-3. Working chart for the Engset formula. Probability of loss=0.01

depends on the blocking probability $\pi_s[n]$. However, if $n = s$, then there will be no blocking in the system and $\pi_s[n] = 0$. In this case, (6-11) becomes

$$a_s = \frac{n\hat{a}}{1 + \hat{a}} \qquad (6\text{-}12)$$

which is known as the intended offered load. Furthermore, if $n = s$, the denominator of (6-5) reduces to unity and (6-5) becomes the binomial distribution

$$p_k[n] = \binom{n}{k} p^k (1 - p)^{n-k}, \ k = 0, 1, ..., n \qquad (6\text{-}13)$$

Example 6-1. Suppose that there are 200 sources to be accommodated by a number of switches. Each switch is capable of accessing a group of 10 trunks. Assume that on average, each of these sources during a busy-hour originates two calls of an average holding time of 140 seconds. How many trunk groups are required in order to serve those sources for a 0.001 grade of service?

Solution. We know that $\gamma = 2$ *calls/hour*, and $\tau = 140$ sec. In telephone systems, the calling rate per source γ is expressed in terms of the number of calls per hour, and the mean holding time $\tau = 1/\mu$ in seconds. Thus, we have the offered load per idle source:

$$\hat{a} = \gamma\tau = 2 \times 140 = 280 \; call-seconds \; per \; source \; per \; hour$$

From the Engset loss curves in Figure 6-2, we find that for 10 trunks with an offered traffic per source of 280 *call–seconds* at a 0.001 grade of service, a group of 43 sources can be served. Therefore, the 200 sources can be divided into $200/43 = 4 + 28/43$. That is, 4 full trunk groups are required and the remaining 28 sources will then be accommodated by the odd group as follows.

The point on the chart which corresponds to 28 sources and an offered load of 280 call-seconds per source per hour lies between 7 and 8. Hence, the odd group will require 8 trunks to carry the traffic. The grouping of the sources will therefore be four full groups, each with 43 sources and one group with 28 sources. There will also be 4 trunk groups, each with 10 trunks and one group with 8 trunks. In total, 48 trunks are required.

6-5. THE ENGSET DELAY SYSTEM

If blocked calls in an Engset system wait as long as necessary for service, there will be no loss of traffic in the system. In this case, we can modify the birth and death rates to take into account the possible $n-s$ waiting calls. Now we choose the birth rate

$$\lambda_k = (n-k)\,\gamma, \;\; k = 0, 1, \ldots, n$$

and the death rate

$$\mu_k = \begin{cases} k\mu, \, k = 0, \, 1, \, \ldots, \, s - 1 \\ s\mu, \, s \leq k \leq n \end{cases}$$

Under equilibrium conditions, substitution of these birth and death rates into (4-24) gives the state probability distribution $p_k[n]$.

Note that for $k \leq s$, the situation is exactly the same as with the Engset loss system. However, for $k \geq s$, the death rate remains constant with rate $s\mu$. Therefore, we have

$$p_k[n] = \begin{cases} \begin{pmatrix} n \\ k \end{pmatrix} \hat{a}^k \, p_0[n] \quad , k = 0, \, 1, \, \cdots, \, s \\ \\ \dfrac{n! \, \hat{a}^k \, p_0[n]}{(n - k)! \, s! \, s^{k-s}} \, , k = s, \, s + 1, \, \cdots \, n \end{cases} \qquad (6\text{-}14)$$

where

$$\hat{a} = \frac{\gamma}{\mu},$$

and $p_0[n]$ is given by

$$\frac{1}{p_0[n]} = \sum_{k=0}^{s} \begin{pmatrix} n \\ k \end{pmatrix} \hat{a}^k + \sum_{k=s+1}^{n} \frac{n! \, \hat{a}^k}{(n-k)! \, s! \, s^{k-s}} \qquad (6\text{-}15)$$

Note that like the Engset loss system, the relation (6-8) remains true for the Engset delay system; that is,

$$\pi_k[n] = p_k[n-1] \ , \ k = 0, 1, \cdots, n \tag{6-16}$$

The probability $P\{W > 0\}$ that a call must wait for service is the probability that an incoming call finds all s lines occupied. Therefore,

$$P\{W>0\} = \sum_{k=s}^{n-1} \pi_k[n] = \sum_{k=s}^{n-1} p_k[n-1]$$

$$= \frac{(n-1)!}{s!} \, \hat{a}^s \left[\frac{\hat{a}}{s} \right]^{n-1-s} p_0[n-1] \sum_{r=0}^{n-1-s} \frac{1}{r!} \left[\frac{s}{\hat{a}} \right]^r . \tag{6-17}$$

6-6. THE WAITING TIME DISTRIBUTION FUNCTION FOR THE ENGSET DELAY SYSTEM WITH SERVICE IN ORDER OF ARRIVAL

To calculate the waiting time distribution, using (5-31), we write

$$P\{W>t\} = \sum_{k=s}^{n-1} \pi_k[n] \, P_k\{W>t\}$$

$$= \sum_{k=s}^{n-1} p_k[n-1] \, P_k\{W>t\}$$

where the probability $p_k[n-1]$ is given by (6-15) and (6-16) by replacing n by $n - 1$. The conditional waiting probability is given by (5-33):

$$P_k\{W>t\} = e^{-s\mu t} \sum_{r=0}^{k-s} \frac{(s\mu t)^r}{r!} \ , \ s \le k \le n - 1$$

It follows then that

$$P_k\{W>t\} = e^{-s\mu t} \sum_{r=0}^{k-s} \frac{(s\mu t)^r}{r!} \quad , \quad s \le k \le n-1$$

It follows then that

$$P\{W>t\} = \sum_{k=s}^{n-1} \frac{(n-1)! \; \hat{a}^k}{(n-1-k)! \; s! \; s^{k-s}} \; p_0[n-1] \; e^{-s\mu t} \sum_{r=0}^{k-s} \frac{(s\mu t)^r}{r!}$$

$$= p_0[n-1] \frac{(n-1)!}{s!} \; \hat{a}^s \; e^{-s\mu t} \sum_{j=0}^{n-1-s} \sum_{r=0}^{j} \left(\frac{\hat{a}}{s}\right)^j \frac{1}{(n-s-1-j)!} \frac{(s\mu t)^r}{r!}$$

$$= p_0[n-1] \frac{(n-1)!}{s!} \; \hat{a}^s \; e^{-s\mu t} \sum_{m=0}^{n-1-s} \sum_{r=0}^{n-1-s-m} \left(\frac{\hat{a}}{s}\right)^{n-s-1-m} \frac{1}{m!} \frac{(s\mu t)^r}{r!}$$

$$= p_0[n-1] \frac{(n-1)!}{s!} \; \hat{a}^s \; e^{-s\mu t} \left(\frac{\hat{a}}{s}\right)^{n-1-s}$$

$$\times \left[\sum_{m=0}^{n-1-s} \sum_{r=m}^{n-1-s} \frac{(s\mu t)^{r-m}}{(r-m)!} \left(\frac{s}{\hat{a}}\right)^m \frac{1}{m!} \right]$$

The quantity in the square brackets can be written in the form

$$\sum_{r=0}^{n-1-s} \frac{1}{r!} \sum_{m=0}^{r} \binom{r}{m}(s\mu t)^{r-m} \left(\frac{s}{\hat{a}}\right)^m = \sum_{r=0}^{n-1-s} \frac{1}{r!}\left(s\mu t + \frac{s}{\hat{a}}\right)^r$$

Consequently, we find

$$P\{W>t\} = c[n] \sum_{r=0}^{n-1-s} \frac{[\phi(t)]^r}{r!} \; e^{-s\mu t} \tag{6-18}$$

where

$$\phi(t) = \frac{s\mu}{\gamma} + s\mu t$$

$$\tag{6-19}$$

and $$c[n] = p_0[n-1] \frac{(n-1)!}{s!} \; \hat{a}^s \left(\frac{\hat{a}}{s}\right)^{n-1-s}$$

6-7. THE OFFERED LOAD AND CARRIED LOAD IN THE ENGSET DELAY SYSTEM

For the Engset delay system, we assume that the size of the waiting room is $n - s$ so that no traffic is lost and the offered load and the carried load are equal. Let N denote the mean number of calls in the system in equilibrium and λ be the mean arrival rate. From Little's formula or (4-40), we have

$$N = \lambda W + a$$

But from (6-10)

$$a = (n - N)\hat{a}$$

Eliminating N and then solving for a yields the offered load

$$a = \frac{n\hat{a}}{1 + \gamma(\tau + W)} \tag{6-20}$$

where the mean waiting time W is given in (6-22) and τ is $1/\mu$.

6-8. THE MEAN WAITING TIME IN THE ENGSET DELAY SYSTEM WITH SERVICE IN ORDER OF ARRIVAL

Using (5-33), we can calculate the conditional mean waiting time:

$$E[W \mid N = k] \doteq -\int_0^\infty t \, d \, P_k \, \{W > t\}$$

$$\tag{6-21}$$

$$= -\sum_{r=0}^{k-s} \frac{1}{r! \, s\mu} \int_0^\infty (rt^r - t^{r+1}) \, e^{-t} \, dt$$

$$= - \sum_{r=0}^{k-s} \frac{1}{r! s \mu} [rr! - (r+1)!]$$

$$= \frac{k-s+1}{s\mu} \quad , \quad k = s, s+1, \cdots, n-1$$

The mean waiting time $E[W]$ can then be written as

$$E[W] = \sum_{k=s}^{n-1} E[W \mid N = k] \, \pi_k \, [n]$$

$$= \sum_{k=s}^{n-1} \left[\frac{k-s+1}{s\mu} \right] \pi_k[n] \qquad (6\text{-}22)$$

$$= \sum_{k=0}^{n-s-1} \frac{k+1}{s\mu} \, p_{s+k}[n-1]$$

Using (6-20) for W, the average time delay in the system can be expressed as

$$E[T] = E[W] + \frac{1}{\mu}$$

$$\qquad (6\text{-}23)$$

$$= \frac{n}{\mu a} - \frac{1}{\gamma}$$

This expression is useful for computer system performance evaluation. In this application, n terminals (sources) request the use of a computer to process transactions. The time spent by the user in generating a request is called the think time; that is, in this instance, $1/\gamma$ is the average think time. The time duration from the instant a terminal generates a request until the computer completes the transaction is called the response time; hence, $E[T]$ is the average response time. The rate at which transactions are processed is called the throughput; thus, μa is the throughput.

Example 6-2. Consider a single-processor, time-sharing computer system with n user-terminals. Assume that the processor times and the think times are independent, identical and exponential random variables with mean τ_p and τ_t, respectively. Determine the average number of requests in the system and the average response time when the scheduling algorithm is FIFO.

Solution. Figure 6-4 shows the model of the computer system in consideration. This model has a finite-source of n terminals.

Let N be the number of requests for service in the system, and let the state probabilities be

$$p_k[n] = P\{N = k\} , k = 0, 1, 2, \cdots , n$$

Then,

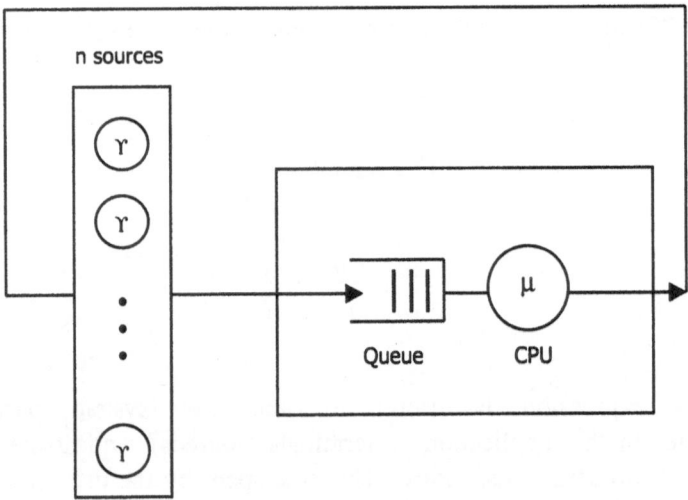

Figure 6-4. Single-processor, time-sharing system.

$$\gamma = \frac{1}{\tau_t}$$

and

$$\mu = \frac{1}{\tau_p}$$

It follows that the offered traffic per idle terminal is

$$\hat{a} = \frac{\gamma}{\mu} = \frac{\tau_p}{\tau_t}$$

Now we may apply the Engset delay model to this system with the state probability given by (6-14) for $s = 1$:

$$p_k[n] = \frac{\hat{a}^k}{(n-k)! \sum_{j=0}^{n} \frac{\hat{a}^j}{(n-j)!}} , \quad k = 0, 1, 2, \cdots, n .$$
(6-24)

Now we may calculate the average response time $E[T]$ by (6-23):

$$E[T] = \frac{n}{\mu a} - \frac{1}{\gamma}$$

where a is the offered load of the system, which remains to be determined. For a single-server queue with a quasi-random input, we have

$$a = \frac{n\hat{a} \, p_0[n]}{p_0[n-1]}$$
(6-25)

where $p_0[n]$ and $p_0[n-1]$ are given by (6-24). Therefore, we obtain

$$E[T] = \frac{1}{\gamma} \left[\frac{p_0[n-1]}{p_0[n]} - 1 \right] \qquad (6\text{-}26)$$

Now we can apply Little's formula (4-35), (6-25) and (6-26) to find the average number of requests in the system.

$$E[N] = a \mu E[T] = n\left(1 - \frac{p_0[n]}{p_0[n-1]}\right) \qquad (6\text{-}27)$$

Expressions (6-26) and (6-27) can be used for calculating the quantities $E[T]$ and $E[N]$ for a single-processor, time-sharing computer system.

6-9. SUMMARY

The essential difference between the Engset systems and the Erlang systems is found in their input processes. The Engset systems have a finite number of independent sources which generate a quasi-random input. The Erlang systems have an infinite source which generates a Poisson input with a constant rate.

Both the Engset loss system and the Engset delay system can be modeled by the birth and death process. An expression for the calculation of the probability of loss, known as the Engset loss formula, is derived. In the Engset loss system, the arriving customer's distribution with n sources turns out to be equal to the outside observer's distribution with $n-1$ sources. Unlike the Erlang systems, the total offered traffic can only be expressed in terms of the number of sources, the offered load per idle source, the number of servers (or trunk lines), and the probability $p_s[n-1]$ and $p_s[n]$.

For the Engset delay system, expressions for the probability of waiting and the waiting time distribution are derived. These expressions are relatively complicated. However, the expression for the mean delay time in the system turns out to be simple.

REFERENCES

[1]   Cooper, R.B., Introduction to Queueing Theory, 2nd ed., New York: Elsevier Science, 1981.

[2]   Syski, R., Introduction to Congestion Theory in Telephone Systems, 2nd ed., New York: Elsevier Science, 1981.

[3]   Fry, T.C., Probability and Its Engineering Uses, 2nd ed., New York: Van Norstrand, 1965.

[4]   Boucher, J.R., Traffic System Design Handbook, Piscataway, N.J.: IEEE, 1992.

[5]   Akimaru, Haruo and Kawashima, Knosuke, Teletraffic Theory and Applications, New York: Springer-Verlag, 1993.

PROBLEMS

6-1.   Thirty sources share access to trunk groups, each with five trunks. The input process is assumed to be quasi-random, and each source generates 2 calls per hour with a mean holding time of 3 minutes. Blocked calls are cleared.

   (a)   If the grade of service is 0.001, how many trunks are required?

   (b)   If the grade of service is 0.01, how many trunks are needed?

6-2.   Consider an Engset loss system in which there are two sources and a single server. An idle source can generate calls at a rate of 1/5 calls per hour with a mean holding time of 2 minutes.

   (a)   Find the offered load.

   (b)   Find the probability of loss.

   (c)   Find the carried load.

6-3.   An Engset delay system consists of four sources and two servers. An idle source generates calls at a rate of 2 calls per hour. The service time is exponentially distributed with a mean of 3 minutes.

(a)   Find the probability of waiting.

(b)   Calculate the server occupancy.

(c)   Calculate the mean waiting time.

(d)   What is the probability of waiting longer than 10 seconds?

6-4.   Show that for an Engset delay system with n sources each generating a quasi-random input, the following relationship is valid:

$$\frac{p_0[n]}{p_0[n-1]} = \frac{1}{n}\left[n - \sum_{k=0}^{n} kp_k[n]\right]$$

6-5.   Let

$$F(n,s) = \sum_{k=0}^{s} \binom{n}{k} \hat{a}^k.$$

Show that

$$p_s[n] = 1 - \frac{F(n, s-1)}{F(n, s)}$$

and

$$\pi_s[n] = 1 - \frac{F(n-1, s-1)}{F(n-1, s)},$$

where the system is an Engset loss system with n sources and s servers, $s < n$.

6-6.   Consider the Engset loss system in Section 6-2 under steady-state conditions. If j particular servers are observed, show that the probability that they are all busy is

$$B(j) = \frac{\pi_s [n + 1]}{\pi_{s-j} [n - j + 1]} \quad .$$

CHAPTER 7

INTRODUCTION TO LOCAL AREA NETWORKS

7-1. INTRODUCTION

A computer network is a communication network that provides interconnections of computers and other terminal devices (such as word processors, storage disks, printers, and graphic terminals) and a means for communication between computers and terminal devices. It is made up of communication links which convey messages or data between computers and devices connected to the network. The communication links may be implemented with twisted-pair wires, coaxial cables, optical fibers, or some other form of communication media.

Depending on the geographical extent covered by the network, computer networks can be classified into three categories: the local area network (LAN), the metropolitan area network (MAN), and the wide area network (WAN) or long-haul network. Local area networks cover only a short distance of up to 25 kilometers. Wide area networks, on the other hand, extend over large wide geographical areas which may be hundreds or thousands of kilometers in distance.

There are significant differences in the design of long-haul and local area networks because of the different requirements for the node-to-node links and network operations. For local area networks, it is possible to install high-speed, low-noise links. However, long-haul networks require the use of shared links for which large bit rate capacities are expensive. Furthermore, links for long-haul networks can be noisy.

Design requirements for long-haul networks may include:

1. Bit rate capacity assignments for connecting links;

2. Choice of network topology;

3. Routing algorithms;

4. Congestion control; and

5.   Network architectures and protocols.

Local area networks are usually designed so that all nodes are connected by a single high-speed shared channel. Data or messages are packetized and then transmitted through the common channel such that a given packet can be received by all nodes on the network. Thus, routing is not required. The primary issues for local area network design are congestion control, channel access protocol, and network architecture.

7-2. DESCRIPTION OF LOCAL AREA NETWORKS

Local area networks are used to interconnect computers, word processors, terminals, graphics terminals, local storage, line printers, telephones, and data files. One reason for utilizing a local area network is to share expensive resources; such as central data files. Another reason is to exchange data and information among systems.

Local Area Network. A local area network is defined as a communication network that provides the interconnection of a variety of data communicating devices within a small area. The phrase "data communicating devices" includes any device that communicates over a transmission medium such as computers, terminals, sensors, telephones, facsimile, and peripheral devices.

The data rates of local area networks are in the range between 0.1 to 100 Mbps and the distances in the range between 0.1 to 25 kilometers. The error rate is typically low, from 10^{-8} to 10^{-11}.

Figure 7-1 shows an example of a local area network which is a single shared communication link. To regulate the use of the single communication link, a protocol to control the access of the link must be developed. Local area networks are used to provide communications between computers and devices, distributed processing, rapid access to data banks, and sharing of expensive devices and resources. Data and messages transmitted on the network are usually in digital form and are packetized. Each packet must carry with it a certain overhead, which includes source and destination addresses and redundant bits for error control.

In the performance analysis of local area networks, network topology, transmission media, network access techniques, and network interfaces are given. From the user point of view, an important performance measure is

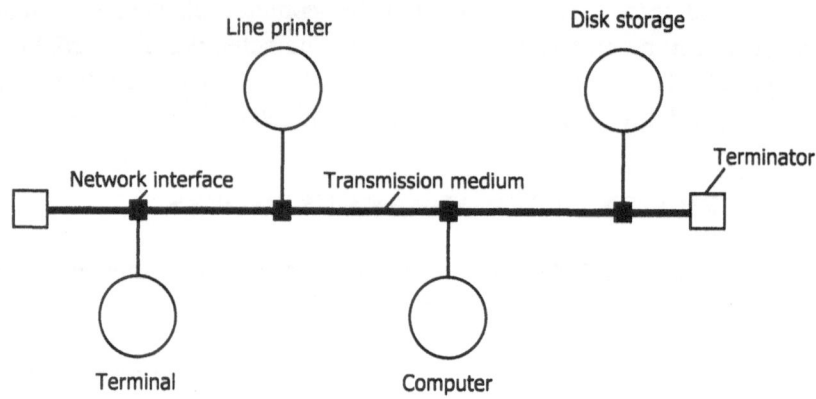

Figure 7-1. Local area network

the response time. Response time is defined as the time required to transmit a packet from the source to the destination and to receive a response, which can be a single acknowledgement.

Response time depends on the state of the network and is thus a random variable. In applications, most performance measures are expressed as average values. Thus, the most common performance measure is simply the average or mean response time.

There are two components involved in the calculation of the average response time:

1. The average one-way packet delay through the network and its interface; and

2. The delay for the user-station links.

The average packet delay through the network is a function of load and the packet size. The delay for the user-station links is relatively load independent and is dependent on the access technique.

At the network level, throughput, which is a measure of the average number of bits per second or packets per second that can pass through the network, becomes significant.

Note that maximizing throughput and minimizing delay for a given network are conflicting requirements. In designing local area networks, we therefore find it necessary to make the most effective compromise between these two requirements. Thus, an important characteristic of many local area networks is the delay-throughput tradeoff curve.

Throughput. Throughput, in bits per second, is a measure of the amount of data or messages passing through a network. It can be defined as the average number of bits passing a given point in a network per second. When calculating throughput, it is usual to include only the data bits of error-free packets. Since the average input rate and the average output rate are equal for a given network in the steady state, throughput is the average number of bits per second either entering or leaving the network. Frequently, throughput is normalized by dividing the average input rate by the channel transmission rate. Normalized throughput is denoted by the symbol S.

Let λ, in packets per second, be the average input rate to a network, $\bar{L}$ be the average packet length in bits, C in bits per second be the channel transmission rate, and $\bar{X} = \bar{L}/C$ seconds be the average packet transmission time. Then the normalized throughput, or simply throughput, is just the product $\lambda \bar{X}$. Thus, throughput, denoted by S, is a dimensionless quantity and is given by

$$S = \lambda \bar{X} \tag{7-1}$$

Channel Utilization or Efficiency. Channel utilization is defined as the fraction of time a channel is busy transmitting packets. The symbol ρ is used for channel utilization. If D is the number of data bits in the packet and H is the number of overhead bits in the packet, then channel utilization is given by

$$\rho = \frac{D}{D + H} S \tag{7-2}$$

In many applications, when the number of overhead bits is much smaller than the number of data bits, the throughput and channel utilization are assumed to be equal.

Average Transfer Delay. Average transfer delay is defined as the average time from the arrival of the last bit of a packet into the sending station of a network until the last bit of this packet is delivered through the network to its receiving station.

As with throughput, it is often convenient to normalize the average transfer delay T by dividing by the average packet transmission time $\bar{X}$, so that the normalized average transfer delay $\hat{T}$ is given by

$$\hat{T} \;=\; T/\bar{X} \qquad\qquad (7\text{-}3)$$

Local Area Network Traffic. The commonly used input traffic in the performance analysis of telephone networks is the Poisson input process. The Poisson process is mathematically attractive because the interarrival times of calls possess the Markov property (or lack of memory).

In local area networks, the data or message traffic is considered to be bursty. A bursty traffic has the characteristic that the demand for service is rare, but when service is needed it requires a high bit rate for both communication and processing. Since the nature of bursty traffic is not yet well understood, to facilitate the performance analysis of local area networks, the arrival process of packets to the network is assumed to be Poisson as an approximation and the packet length distributions are either exponential or constant.

In the analysis of local area networks, we assume that the arrival process of packets at each station is an independent Poisson process with rate λ packets per second, and the packet transmission times are either identical and exponentially distributed with an average length of $\bar{X} = 1/\mu$ seconds or constant equal to $\bar{X}$ seconds.

7-3. CHANNEL ACCESS TECHNIQUES

A characteristic of local area networks is that the transmission medium is shared by many users. Thus, it is possible for two or more user stations in the network to transmit simultaneously, thereby causing their signals to interfere and become garbled. To resolve these conflicts, some means of controlling access to the transmission medium is needed.

There are three major multiaccess techniques for controlling the access to the transmission medium on local area networks: fixed assignment, random assignment, and demand assignment.

In local area networks, the capacity of the transmission medium is usually much greater than that required for the transmission of a single signal. In this case, multiplexing schemes can be used with fixed assignment. Two techniques are in common use: the frequency-division multiple access (FDMA) and the time-division multiple access (TDMA). Another well known technique is the code-division multiple access (CDMA).

Random assignments permit any user station to transmit a message at any time. There is no control mechanism for determining who may have access to the transmission medium and no need for coordination among user stations. However, if two or more stations transmit at the same time so that their packets overlap in time, interference results, and errors are produced in the overlapping portion of the packets. In this case, a collision has occurred and retransmissions of the collided packets must be scheduled at later times. To avoid repeat collisions of the collided packets, the retransmission times are usually chosen at random by the stations involved.

The demand assignment method requires a central controller or control computer to determine the access to the transmission medium at any time. The operation of a central control network also requires a two-way communication between the controller and the stations.

7-4. FIXED ASSIGNMENT ACCESS METHODS

Consider the local area network with M user stations, which share a common communication medium as shown in Figure 7-2.

Let the channel capacity of the common transmission medium be C bits per second. Packets of fixed length $L = \bar{L}$ bits arrive at each station at a rate of λ packets per second according to independent Poisson process and a buffer for storing the packets at each station is permitted. Thus, the packet transmission time is $\bar{X} = \bar{L}/C$ seconds.

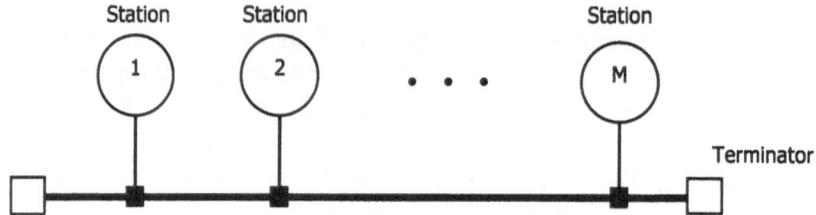

Figure 7-2. Local area network with a bus structure.

7-4-1. Frequency-Division Multiple Access. The concept of FDMA is to divide the channel capacity C into a number M of smaller independent channels, each having a capacity of C/M bits per second. In this case, all stations can transmit their packets simultaneously on the network and each signal is modulated onto a different carrier frequency. The carrier frequencies are sufficiently separated with guard bands so that the bandwidths of the signals do not overlap. The bandwidth of each signal with its carrier frequency at the center is called a channel.

Since each of the M channels operates independently with the same capacity C/M bps, the local area network with a FDMA method can be regarded as M independent M/D/1 queues. Each of these M/D/1 queues has the effective bit rate of C/M bps. Thus, the average transfer delay for a FDMA local area network is equal to the sum of the packet transmission time and the average queueing delay at the station. That is, using (4-62) with $\tau = M\bar{X}$, we can write the average transfer delay

$$T = M\bar{X} + \frac{\rho M \bar{X}}{2(1-\rho)} \qquad (7\text{-}4)$$

where the utilization factor ρ of each queue is

$$\rho = \lambda M \bar{X} \qquad (7\text{-}5)$$

For the FDMA local area network, the network throughput S in the steady state is equal to the total input rate, $M \lambda$ packets per second, multiplied by the packet transmission time. Thus, S is given by

$$S = M \lambda \bar{X} \tag{7-6}$$

From (7-5) and (7-6), we obtain

$$S = \rho$$

Thus, (7-4) can be written as

$$T = M \bar{X} + \frac{M \bar{X} S}{2(1 - S)} \tag{7-7}$$

Dividing this expression by the average packet transmission time $\bar{X}$ yields the normalized average transfer delay

$$\hat{T} = M + \frac{M S}{2(1 - S)} = \frac{M(2 - S)}{2(1 - S)} \tag{7-8}$$

7-4-2. Time-Division Multiple Access. In TDMA networks, all signals use the same carrier frequency and bandwidth but operate at different times. Like the guard bands in FDMA networks, TDMA networks make use of small time slots to separate signals from adjacent stations in order to prevent cross-talks. Along with the signal, one synchronization bit is transmitted with each cycle of signals from the M stations. This cycle plus the controlling synchronization bit is known as a frame. Thus, there are M equal length time slots in a frame, one for each station, as shown in Figure 7-3. The operation of TDMA is also known as synchronous time-division multiplexing (STDM). Packets arriving at a station are stored in the station buffer and wait for their turn (or time slot) to be transmitted.

For simplicity, we assume that each time slot is one packet transmission time $\bar{X}$ long. Thus, with M stations, each station can transmit one packet per frame. The length of a frame is $M \bar{X}$ seconds.

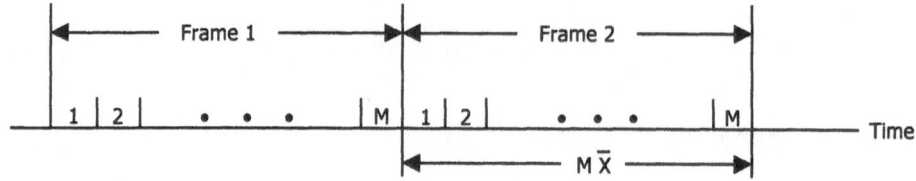

Figure 7-3. Time slots for TDMA network.

Now consider a packet arriving at an arbitrary station. The average transfer delay consists of three components:

1. An average slot synchronization delay before the station gets its turn to transmit the packet on the network;

2. The average delay in the station buffer; and

3. The actual transmission time, $\bar{X}$ seconds.

Here we have neglected the propagation delay on the channel. Since the arrival process is a purely random or Poisson process, the average slot synchronization delay is one-half of the frame time or $M\,\bar{X}/2$.

To calculate the average delay in the station buffer, the network can be modeled as M independent identical M/D/1 queues, each with arrival rate λ packets per second. For each such queue, the effective packet transmission time is $M\,\bar{X}$. From (4-62), replacing τ by $M\,\bar{X}$, we obtain the average delay in the buffer as

$$W = \frac{\rho\,M\,\bar{X}}{2(1-\rho)} \tag{7-9}$$

Note that for TDMA networks, the relations in (7-5) and (7-6) are also valid. Thus, (7-9) can be written as

$$W = \frac{S \, M \, \bar{X}}{2(1 - S)} \tag{7-10}$$

The average transfer delay T is the sum of the average synchronization delay, the average waiting time W in the buffer, and the packet transmission time, so that

$$T = \frac{M \, \bar{X}}{2} + \frac{M \, S \, \bar{X}}{2(1 - S)} + \bar{X} = \frac{M\bar{X}}{2(1 - S)} + \bar{X} \tag{7-11}$$

Dividing this expression by $\bar{X}$ gives the normalized transfer delay

$$\hat{T} = \frac{M}{2(1 - S)} + 1 \tag{7-12}$$

By comparison of (7-8) and (7-12), we see that

$$\hat{T}_{FDMA} = \hat{T}_{TDMA} + \frac{M}{2} - 1 \tag{7-13}$$

Since the number of stations must be at least two, we conclude that $\hat{T}_{FDMA} \geq \hat{T}_{TDMA}$. Thus, for a network with more than two stations, FDMA always yields greater average transfer delay than TDMA.

The average normalized delays for both FDMA and TDMA are plotted against network throughput S in Figure 7-4 (see p. 314) for different values of M.

Example 7-1. Consider a local area network with 100 stations which are connected to a multiaccess transmission medium. The packet arrival process to each station is an independent Poisson process with a rate of 1 packet per minute. The constant packet length is 200 bits. The network has a capacity of 1 M bps. Calculate the average transfer delay for the network with (a) FDMA and (b) TDMA.

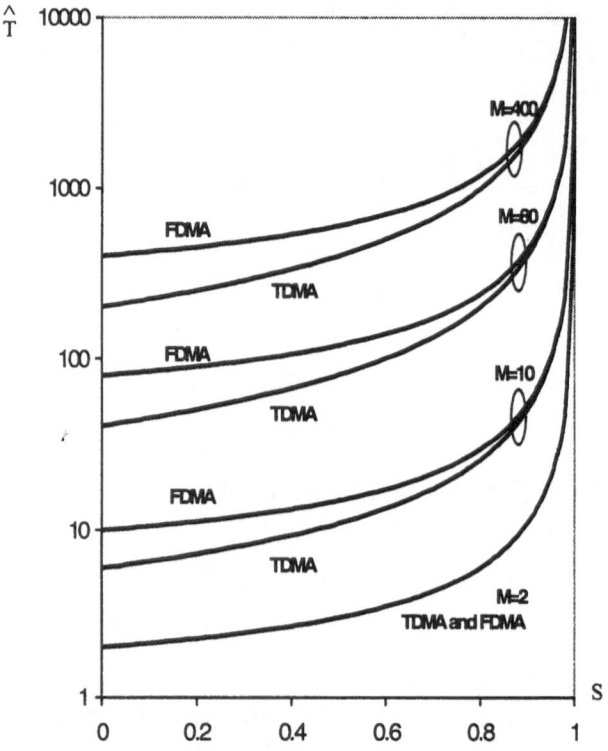

Figure 7-4. Average normalized delay versus throughput for TDMA and FDMA with the number of stations, M, as a parameter.

Solution. The network data are:

$\lambda = 1/60$ packet/second,

$M = 100$ stations,

$C = 10^6$ bps, and

$\bar{X} = 200/10^6 = 200 \ \mu s$

Using (7-6), we obtain the throughput

$$S = M \lambda \bar{X} = 10^{-3}/3$$

(a) Using (7-7), the average transfer delay on a FDMA network is

$$T = \frac{M\bar{X}}{2} \left[\frac{2-S}{1-S} \right] = 20 \ ms.$$

(b) Using (7-11), the average transfer delay on a TDMA network is

$$T = \bar{X} \left[\frac{M}{2} + \frac{MS}{2(1-S)} + 1 \right] = 10.2 \ ms.$$

7-4-3. Code-Division Multiple Access. As mentioned in Section 7-3, the two commonly used fixed assignment access methods are FDMA and TDMA. In FDMA, all M user stations can transmit simultaneously but using only $1/M$ of the channel capacity C (that is, C/M bps). In TDMA, all M users can use the full channel capacity C for transmission, but they are permitted to transmit one packet in a slot of a frame of M slots, one station to a slot. The slot length is $\bar{X}$ seconds. Furthermore, TDMA requires synchronization of all stations.

Both FDMA and TDMA work well under heavy traffic conditions and with a small and fixed number of stations. If the number of stations is large and continuously changing (stations are added and deleted), and if the traffic is bursty, FDMA and TDMA may become inefficient. The basic problem is that the allocated bandwidths or time slots to the stations will be wasted when they have no packets to transmit.

If all M stations are permitted to transmit simultaneously, channel utilization can be improved by using some other method known as the code-division multiple access (CDMA). Since all stations are using the channel simultaneously, each station is assigned a unique code. The pattern of the code specifies how a station to transmit its packet by spreading the signal over a length of $M\bar{X}$ seconds, where $\bar{X} = L/C$, is the packet transmission time, L is the packet length in bits and C is the channel capacity in bits per second.

If the codes assigned to the different stations do not overlap in frequency at any time, then CDMA is equivalent to FDMA or TDMA. However, in CDMA, the codes allow transmitted signals to overlap causing collisions. A main feature of a CDMA network is that collisions are not destructive; each of the signals involved in a collision would be received with only a slight increase in error rate. The use of long codes and extensive error control procedures make it possible to allow several simultaneously transmitted signals without serious performance degradation. Thus, the effect of collision can be minimized by the ability of the receiver to lock on one packet (signal) while all other overlapping packets appear as noise.

To accomplish CDMA, a class of signal techniques known collectively as spread-spectrum modulation is used. In particular, each station is assigned a unique code to perform a spread-spectrum modulation. Each code is approximately orthogonal with all the other codes. The CDMA network operates asynchronously so that the transmission time of a station's packet does not have to coincide with those of the other stations.

The use of CDMA provides three attractive features over TDMA:

•   CDMA provides an external interference rejection capability;

•   CDMA does not require synchronization of stations, which is essential for TDMA; and

•   CDMA offers a gradual degradation in performance as the number of stations is increased.

The transmission of signals on a CDMA network is illustrated in Figure 7-5.

Because of the spread-spectrum modulation, we assume that each station has a mean effective (or equivalent) packet transmission time equal to $M\bar{X}$ seconds. This average effective packet transmission time is difficult to estimate because it depends on many factors such as the packet length L, the chip width of the code, the channel capacity C and the number of simultaneously transmitted signals. The bit error probability is neglected in the following approximate analysis. In the CDMA network, each station

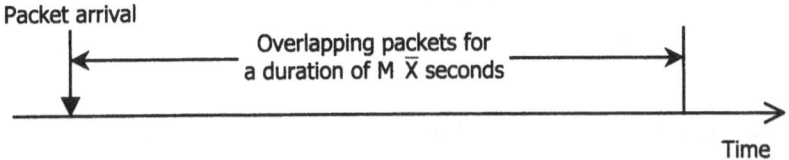

Figure 7-5. Transmission of overlapping packets for a duration of $M\bar{X}$ seconds in a CDMA network

can be modeled as an $M/D/1$ queue with Poisson packet arrival rate of λ packets per second and mean effective packet transmission time $M\bar{X}$, where $\bar{X} = L/C$. Using (4-62), the mean packet transfer delay can be written as the sum of the mean waiting time in the queue and the mean effective packet transmission time $M\bar{X}$,

$$T = \frac{\rho M\bar{X}}{2(1 - \rho)} + M\bar{X} \tag{7-14}$$

Since $\rho = \lambda M\bar{X} = S$, (7-14) can be written as

$$T = \frac{SM\bar{X}}{2(1 - S)} + M\bar{X} \tag{7-15}$$

The normalized mean transfer delay is given by

$$\hat{T} = \frac{SM}{2(1 - S)} + M \tag{7-16}$$

which is identical to (7-8). It follows from (7-13) that we can write

$$\hat{T}_{CDMA} = \hat{T}_{TDMA} + \frac{M}{2} - 1 \tag{7-17}$$

and conclude that for $M \geq 2$,

$$\hat{T}_{CDMA} \geq \hat{T}_{TDMA}$$

Comparison of (7-8) and (7-16) shows that CDMA and FDMA have the same mean transfer delay under the assumption that the mean effective packet transmission time equals $M\bar{X}$ seconds. If this assumption is not true, then the result will be different.

7-5 RANDOM ACCESS METHODS

Consider the local area network as shown in Figure 7-2. Now the access method used is completely random in the sense that station transmissions occur randomly. The earliest and most basic of random access methods, known as the pure ALOHA, was developed at the University of Hawaii for ground based packet radio broadcasting networks.

On a random access network, whenever a station has a packet to send, it will send immediately. Then the station listens for a length of time equal to the maximum possible round-trip propagation time on the network (twice the time for sending a packet between the two most widely separated stations). If the station receives an acknowledgement during this time, the transmission is successful; otherwise, it will resend the packet at an appropriate time later. A receiving station will check the message. If the message is correct, the station acknowledges immediately. If the message is in error due to noise on the network or because more than one station transmits at the same time, interference results. This phenomenon is known as a collision. In this case, no acknowledgement is sent by the receiving station. Because of collisions, a pure ALOHA network has a maximum utilization factor of about 18%.

7-5-1. Pure ALOHA Networks. To calculate the transfer delay in a pure ALOHA network, we assume that P seconds are required for the transmission of a packet. This time includes the time for transmitting all the bits in the packet and the maximum propagation time on the network. For local area networks, the maximum propagation time is usually much smaller than the packet transmission time. We also assume that the arrivals of new packets as well as retransmitted packets that collided earlier for transmission on the network follow a Poisson process with a rate of Λ packets per second, so that the event of k arrivals in t seconds has the probability distribution:

$$P_k\ (t)\ =\ \frac{(\Lambda t)^k}{k!}\ e^{-\Lambda t}\ ,\ k\ =0\,,1\,,\ldots \qquad (7\text{-}18)$$

For analysis of the pure ALOHA network, we let G be the average number of attempted transmissions in a packet transmission time, P seconds. Then

$$G\ =\ \Lambda P \qquad (7\text{-}19)$$

is the total traffic offered to the network per packet transmission time, which is dimensionless. Furthermore, we define throughput S as the average number of successful transmissions per packet transmission time, P seconds. Since G is the average number of attempted transmissions in a time interval P, and S is the average number of successful transmissions in P, the probability of a successful transmission is then

$$P\ \{\ a\ \ successful\ \ transmission\ \}\ =\ S/G \qquad (7\text{-}20)$$

Now consider a test packet which requires a transmission time of P seconds as shown in Figure 7-6.

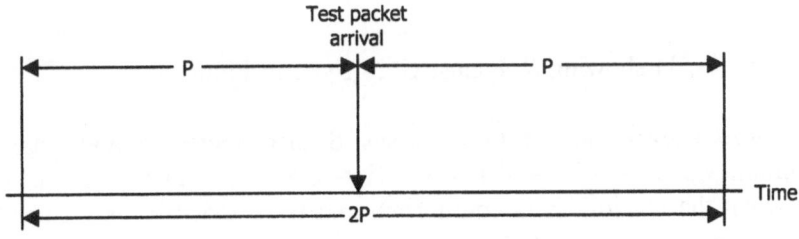

Vulnerable period for the test packet

Figure 7-6. ALOHA vulnerable period for a test packet transmission

Observe that any other transmissions within P seconds before and after the test packet will result in a collision. Thus, the vulnerable period can be seen to consist of two adjacent time intervals, each with length P seconds. The probability of a successful transmission is then the same as the probability of no other transmissions or arrivals in this vulnerable period of length $2P$ centered at the arrival epoch of the test packet.

For a Poisson arrival process, the number of arrivals in a given time interval depends only on the length of the time interval and is independent of the starting instant of transmission of the test packet. Thus, we obtain from (7-18) and (7-19)

$$P \{ a \ successful \ transmission \ \} \ = \ P \{ no \ arrivals \ in \ P \ seconds$$

$$before \ the \ test \ packet \ \} \times P \{ no$$

$$P \ arrivals \ in \ seconds \ after$$

$$the \ test \ packet \ \}$$

$$= \ e^{-\Lambda P} \ \times \ e^{-\Lambda P} \ = \ e^{-2G}$$

It follows from (7-20) that throughput S is given by

$$S \ = \ G \ e^{-2G} \tag{7-21}$$

A plot of (7-21) showing S against G is given in Figure 7-7.

The curve shows that as G increases S also increases and attains a maximum value of $1/2e = 0.184$ for G = 0.5. Note that (7-21) is valid only when the network is in equilibrium and $G < 0.5$. If G is greater than 0.5, an increase in G will lead to a decrease in S due to an increasing number of collisions. As a result, the throughput decreases further and eventually goes to zero. This phenomenon is known as channel (or network) saturation. Thus, it is not possible for a pure ALOHA network to have a stable operation for values of G greater than 0.5.

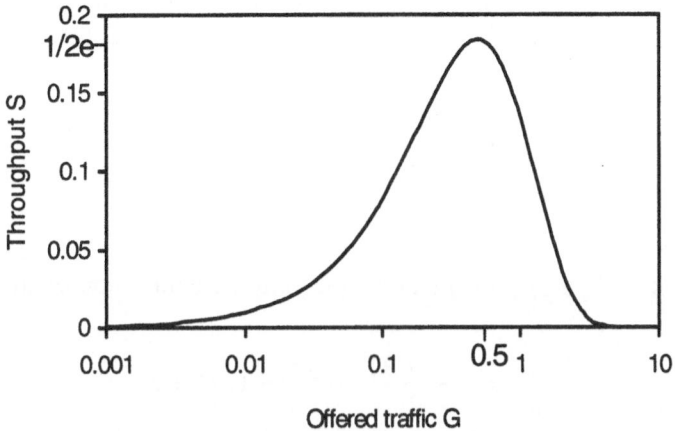

Figure 7-7. Throughput versus offered traffic for the pure ALOHA network

To determine the average transfer delay excluding the packet propagation time, we observe that the average number of attempted transmissions per successful transmission is given by

$$G/S = e^{2G} \tag{7-22}$$

Thus, the average number N_r of unsuccessful attempts per successful transmission is

$$N_r = \frac{G}{S} - 1 = e^{2G} - 1 \tag{7-23}$$

If a collision occurs, the collided packets will be retransmitted at some randomly chosen time. This causes a delay during which the packet is said to be in a state of backoff. Let the average backoff delay be denoted by $\bar{B}$. Note that each unsuccessful attempt to transmit a packet requires $(P + \bar{B})$ seconds, on the average. Since, for a packet to complete a transmission, it requires an average of $(e^{2G} - 1)$ unsuccessful attempts and one successful attempt. Thus, the average transfer delay is given by

$$T = (e^{2G} - 1)(P + \bar{B}) + P \qquad (7\text{-}24)$$

The backoff strategy is to select an integer from the integers $0, 1, 2, \ldots,$ K-1 with equal probabilities. The backoff delay for a chosen integer k is then set at a value of kP seconds. Thus, the average backoff delay is given by

$$\bar{B} = \sum_{k=0}^{K-1} kP\ (1/K) = (K - 1)\ P/2 \qquad (7\text{-}25)$$

Substitution of $\bar{B}$ into (7-24) and simplifying the expression yields

$$T = \frac{(K + 1)}{2}\ (e^{2G} - 1)\ P + P \qquad (7\text{-}26)$$

The average transfer delay in (7-26) can be normalized by dividing both sides by P to yield

$$\hat{T} = \frac{(K + 1)}{2}\ (e^{2G} - 1) + 1 \qquad (7\text{-}27)$$

Example 7-2. Consider a pure ALOHA network using a random access protocol. The packet length is 1000 bits and the channel capacity is 10^6 bps. Input process to each station is Poisson with a rate of λ packets/second.

(a) If the network has 100 stations, what is the permissible maximum average input rate for each station?

(b) What is the maximum total network input rate?

(c) Under maximum throughput conditions, what is the average number of retransmissions per successful transmission?

(d) What is the average backoff delay if the backoff strategy is to select an integer from the set of integers $\{0, 1, 2, \cdots, 19\}$ with equal proabilities?

(e) What is the average transfer delay?

Solution. The number of stations on the network, $M = 100$.

The packet length, $\bar{L} = 1000$ bits.

Channel capacity, $C = 10^6$ bps.

The packet transmission time, $P = \bar{X} = \bar{L}/C = 10^{-3}$
seconds

(a) Since maximum throughput occurs at $G = 0.5$, then

$$S_{max} = G\,e^{-2G} = 1/2e = 0.1839$$

Since the network has M stations, each with throughput $\lambda\bar{X}$,

$$S_{max} = M\,\lambda_{max}\,\bar{X}$$

Thus, the permissible maximum average input rate per station is given by

$$\lambda_{max} = S_{max}/M\,\bar{X}$$
$$= 0.1839\,/(100 \times 10^{-3})$$
$$= 1.839\ packet/s$$

(b) The maximum total network input rate is given by

$$M\,\lambda_{max} = 100 \times 1.839 = 183.9\ packets/s.$$

(c) Application of (7-23) gives the average number of retransmissions per successful transmission at maximum throughput

$$N_r = \frac{G}{S_{max}} - 1 = e^{2G} - 1 = e^1 - 1 = 1.7183$$

(d) Using (7-25), we find the average backoff delay

$$\bar{B} = \frac{20 - 1}{2} \times 10^{-3} \; seconds = 9.5 \; ms$$

(e)   Using (7-24), and the results of (c) and (d), we obtain

$$T = 1.7183(1 + 9.5) + 1$$

$$= 19.0422 \; ms$$

7-5-2. Pure ALOHA Networks with Captures. The throughput of the ALOHA network can be improved by dividing stations into two groups: one group transmits packets at high power and the other group at low power. High-power packets are not affected by low-power packets. If there is only one high-power packet and some low-power packets transmitted within the vulnerable period, the high-power packet can always be captured by the intended receiver station. Low-power packets transmitted within the vulnerable period are unsuccessful and will be retransmitted at a later time. This procedure is known as ALOHA with captures.

Consider the ALOHA network with capture capability. Suppose that packet arrival processes (new and collided packets) for both high-power and low-power groups are independent Poisson processes. Let G_H be the offered traffic (new plus collided packets) for the high-power group. Let S_H be the portion of successfully transmitted traffic. Let G_L and S_L be the corresponding traffic for the low-power group. We shall show that the maximum throughput for the pure ALOHA network with captures can be increased to 27%.

Applying (7-22) to the high-power packet yields

$$S_H = G_H e^{-2G_H} \tag{7-28}$$

Note that low-power packet transmissions have no effect on the high-power ones, whereas the opposite is not true. For the low-power packet, note that

P {a successful transmission} = P {no arrivals of both the

low–power and *high–power packets}*

$$= e^{-2(G_L+G_H)}$$

Thus
$$S_L = G_L e^{-2(G_L + G_H)}$$
(7-29)

The total throughput is then given by the sum of S_H and S_L,

$$S = G_H e^{-2G_H} + G_L e^{-2(G_L + G_H)}$$
(7-30)

To determine the maximum throughput, we solve the following two equations:

$$\frac{\partial S}{\partial G_L} = 0 \quad \text{gives} \quad G_L = 0.5$$

and

$$\frac{\partial S}{\partial G_H} = 0 \quad \text{gives} \quad G_H = 0.3161$$

Substitution of these values for G_L and G_H in (7-30) yields

$$S_{max} = 0.2657$$

7-5-3. Slotted ALOHA Networks. For pure ALOHA networks, a station can transmit whenever it has a packet to send. Random access methods are well suited for bursty users. Under light load conditions, a user can, on average, successfully have access to the channel after only a few attempts. As shown in Figure 7-7, maximum throughput for a pure ALOHA network is only 1/2e or about 18% at an offered traffic of 0.5. This low throughput is essentially due to the number of collisions in the transmission.

To improve the performance of pure ALOHA networks, more sophisticated techniques are developed to reduce the number of collisions and the length of collision intervals. Collision interval is defined as the time interval during which the channel is occupied by corrupted or partially corrupted packets. Slotted ALOHA is such an improvement. For a slotted ALOHA network, all stations are synchronized and packet lengths are fixed. Thus, the channel time is divided into intervals of length equal to the packet transmission time P seconds. We shall show that these modifications can lead to a decrease of the vulnerable period from $2P$ to P. As a result, maximum throughput can double that of the pure ALOHA network because packets can collide either completely or not at all.

Figure 7-8 shows the channel time which is segmented into slots of P seconds (the packet transmission time). The operation of slotted ALOHA is such that a station can start to transmit only at the beginning of a time slot. Strictly speaking, a slotted ALOHA is not a random access method but only a restricted random access method.

Consider an arbitrary arriving test packet at a station. Since the channel time is slotted, the test packet can not be transmitted immediately. It has to wait for transmission at the beginning of the next slot. If there are no other transmissions during the transmission of the test packet, the test packet has a successful transmission. Upon receipt of the test packet, the receiving station sends a positive acknowledgement over a separate error-free channel to the sending station. After transmitting the test packet,

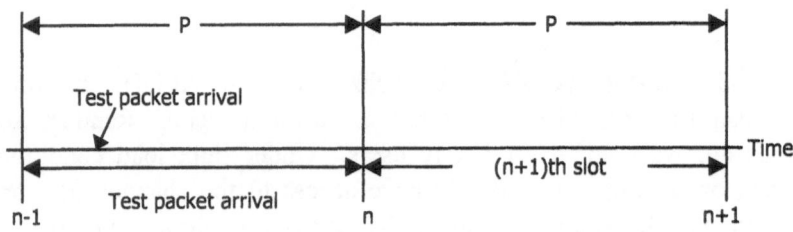

Figure 7-8. Slotted ALOHA vulnerable period for a test packet

the sending station will wait for a period equal to the two-way propagation time of the most widely separated stations. If no acknowledgement is received, a collision has occurred. The sending stations then select a backoff time for retransmitting the collided packets. This procedure is repeated until the transmissions become successful. From Figure 7-8, we see that the vulnerable period for a slotted ALOHA network now becomes P seconds.

Like the pure ALOHA network, the arrivals to the network consist of both new arrivals and unsuccessful transmissions. Thus, the total arrival process to the network is assumed to be a Poisson process with a rate of Λ packets per second. Since P is the packet transmission time, the actual offered load to the network in a packet transmission time of P seconds is given by

$$G = \Lambda P$$

Note that the probability of a successful transmission is the same as the probability of no other arrivals in the vulnerable period of length P. Thus, from (7-18) to (7-20), we obtain:

$$P \ \{ a \ successful \ transmission \ \} \ = \ e^{-\Lambda P} \ = \ e^{-G} \ = \ S/G$$

Hence, we have

$$S \ = \ G \ e^{-G} \tag{7-31}$$

This result is plotted in Figure 7-9.

Note that maximum throughput for slotted ALOHA occurs at $G = 1$ and is $1/e$ or 36.8%, twice that of pure ALOHA network.

The assumption of Poisson arrival process implies that the number of users at each station is infinite. To calculate the transfer delay for slotted ALOHA networks, we assume that the maximum propagation delay between two stations is τ_p seconds. A transmitting station learns whether or not its transmission is successful after a time of rP seconds, where r is the smallest integer greater than $2\tau_p /P$. The processing time of the

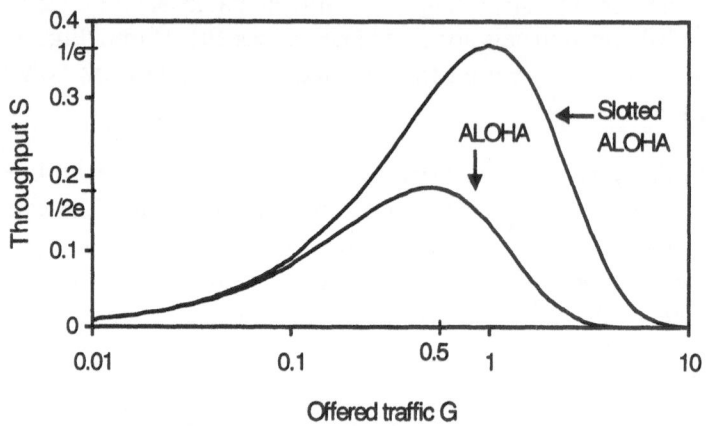

Figure 7-9. Throughput versus offered traffic for a slotted ALOHA network

receiver is neglected.

The average transfer delay in a slotted ALOHA network has four components:

1. The time delay after arrival until the beginning of the next slot. For Poisson input process the arrival time in an arbitrary time slot has a uniform distribution with probability $1/P$. Thus, the average delay between the arrival and the beginning of the next slot is $P/2$;

2. The packet transmission time, P seconds;

3. The average delay due to retransmissions. Suppose that the average backoff delay $\bar{B}$ is given by (7-25). Then an average backoff cycle (defined as the sum of the packet transmission time P, time to determine the packet has collided, rP, and the average backoff time) is given by

$$P + rP + (K - 1)\,P/2 = rP + \left[\frac{K + 1}{2}\right] P$$

If N_r is the average number of retransmissions per successful transmission, then the average time per successful transmission is given by

$$N_r \left[rP + \left(\frac{K+1}{2} \right) P \right] \qquad (7\text{-}32)$$

where

$$N_r = G/S - 1 = e^G - 1 \qquad (7\text{-}33)$$

4. The average propagation delay. If the stations are uniformly distributed along the bus in a local area network, the average distance between two stations is approximately equal to one-third of the bus length. Thus, the average propagation delay is approximately τ_p /3 seconds.

Now, by adding up the four components, we obtain the average transfer delay,

$$T = \frac{3P}{2} + N_r \left[rP + \left(\frac{K+1}{2} \right) P \right] + \frac{\tau_p}{3} \qquad (7\text{-}34)$$

Dividing T by the packet transmission time P yields the normalized transfer delay

$$\hat{T} = 1.5 + N_r \left[r + \left(\frac{K+1}{2} \right) \right] + \frac{\hat{\tau}_p}{3} \qquad (7\text{-}35)$$

where $\hat{\tau}_p = \tau_p /P$ is the normalized end-to-end propagation delay and N_r is given by (7-33).

Example 7-3. A slotted ALOHA network has a channel of capacity 1 Mbps. Packets are 1000 bits long. There are 150 stations, each having a Poisson input of an average rate of 0.1 packets per second. End-to-end propagation delay is 0.125 ms. The backoff algorithm selects an integer k at random from 0, 1, . . . , 19 so that the backoff time is set at k times of the packet transmission time. Under maximum throughput conditions, determine the average transfer delay.

Solution. The packet transmission time is

$$P = \bar{X} = \frac{10^3}{10^6} = 1 \ ms.$$

The end-to-end propagation delay is

$$\tau_p = 0.125 \ ms.$$

The time delay to determine that the packet has collided is

$$rP = \text{the smallest integer greater than } (2 \ \tau_p/P \ )\times P$$

$$= \left[\frac{2 \times 0.125}{1} \right] \times 1$$

$$= 1 \ ms.$$

Under maximum throughput conditions, we find

$$S_{max} = 0.368 \quad \text{and} \quad G = 1$$

Thus, the average number of retransmissions per successful transmission is

$$N_r = \frac{G}{S} - 1 = \frac{1}{0.368} - 1 = 1.7174$$

Therefore, using (7-34), we find the average transfer delay

$$T = \frac{3}{2} + 1.7174 \ [\ 1 + (\ \frac{20 + 1}{2} \ ) \times 1 \ ] + \frac{0.125}{3}$$

$$= 21.2918 \ ms.$$

7-5-4. Slotted ALOHA Networks with Captures. For the slotted ALOHA network, applying (7-28) to the high-power packet yields

$$S_H = G_H e^{-G_H} \tag{7-36}$$

For the low-power packet, the probability of a successful transmission becomes $e^{-(G_L + G_H)}$.

Thus, the throughput is given by

$$S_L = G_L e^{-(G_L + G_H)} \tag{7-37}$$

The total throughput is

$$S = G_H e^{-G_H} + G_L e^{-(G_L + G_H)} \tag{7-38}$$

By solving the following two equations:

$$\frac{\partial S}{\partial G_L} = 0 \quad \text{gives} \quad G_L = 1$$

and

$$\frac{\partial S}{\partial G_H} = 0 \quad \text{gives} \quad G_H = 1 - 1/e$$

we obtain the maximum throughput by substituting these values for G_L and G_H in (7-38),

$$S_{max} = e^{-(1 - 1/e)} = 0.5315$$

Hence, the maximum throughput is increased to about 53%.

7-6. CENTRAL CONTROL ACCESS METHODS

Consider a central control network shown in Figure 7-10, where the control computer determines which of the M stations can have access to the channel at a given time.

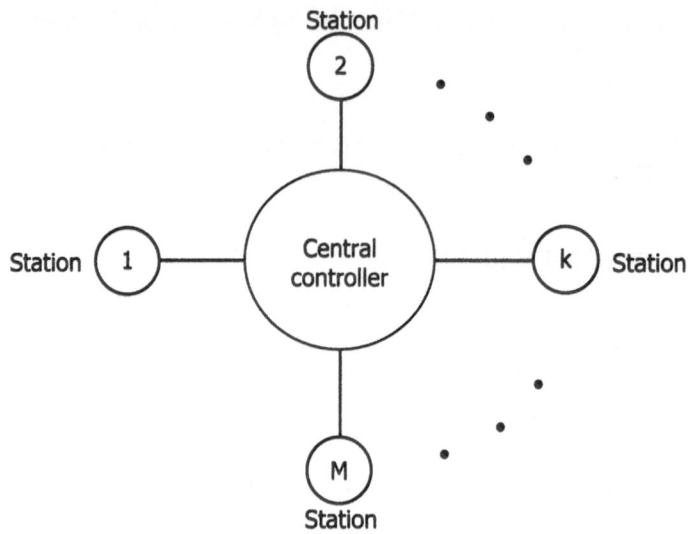

Figure 7-10. Central control local network

For simplicity, we assume that transfer of the channel from one station to another by the control computer is done instantaneously, so that the control computer can assign the channel (when idle) to a station which has data to transmit with negligible delay. Waiting packets can queue in the buffers at their stations. Stations are served in order of the first arrivals. The arrival process at each of the M stations is Poisson with a rate of λ packets per second. Packets have a constant transmission time $\bar{X}$ seconds. We also assume that the propagation time is negligible.

Since packets do not interfere with each other and since no delay is associated with transferring the use of the channel from one station to another, the whole distributed network behaves as a single $M/D/1$ queue with a total input rate $M\lambda$ packets/second. Using (4-62) with the constant packet transmission time $\bar{X}$, we obtain the average transfer delay:

$$T = \bar{X} + \frac{\rho}{2(1-\rho)} \, \bar{X} \qquad (7\text{-}39)$$

where the utilization factor $\rho = M\lambda \, \bar{X}$.

When the channel is busy, it always transmits good packets, and hence, throughput S and utilization factor ρ are equal. Substitution of S for ρ into (7-39) yields

$$T = \bar{X} + \frac{S}{2(1-S)} \bar{X} \qquad (7\text{-}40)$$

Dividing this expression by $\bar{X}$ gives the normalized average transfer delay

$$\hat{T} = 1 + \frac{S}{2(1-S)} = \frac{2-S}{2(1-S)} \qquad (7\text{-}41)$$

Example 7-4. A central control network with channels capacity of 64 K bps has 50 stations, each having a Poisson input with a rate of 40 packets per minute. Packets have a constant length of 500 bits. Assume that the propagation time is negligible and the control of use of the channel is instantaneous. What is the average transfer delay on the network?

Solution. The throughput S is given by

$$S = \rho = M \lambda \bar{X} = 50 \times \frac{40}{60} \times \frac{500}{64 \times 10^3} = 0.26$$

Since $\bar{X} = 500/64 \times 10^3 = 7.8125ms$, using (7-40), the average transfer delay is

$$T = \left[1 + \frac{0.26}{2(1-0.26)} \right] 7.8125 = 9.185 \; ms.$$

7-7. SUMMARY

In local area networks, the transmission medium for communication is shared by many users. To avoid conflicts of access to the transmission medium, TDMA, FDMA, and demand assignment techniques may be used. However, for bursty input traffic, random access techniques have been found to be more efficient.

Expressions for determining the average transfer delay in local area networks using TDMA, FDMA and CDMA are derived. In general, the average transfer delay using TDMA is always smaller than that of using FDMA or CDMA. It is shown that CDMA and FDMA have the same average transfer delay.

For random access networks, expressions for the throughput and the average transfer delay for both the pure ALOHA and the slotted ALOHA are obtained. It is shown that the slotted ALOHA has twice the maximum throughput as that of the pure ALOHA. By introducing capture capability to the ALOHA network, the maximum throughput can be increased further.

For demand assignment with idealized central control access, expressions for the throughput and the average transfer delay are also determined.

REFERENCES

[1]   Hammond, J.L. and O'Reilly, P.J.P., Performance Analysis of Local Computer Networks, Reading Mass.: Addison-Wesley, 1986.

[2]   Stallings, W., Local Networks, 3rd ed., New York: Macmillan, 1990.

[3]   Hopper, A., Temple, S. and Williamson, R., Local Area Network Design, Reading, Mass.: Addison-Wesley, 1986.

[4]   Schwartz, M., Telecommunication Networks: Protocols, Modeling and Analysis, Reading, Mass.: Addison-Welsey, 1987.

[5]   Lam, S.S., "Multiple Access Protocols, " in Computer Communications, Vol. 1, Principles, ed. Chou, W., Englewood, N.J.: Prentice Hall, 1983.

[6]   Frost, V.S. and Melamed, B., "Traffic Modeling for Telecommunications Networks", IEEE Communications Magazine, Vol. 32, No. 3, March, 1994, pp. 70-81.

[7]   Leland, W.F., Taqqu, M.S., Willinger, W. and Wilson, D.V., "On the Self-Similar Nature of Ethernet Traffic", IEEE/ACM Trans. on Networking, Vol. 2, No. 1, February 1994, pp. 1-15.

PROBLEMS

7-1. Each station of a local computer network has an average input rate of 0.25 messages per second. Messages are 10^4 bits in length with no overhead, no retransmissions, and no transmission errors. The channel of the network has a bit rate of 1 Mbps.

    (a) How many stations in the network will generate a steady-state unnormalized throughput of 10 messages per second?

    (b) What is the normalized throughput S that corresponds to 10 messages per second?

7-2. Consider the following two different transmission schemes under error-free conditions.

    Scheme 1: Fifty overhead bits are used for error-correction in each 250-bit packet; the remaining 200 bits carry data. Packets are transmitted in a continuous sequence.

    Scheme 2: Ten overhead bits are used for error-correction in each 250-bit packet, the remaining 240 bits carry data. When each packet is received, a 5-bit acknowledgement message is sent back to the sending station. A new packet is then transmitted after receipt of the acknowledgement.

    Calculate the transmission efficiency for the two schemes.

7-3. A local area network with 120 stations uses a cable with a capacity of 1 Mbps. The network processes packets of 12-byte characters. What is .the limit on the average input rate λ in packets/second of each station if the average transfer delay is less than 0.1 second? We assume that each station has the same average input rate. Obtain a result for a local area network with TDMA protocol and for an idealized central control network.

7-4. A local area network uses a 10 Mbps bus for transmission medium. Access to the bus is a fixed assignment protocol.

Each station has a Poisson input with a rate of 5 packets per second. The packet length is 1000 bits. There are 500 stations on the network. Calculate

(a) The mean transfer delay for a FDMA protocol and

(b) The mean transfer delay for a TDMA protocol.

7-5.   A local area network supports a number of stations, each of which has a Poisson input with a rate of 0.2 messages per second. Messages are 2000 bits in length. All stations share a common channel of bit rate 1 Mbps. If the average transfer delay on the network is limited to 0.1 seconds, what is the maximum number of stations that the network can support by using (a) a FDMA protocol, and (b) a TDMA protocol.

7-6.   An idealized central control network with 100 stations is constructed with a line of bit rate C = 1 Mbps.

   (a)   At a throughput of S = 0.4, what is the average message delay for 1000-bit messages?

   (b)   What is the average waiting time in the buffer?

7-7.   (a)   Assuming that a guardband of g Hz is necessary for each channel in an FDMA system, obtain an expression for the average transfer delay with 1 bps/Hz modulation. Give expressions for the maximum throughput that can be obtained with this system and the fractional increase in the minimum transfer delay due to the guardbands.

   (b)   For a channel of total bandwidth 1 MHz and 100 stations, calculate the average transfer delay for an average arrival rate of 5 packets per second, a guardband of 1 KHz, and 1000-bit packet length.

7-8.   (a)   A TDMA system makes use of extra bits for frame synchronization. Let the fractional overhead time per frame be y. Obtain an expression for the average transfer delay for a TDMA system in terms of the number of stations M, the throughput S, the average packet transmission time $\bar{X}$, and the overhead factor y. What is the maximum throughput when the overhead factor is included in the analysis?

(b) For a TDMA network with 100 stations and a normalized throughput of 0.5, calculate the percentage error introduced in the average transfer delay by neglecting the frame overhead when the overhead factor is 0.01.

7-9.   A random access network uses the ALOHA access protocol. The average input rate to each of the 150 stations is 1 packet per second and the packet is of constant length of 1000 bits. The channel capacity is 1 Mbps.

(a) If the network is operated at this given throughput, what is the total traffic offered to the network per packet transmission time?

(b) What is the average rate of retransmission per station?

(c) What is the average number of retransmissions per successful transmission?

(d) What is the average transfer delay if the backoff strategy is to select an integer from the set $\{0, 1, 2, \cdots, 19\}$ with equal probabilities?

7-10.   A slotted ALOHA network has the throughput $S = Ge^{-G}$ as given by (7-27). Let the state of the network be defined by the number of stations with a collided packet waiting for retransmission. Assume that stations with a collided packet can not generate new packets. When the network is in state k, the throughput is equal to $S = (M - k)\sigma$, where σ is the probability that an idle station generates a new packet and M is the number of stations. Suppose that we choose $\sigma = \dfrac{1}{Me}$ and that a station with a collided packet will transmit with probability $p = \dfrac{1}{M}$. The average number of attempted transmissions in a slot when the network is in state k is then equal to $G = (M - k)\sigma + kp$.

(a)   Draw the curve S versus G and the load line; and

(b)   Find the intersection of the curve and the load line.

CHAPTER 8

POLLING NETWORKS

8.1. INTRODUCTION

Polling is a method of controlling the access to a transmission medium which is shared by a number of stations. A station is a user requesting service. Polling techniques have been widely used in inquiry - response networks. In polling networks, a controller is required to either initiate or to carry out the polling.

The basic feature of a polling network lies in the action of the central control computer (controller) in polling each of the stations on the network in a prespecified cyclic order to provide access to the communication channel. Polling networks fall into the class of central control networks discussed in Chapter 7.

Messages arriving at each station are first packetized, and then stored in the transmit buffer, where they wait for the poll. As each station is polled, those stations with packets to transmit can use the full capacity of the channel to send their packets to the central computer.

Transmissions between stations take place through the central computer, which receives packets from each station and then transmits them to the appropriate stations. The lines connecting the stations and the central computer are usually high-speed lines.

Polling networks can operate in either one of two modes: roll-call polling and hub polling.

For roll-call polling, the central computer initiates the polling sequence by sending a poll (or polling packet) to a chosen station. After this station has transmitted its packets to the central computer, it notifies the central computer with a suffix called go-ahead which is added to its last packet. After receiving this suffixed packet, the central computer sends a poll to the next station in the polling sequence, and the process is continued.

For hub polling, the central computer sends out a poll to an initial station in the same manner as for roll-call polling. When this station has completed its transmission, it suffixes a go-ahead to the end of its last packet as before, but in addition, a next-station address is specified. With hub polling, all stations are continually monitoring the incoming traffic to the central computer. When the next station detects the go-ahead and recognizes its own address, it starts transmitting immediately.

Polling networks can have bus, ring, or star (or tree) configurations as shown in Figure 8-1.

The dashed lines in Figure 8-1(a) are used for hub polling only. Note that the total time required for roll-call polling is longer than for hub polling since successive polls must be sent back and forth between the central computer and the stations. Hub polling is generally not used for large networks with bus configuration because extra wiring for monitoring

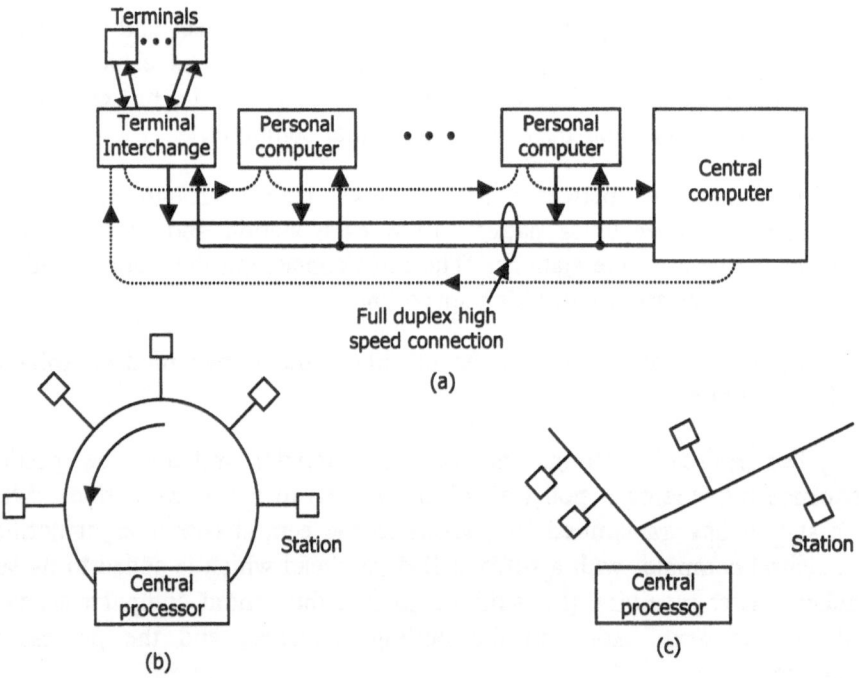

Figure 8-1. Polling networks (a) bus (b) ring (c) tree

between the stations and the central computer is required. However, hub polling is simpler to implement on local area networks with ring configuration.

8-2. OPERATION OF POLLING NETWORKS

This section presents the details for the operation of a polling network. Both roll-call polling and hub polling are examined as follows.

For roll-call polling, the following sequence of operations is typical:

- The central computer sends out a poll to station i in the polling sequence;

- Station i receives the poll and reads the station address;

- Station i transmits all of its packets to the central computer;

- Station i appends a go-ahead, and possibly a next-station address, to its last packet;

- The central computer receives all of the packets, including the go-ahead and the next-station address;

- The central computer transmits all of the packets received from station i to the destination station; and

- The central computer sends out a new poll to station i + 1 in the polling sequence.

The above procedures are repeated for each station on the polling list until all of the M stations have been polled. The whole cycle is then repeated again, starting with station i. If a station receives a poll and has no packet to send, it either ignores the poll or sends back a go-ahead with the next-station address to the central computer.

For hub polling, each station with packets to send must add the next-station address following the go-ahead to the last packet of its transmission. If a station has no packet to send, it must send out the go-ahead plus the next-station address. With this arrangement, the central computer is no longer needed in transferring the poll between stations. Now each station is continually monitoring the poll.

Note that for both roll-call polling and hub polling, the go-ahead suffix added to the last packet is necessary, whereas the next-station address is optional for roll-call polling and is required for hub polling.

A polling network may be regarded as a multiple-queue system with cyclic service. Figure 8-2 shows a queueing model for a polling network with M stations.

Important measures of performance include the average waiting time, average cycle time, average number of packets in a station queue, and average transfer delay.

8-3. PERFORMANCE ANALYSIS OF POLLING NETWORKS

Consider a polling network, where a communication channel is shared by a number of M stations. At any given time, only one station can transmit on the network. Because of polling, the channel time is comprised of two different intervals: the polling interval and the data

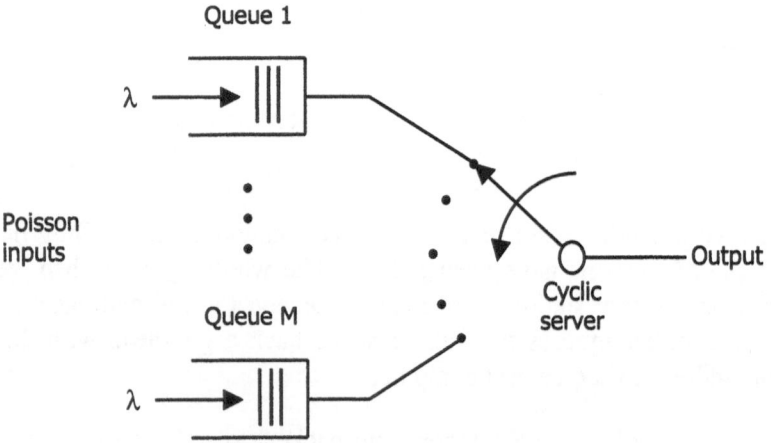

Figure 8-2. Queueing model for a polling network

interval, as shown in Figure 8-3. The polling interval includes all of the time required to transfer the poll from one station to the next station.

According to the rules for specifying which packets in a given station are sent during a data interval, services in polling networks may be divided into three types: (a) In the exhaustive network, all packets that arrive in a station are sent until the buffer is empty; (b) . In the fully gated network, only those packets that arrived prior to the station's preceding polling interval are transmitted; and (c) In the partially gated network, only those packets that arrived prior to the station's data interval are sent.

The arrival intervals for station 0 in the exhaustive, fully gated and partially gated networks are shown in Figure 8-4.

Average Waiting Time. We shall derive expressions for both the average waiting time of a packet before it is transmitted to the central computer, and the average delay time, which is defined as the average time between the packet arrival instant at a station and its delivery to the central computer for all the three types of services.

To simplify the analysis, we shall make the following assumptions:

• The network serves M stations, which are numbered 0, 1, ..., M-1. The packet arrival processes of the stations are independent, identical

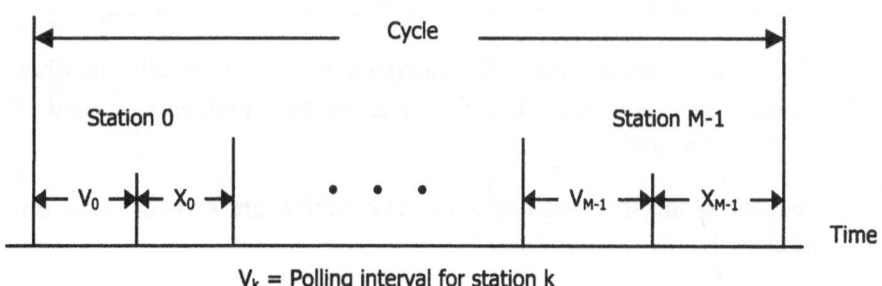

V_k = Polling interval for station k

X_k = Data interval for station k

Figure 8-3. Polling cycle

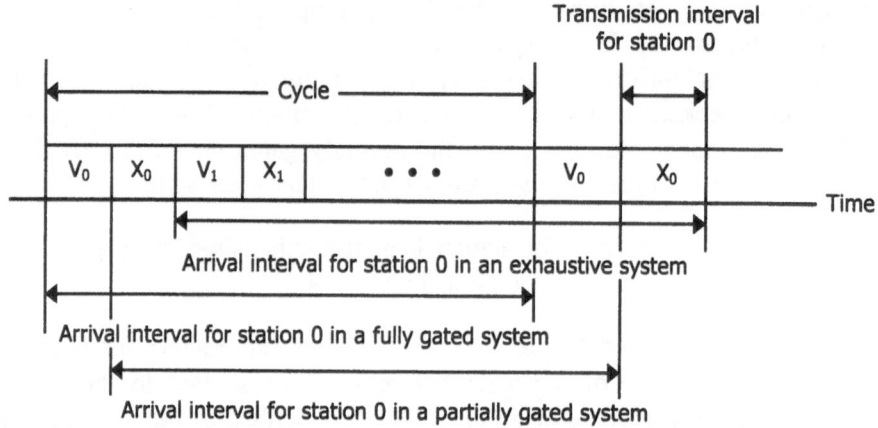

Figure 8-4. Arrival intervals for station 0

Poisson processes with a rate of λ packets per second;

- The kth polling interval is used for station k mod M and the subsequent kth data interval is used to send packets corresponding to those polling intervals;

- The first and second moments of the packet transmission time are $\bar{X} = 1/\mu$ and $\overline{X^2}$, respectively. The throughput is $S = M\lambda/\mu$;

- Polling intervals are mutually independent and identically distributed random variables with the first and second moments $\bar{V}$ and $\overline{V^2}$, respectively; and

- Packets in the transmit buffer of each station are served in the order of arrival.

Consider the ith packet arriving in the polling network (packets are counted in the order of arrival, regardless of station). Let

R_i = the residual service time; this is the residual time either for the packet transmission time or the polling time in progress in the network when the ith packet arrives;

N_i = the number of packets in all the M queues in the network ahead of the ith packet; and

Y_i = the duration of all the M polling intervals up to the starting instant of transmission of the ith packet.

Therefore, the waiting time of the ith packet is given by (see Figure 8-5)

$$W_i = R_i + \sum_{j=1}^{N_i} X_j + Y_i \qquad (8\text{-}1)$$

where X_j is the transmission time of the jth packet.

Since N_i and X_j are mutually independent random variables, by taking expectation of the above equation, we have

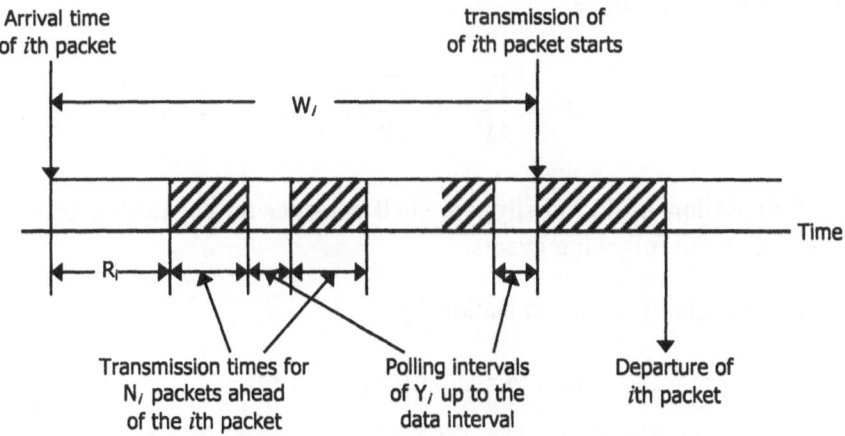

Figure 8-5. Waiting Time W_i of the ith packet

$$E[W_i] = E[R_i] + E[N_i] E[X_j] + E[Y_i]$$

$$= E[R_i] + E[N_i]/\mu + E[Y_i]$$

When the network is in equilibrium, or as $i \to \infty$, we obtain the average waiting time

$$W = R + N_q/\mu + Y \qquad (8\text{-}2)$$

From Little's formula, the second term on the right-hand side is equal to $M\lambda W/\mu = SW$. It follows from (8-2) that

$$W = \frac{R + Y}{1 - S} \qquad (8\text{-}3)$$

where $S = M\lambda/\mu$ is the network throughput. It remains to calculate the average residual service time R and the average polling time Y.

Calculation of R. The average residual service time R can be either a residual packet transmission time $\overline{X^2}/2\overline{X}$ with probability S, or a residual polling time $\overline{V^2}/2\overline{V}$ with probability 1-S. Thus, by using the formula of total probability, we have

$$R = \frac{\overline{X^2}}{2\overline{X}}S + \frac{\overline{V^2}}{2\overline{V}}(1 - S) \qquad (8\text{-}4)$$

Calculation of Y. Firstly, we shall consider an exhaustive network. Let us define the following events:

A_k = packet i arrives in station k;

B_{kj} = packet i belongs to station (k+j) mod M.

Since there are exactly j polling intervals between station k mod M and station $(k + j)$ mod M, the two events, $\{Y_i | A_k B_{kj}\}$ and $\{\sum_{r=1}^{j} V_{(k+r)modM}\}$, are equivalent. It follows that

$$E[Y_i | A_k \, B_{kj}] = E[\sum_{r=1}^{j} V_{(k+r)modM}] = j\bar{V} \tag{8-5}$$

and

$$E[Y_i | A_k] = \sum_{j=0}^{M-1} E[Y_i | A_k \, B_{kj}] \, P(B_{kj})$$

But $P(B_{kj}) = 1/M;$

hence,

$$E[Y_i | A_k] = \sum_{j=0}^{M-1} j\bar{V}/M = \frac{(M-1)\bar{V}}{2}$$

Furthermore, we have

$$E[Y_i] = \sum_{k=0}^{M-1} E[Y_i | A_k] P(A_k)$$

$$= \sum_{k=0}^{M-1} \frac{(M-1) \, \bar{V}}{2} P(A_k)$$

As $i \to \infty$, we obtain

$$Y = \lim_{i \to \infty} E[Y_i] = \frac{(M-1)\bar{V}}{2} \sum_{k=0}^{M-1} P(A_k) \tag{8-6}$$

In the steady state, a packet will arrive at station k with probability $1/M$. Using this fact, we obtain

$$P(A_k) = \frac{1}{M}$$

Substitution of this probability into (8-6) yields

$$Y = \frac{(M-1)\,\bar{V}}{2} \tag{8-7}$$

Combining (8-3), (8-4), and (8-7), we obtain the average packet waiting time:

$$W = \frac{1}{2(1-S)}\,[\frac{S\overline{X^2}}{\bar{X}} + (1-S)\,\frac{\overline{V^2}}{\bar{V}} + (M-1)\,\bar{V}] \tag{8-8}$$

Since $S = M\lambda\bar{X}$ and $\overline{V^2} = \sigma_V^2 + \bar{V}^2$, (8-8) can be written as

$$W = \frac{M\lambda\overline{X^2}}{2(1-S)} + \frac{(M-S)\bar{V}}{2(1-S)} + \frac{\sigma_V^2}{2\bar{V}} \qquad \text{(exhaustive)} \tag{8-9}$$

We shall now consider a partially gated network. If a packet arrives in its own data interval with probability S/M, it will be delayed by an additional duration of $M\bar{V}$, which is the average sum of polling intervals in a cycle. Thus, Y in (8-7) is increased by $S\bar{V}$, and we obtain

$$W = \frac{M\lambda\overline{X^2}}{2(1-S)} + \frac{(M+S)\bar{V}}{2(1-S)} + \frac{\sigma_V^2}{2\bar{V}} \qquad \text{(partially gated)} \tag{8-10}$$

Finally, we will consider the fully gated network. If a packet arrives in its own station data interval, it will be delayed by $M\bar{V}$ with probability S/M. This is the same as the partially gated system. However, if a packet arrives in its own polling interval, it will be delayed by $M\bar{V}$ with probability $(1-S)/M$. Thus, Y in (8-7) is increased by an additional delay, $M\bar{V}\frac{S}{M} + M\bar{V}\frac{(1-S)}{M} = \bar{V}$. We then obtain

$$W = \frac{M\lambda\overline{X^2}}{2(1-S)} + \frac{(M+2-S)\overline{V}}{2(1-S)} + \frac{\sigma_V^2}{2\overline{V}} \qquad \text{(fully gated)} \qquad (8\text{-}11)$$

Having determined the average waiting time, it is straight-forward to find the average number of packets in a station transmit buffer. Using Little's formula directly, the average number of packets in a station buffer is given by

$$N_q = \lambda W \qquad (8\text{-}12)$$

where W is given by (8-9), (8-10), or (8-11).

Example 8-1. Consider a metropolitan area network with a central processor (controller) located at the headend of a broadband CATV (community antenna or cable television) network that has a tree topology. The network has the following characteristics:

- Maximum distance from headend to a subscriber station = 20 km;

- Number of subscribers = 1000;

- Length of go-ahead = 1 byte or 8 bits;

- Length of polling packet = 8 bytes;

- Access technique is roll-call polling;

- Data rate of channel = 56 Kbps;

- Propagation time = 5 μs/km;

- Packet length distribution for packets from subscriber to headend is exponential with mean of 100 bytes; and

- Modem synchronization time = 5 *ms*.

Assume that the network is operated in exhaustive mode.

(a)   If each user generates packets at a rate of one packet per minute, what is the mean waiting time?

(b)   If the number of subscribers is increased to 1500, what is the corresponding mean waiting time?

(c)   Calculate the average number of packets in each station queue for parts (a) and (b).

Solution. For roll-call polling, the mean polling time for each station consists of the modem synchronization time, the transmission times of go-ahead and of the polling packet, and the propagation times from a station to the headend processor and from the headend processor to the next station.

Modem synchronization time = 5 ms.

Transmission times of go-ahead and polling packet = $\dfrac{(8 + 8 \times 8)bits}{56 \times 10^3 bps}$

= 1.2857 ms.

Maximum one-way propagation time = $20km \times 5 \ \mu s/km = 0.1 \ ms$

Thus, the mean polling time is

$$\bar{V} = 5 + 1.2857 + 2 \times 0.1 = 6.49 \ ms.$$

(a)   Mean packet transmission time = $\bar{X} = \dfrac{100 \times 8}{56 \times 10^3} = 14.29 ms$

Mean polling time = $\bar{V} = 6.49 \ ms$

Total throughput = $S = M \lambda \bar{X} = 1000 \times \dfrac{1}{60} \times 14.29 \times 10^{-3} = 0.238$

Since the packet length distribution is exponential and the polling time is constant, we have

$$\overline{X^2} = 2\bar{X}^2 = 2 \times (14.29)^2 = 408.41 \times 10^{-6}(seconds)^2$$

and $\sigma_V^2 = 0$. Using these results in (8-9), we obtain

$$W = 0.0047 + 4.2575 = 4.2622 \text{ seconds}$$

(b) Now $M = 1500$ subscribers;
$$S = 1500 \times \frac{1}{60} \times 14.29 \times 10^{-3} = 0.3573$$

Thus, $W = 0.0079 + 7.572 = 7.58 \text{ seconds}$

(c) For part (a), the average queue length is

$$N_q = \lambda W = 0.077 \text{ packets}$$

For part (b), $N_q = \lambda W = 0.126 \text{packets}$

Average Cycle Time. The total time required to poll all of the stations and to return to the starting station in the polling sequence is called the cycle time. The cycle time depends on the number of packets to be transmitted by each station after the last poll and hence is a random variable. The average cycle time T_c is an important parameter for describing the operation of polling networks.

Let N_c be the average number of packets that arrive at a station in an average cycle time T_c. From Little's formula, we obtain

$$N_c = \lambda T_c \qquad (8\text{-}13)$$

Since the average packet transmission time is $\bar{X}$ and the average polling interval is $\bar{V}$, the average cycle time can be expressed as

$$T_c = M(N_c \bar{X} + \bar{V}) \qquad (8\text{-}14)$$

By combining these two equations, we obtain the average cycle time:

$$T_c = \frac{M\bar{V}}{1 - M\lambda\bar{X}}$$

$$\text{(8-15)}$$

$$= \frac{M\bar{V}}{1 - S} \text{ seconds}$$

Average Transfer Delay. The average transfer delay T is defined as the average total time between the packet arrival instant at a station and its delivery to the central computer. It is the sum of the average waiting time W, the average packet transmission time $\bar{X}$, and the average propagation time τ_{ave} from a station to the central computer. The average transfer delay can be expressed as

$$T = W + \bar{X} + \tau_{ave} \qquad \text{(8-16)}$$

where W is given in (8-9) to (8-11).

Let τ denote (a) the end-to-end propagation time for a bus network, (b) the maximum propagation time from the most remote station to the central computer for a tree network, or (c) the propagation time around the ring for a ring network. The average propagation time can be approximately expressed as

$$\tau_{ave} = \frac{\tau}{2} \qquad \text{(8-17)}$$

Comments. By examining (8-9) to (8-12) and (8-16), we see that both the average number of packets N_q in the queue and the average transfer delay T are linear functions of the average waiting time W. Normally, it is desirable to keep both N_q and T as small as possible. This requires that W be small. However, W is proportional to the number of stations in the network. Also, W is inversely proportional to $1 - S$. As the throughput increases and approaches unity, W grows in an unbounded fashion. This is an effect similar to that of ρ on T in Figure 4-8.

If $\bar{V}$ is large, then the number of stations M should be kept small because of the factor (M-S) $\bar{V}$ being in the second term on the right-hand side of the expression for W. In the first term, the factor $M\bar{X^2}$ would be at a minimum if the packet length is constant and equal to $\bar{X}$, for $\overline{X^2} = \bar{X}^2$.

Example 8-2. For the metropolitan area network of Example 8-1 with 1000 subscribers, calculate (a) the average cycle time, and (b) the average transfer delay.

Solution.

(a) $M = 1000$, $\bar{V} = 6.49$ *ms*. $S = 0.238$, using (8-15), we find the mean cycle time

$$T_c = \frac{1000 \times 6.49}{1 - 0.238} = 8.52 \text{ seconds}$$

(b)   Since $W = 4.2622$ *seconds*,

$$\bar{X} = 14.29 \times 10^{-3} \text{ seconds},$$

$$\tau_p = 1 \times 10^{-4} \text{ seconds},$$

then applying (8-16) yields

$$T = 4.2622 + 0.01429 + \frac{1}{2} \times 10^{-4} = 4.2765 \text{ seconds}$$

8-4. SUMMARY

This chapter presents the operation of roll-call and hub pollings. A polling network may be considered as a multiple-queue with a cyclic server. Because of polling, the channel time for each station is comprised of a polling interval and a data transmission interval.

For performance analysis, we obtain expressions for the mean waiting time of exhaustive, partially gated, and fully gated polling networks. From these mean waiting times, the corresponding average transfer delays are also obtained.

Furthermore, we derive an expression for the average cycle time. It should be pointed out that the queueing model for the analysis of polling networks can also be used for the analysis of ring networks, as will be discussed with modifications in the next chapter.

REFERENCES

[1] Hammond, J.L. and O'Reilly, P.J.P., Performance Analysis of Local Computer Networks, Reading, Mass.: Addison-Wesley, 1986.

[2] Bertsekas, D. and Gallager, R., Data Networks, 2nd ed., Englewood Cliffs, N.J.: Prentice Hall, 1992.

[3] Schwartz, M., Telecommunication Networks: Protocols, Modeling and Analysis, Reading, Mass.: Addison-Wesley, 1987.

[4] Takagi, H., Analysis of Polling Systems, Cambridge, Mass.: MIT Press, 1986.

PROBLEMS

8-1.   Consider the pairs of multidrop telephone lines which connect one primary node to a number of secondary nodes. The signal transmitted by the primary node passes through a pair of wires and is received by all of the secondary nodes. Similarly, there is a return pair of wires which carries the sum of the transmitted signals from all of the secondary nodes to the primary node. The mode of operation for multidrop telephone lines is for the primary node to poll each secondary node in some order. Each secondary node responds to its poll either by sending data back to the primary station or by indicating that it has no data to send. Data for the network are as follows:

- Maximum distance between the primary station and the furthest secondary stations is 10 km;

- Propagation time is 5 μs per km;

- Modem synchronization time is 20 milliseconds;

- Length of go-ahead is 1 byte;

- Length of polling packet is five 8-bit bytes;

- Bit rate of the channel is 96 kbps;

- Exponential packet length with average length of 100 bytes;

- Input to each station is independent Poisson with an average rate of 1 packet per minute; and

- Assume exhaustive mode.

(a) What is the maximum number of stations such that the average transfer delay is at most one second?

(b) What is the maximum number of stations for a normalized throughput equal to 0.5?

8-2.   A polling network is to be constructed with hub polling and the data listed below:

- Go-ahead message is two 8-bit bytes;

- Synchronization time for go-ahead is 5 *ms* ;

- Constant packet length equals 500 bits;

- Independent Poisson input with average rate to each station is 1 packet per minute;

- Number of stations is 200;

- Propagation time is negligible; and

- Bit rate of channel is 2400 bps.

(a) What is the minimum bit rate for the operation of the network to be stable?

(b) Calculate the average transfer delay for the exhaustive, fully-gated and partially-gated operations.

(c) What is the mean cycle time?

8-3.   Design a (roll-call) polling network with 10 stations, 200 meters apart with one station 200 meters from a central computer. Each station is to service 10 terminals, whose inputs are all independent Poisson processes with average rates varying from 1-10 messages per minute. Each message is fifty 8-bit bytes long and the messages are transmitted by the stations to the central computer. The terminals transmit messages into a buffer at a station over a 1 kbps line, where the messages are stored until the station can transmit to the central computer. A specification for the design is that the total average transfer delay for messages from a terminal to the central computer must not exceed one second. Assume that the go-ahead message and the polling message are respectively 1 byte and 3 bytes long, while the propagation time is 5 $\mu s/km$. Assume exhaustive operation.

8-4. M stations 100 meters apart are connected to a bus in tandem with 100 meters between the nearest station and the central computer. The connections are all made with 20 kbps lines. Messages transmitted from stations to the central computer are all 500 bits long. Each station serves 30 terminals, each of which generates independent Poisson traffic with an average rate λ_i of 1 message per minute. Modem synchronization time is 10 milliseconds, propagation delay is 5 μs per kilometer, polling packets are five 8-bit characters, and the go-ahead is two 8-bit characters. Assume exhaustive operation:

   (a)   How many stations can be serviced if the total average transfer delay for messages from the stations to the central computer must be less than 1 second?

   (b)   What is the minimum value of total average transfer delay as the throughput approaches zero? Given the configuration of stations, how can this transfer delay be reduced?

   (c)   What is the absolute limit on the number of stations to be serviced by the network?

8-5.   Show that for both the fully-gated and partially-gated polling networks, the steady-state probability of a polling interval being followed by an empty data interval is $\dfrac{(1 - S - \lambda\bar{V})}{(1 - S)}$. Hint: If p is the required probability, argue that the ratio of the times used for data intervals and for polling intervals is $\dfrac{(1 - p)\bar{X}}{\bar{V}}$.

CHAPTER 9

RING NETWORKS

9-1. INTRODUCTION

A ring network is a local area network which consists of a number of repeaters joined by point-to-point links to form a closed loop. The repeater is a device, capable of receiving data on one link and transmitting it, bit by bit, on the other link as fast as it is received with no buffering. The links are unidirectional so that data are transmitted in one direction only around the ring, and can be read by all stations attached to the ring (Figure 9-1, see p. 358).

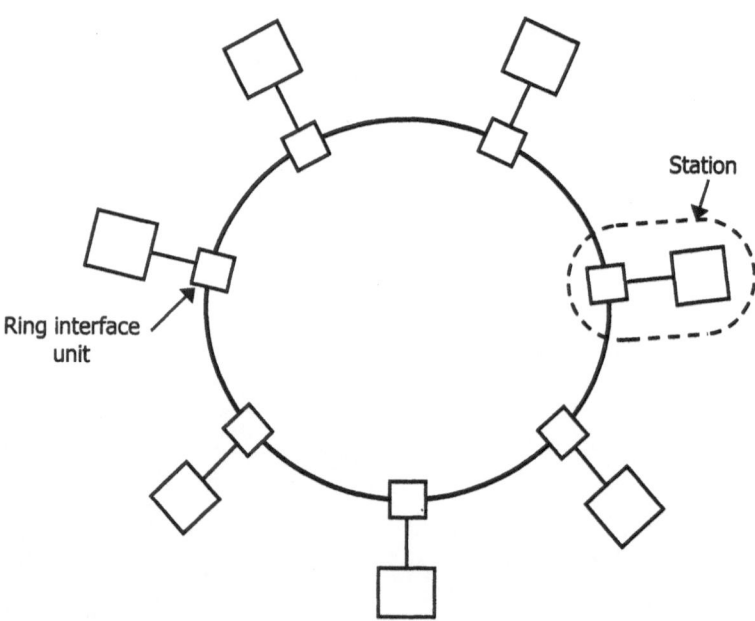

Figure 9-1. Ring network

information, including the destination station address. These packets are stored in the station transmit buffer when awaiting transmission on the ring.

For a ring network to operate as a communication network, three functions are required: data insertion, data reception, and data removal. In addition to serving as an active transceiver on the ring, each repeater also serves as a device for data insertion.

A number of different types of ring networks has been designed and implemented. Of the many different types of rings proposed, three types deserve more attention: token rings, slotted rings, and register insertion rings.

This chapter presents performance analysis of the above mentioned three types of rings in detail. The main performance measure for ring networks is the mean packet delay. Thus, performance analysis consists of determining the mean packet delay, and how it is affected by the various parameters of the network, such as statistics of the packet service time, the network throughput, and the number of stations.

9-2. TOKEN RING NETWORKS

When a station has data to send, it first carries out the packetization, stores the packets in the transmit buffer and waits for the free token which controls the access to the ring for transmission. When this station's repeater receives the free token, the repeater retransmits the bits received from the station on the outgoing link. During this period of transmission, bits may arrive on the incoming link. In this case, there are two possibilities that are treated differently:

1. If the packet length is longer than the bit length of the ring (the round-trip delay in bit times often called the ring latency), the bits could be from the same packet that the repeater is still sending. The repeater passes the bits back to the station, which can check them as a form of acknowledgement.

2. For multiple-token rings, more than one packet could be on the ring at the same time. If the repeater, while transmitting, receives bits from a packet it did not originate, it must store them in a buffer for later transmission.

A packet may be removed by the repeater of the destination station or by the transmitting station's repeater after it has made a round trip on the ring. The latter approach is more desirable because it permits automatic acknowledgement and multicast addressing (one packet being sent to multiple stations).

The repeater on the ring network serves two main purposes: to maintain the proper functioning of the ring by passing on all the data coming from the incoming link and to provide an access point for the attached stations to send and receive data.

The most important feature of the ring structure is that it uses point-to-point communication links. This fact brings a number of advantages:

- Since the transmitted signal is regenerated at each node, greater distances can be covered;

- The ring structure can accommodate optical fibre links;

- Implementation is simple because transmission paths are fixed;

- Routing algorithms are not required;

- Stations introduce only a small delay to a packet circulating on the ring; and

- Multiple addressing is straightforward.

Disadvantages, on the other hand, are:

- Reliability is not good because a failure of an interface unit or a line segment can cause a breakdown on the ring;

- The ring must be broken to add or delete stations so that it is inflexible;

- Propagation delay is proportional to the number of stations on the ring; and

- There is a practical limit to the number of repeaters on a ring.

9-3. OPERATION OF TOKEN RING NETWORKS

Token ring is a medium access control technique for ring networks originally proposed by W.D. Farmer and E.E. Newhall in 1969, and referred to as the Newhall ring. It is based on the use of a small frame, called a token, which is a distinctive bit pattern. The token circulates around the ring from station to station and can be in one of two possible states: idle or busy. When the ring is first activated or initialized, an idle (or free) token is inserted in the ring. A station with data to transmit must wait until it seizes the idle token and changes it to busy state before transmitting. The busy token then becomes part of the header of data transmitted on the ring by the station repeater. Thus, other stations on the ring can read the header, detect the busy token, and refrain from transmitting. The busy token on the ring will make a round trip and be removed by the transmitting repeater. When the station has completed transmission of its data and the first bit of its transmitted data has returned to the station (after a round trip), the transmitting station inserts a new free (or idle) token on the ring. Thus, the use of a token guarantees that only one station may transmit at a time.

In an elementary ring network, the token can be a distinctive pattern of several bits. In more sophisticated ring networks, such as the IEEE 802 token ring, the token can be several bytes long. For a simple token which could be a sequence of eight 1's, to prevent the occurrence of this bit pattern in the data, it is necessary to use the bit stuffing technique. When bit stuffing has been used, the receiver must be able to identify the added (stuffed) bits, so that they can be removed before error detection.

A long dedicated bit pattern with bit stuffing can be inefficient. An alternative is to use special bits for the token, or special bits to mark the beginning and ending of a control field. For example, Manchester coding may be used for data with violations for control bits.

Logical operation of the repeater (interface unit) is shown in Figure 9-2. There are two modes: the listen mode and the transmit mode. Stations are normally in the listen mode, in which the station can read the address of the data circulating on the ring and copy the contents of packets addressed to it. Otherwise, the station retransmits each bit with a small delay. When the station has data to send and when its repeater seizes a free token, the repeater receives bits from the station and retransmits them on its outgoing link to the ring after changing the free token to busy state.

There are two types of transmission schemes: the exhaustive service and nonexhaustive service. For the exhaustive service, the station transmits all the data stored in the transmit buffer and then places a new free token on to the ring. For nonexhaustive service, the station is allowed to transmit only a specified number of packets each time it captures a free token. In both service schemes, the transmitting station repeater is responsible for removing its data from the ring. In the sequel, the exhaustive type of operation is assumed.

Station Latency. In a ring network, bits are transmitted serially out of each station interface. Normally, the input to each repeater is the output of the previous repeater delayed by the propagation time between the two stations. In the listen mode, each repeater passes on the data received from its incoming link at the input after a delay known as the station latency. The station latency can be a minimum of one bit time or a dozen of bit times. Let this delay be b bit times and the channel capacity be C bps. The station latency is given by

$$\tau_s = b/C \ seconds \qquad\qquad (9\text{-}1)$$

Ring Latency. Bits circulate around the ring in a synchronous fashion. If all the repeaters are in the listen mode, the token circulates around the ring in a time equal to the sum of the propagation delays plus the sum of the station latencies. This composite time is called the ring latency.

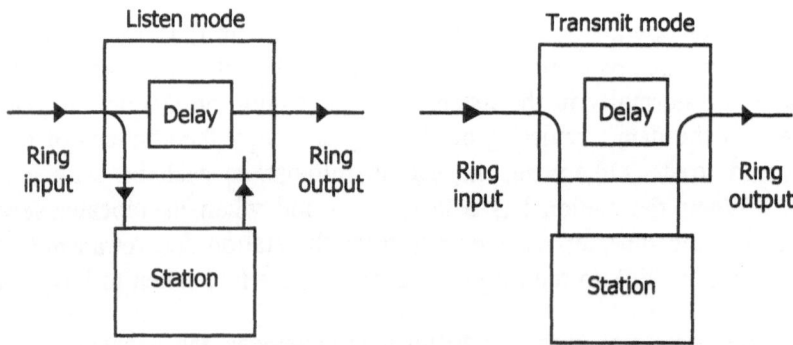

Figure 9-2. Repeater/station logical operation

Let τ_p be the round-trip propagation delay for the ring and M be the number of stations on the ring. The ring latency τ_r is given by

$$\tau_r = \tau_p + M\tau_s \ \text{seconds} \tag{9-2}$$

Depending on when a new free token can be generated, there are three operational modes which are referred to as multiple-token operation, single-token operation, and single-packet operation.

Multiple-Token Operation. For multiple-token operation, the transmitting station repeater places a new free token on the ring immediately following the last bit of the transmitted packet. This type of operation permits several busy tokens and one free token on the ring at any given time. With more than one token on the ring, packet lengths and station latencies must be related, so that busy tokens do not progress far enough around the ring to be removed by their originating station before the new free token is generated.

Single-Token Operation. For single-token operation, a transmitting station generates a new free token only after it has removed its own busy token. If a packet length is longer than the ring latency in bit times, the transmitting station will remove its busy token before it finishes transmitting data. In this case, the station must continue transmitting data and generate a new free token only after the last data bit has been transmitted. This situation is exactly the same as the multiple-token operation. Thus, single-token and multiple-token operations differ only in cases of packet lengths being shorter than the ring latency.

Single-Packet Operation. For single-packet operation, a transmitting station generates a new free token only after both its busy token and all of its transmitted data have circulated a round-trip on the ring and are completely removed from the ring. This type of operation ensures that two transmissions will not interfere each other.

To illustrate the three types of operations, consider a four-station token ring network with 1 bit time delay per station and negligible propagation delay (Figure 9-3). We assume that initially station 1 receives a free token at $t = 1$ and outputs a busy token at $t = 2$. Note that the ring latency is four bit times.

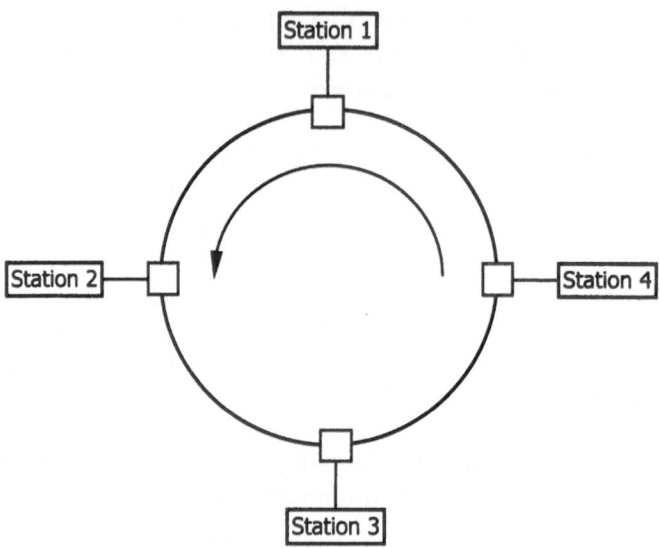

Figure 9-3. Four-station ring

At $t = 5$, after traversing the ring with a ring latency of four bit times, the busy token and the transmitted data of station 1 begin to return to station 1 and hence are removed. As shown in Table 9-1, in all three types of operations, station 1 initially transmits a packet of total length six bit times at $t = 1$ and station 4 transmits a packet of total length three bit times.

Note that the time at which a free token is subsequently generated depends on the mode of operation. Specifically, a new free token is generated by station 1 at $t = 8$ and by station 4 at $t = 14$ for multiple-token operation; by station 1 at $t = 8$ and by station 4 at $t = 15$ for single-token operation; and by station 1 at $t = 11$ and by station 4 at $t=20$ for single-packet operation.

Table 9-1. Three operational modes for a four-station ring

Discrete time	Multiple token					Single token					Single packet					Discrete time
	In #1	Out #1	Out #2	Out #3	Out #4	In #1	Out #1	Out #2	Out #3	Out #4	In #1	Out #1	Out #2	Out #3	Out #4	
0																0
	□				□	□				□	□				□	
2		■					■					■				2
		d	■				d	■				d	■			
4			d	d	■			d	d	■			d	d	■	4
	■	d	d	d	■	■	d	d	d	■	■	d	d	d	■	
6	d	d	d	d	d	d	d	d	d	d	d	d	d	d	d	6
	d	d	d	d	d	d	d	d	d	d	d	d	d	d		
8	d	□	d	d	d	d	□	d	d	d	d		d	d	d	8
	d		□	d	d	d		□	d	d	d			d	d	
10	d			□	d	d			□	d	d				d	10
	■				■	■				■		□				
12	d	■			d	d	■			d			□			12
	d	d	■		d	d	d	■		d				□		
14	□	d	d	■	□		d	d	■		■				■	14
		□	d	d		□		d	d	□	d	■				
16				□	d	□			d		d	d	■			16
					□				□		d	d	d	■		
18	□				□				□			d	d	d		18
						□				□			d	d	d	
											□				□	20
												□				
													□			22
														□		
											□				□	24
																26

□ = Free token
■ = Busy token
d = Data bit

9-4. PERFORMANCE ANALYSIS FOR TOKEN RING NETWORKS

Packets generated at a station are stored in the transmit buffer and wait for the free token. In addition to the waiting time, the packet has three more delays before it arrives at its destination station:

1. The delay due to station latency;

2. The propagation delay associated with each link segment of the ring the packet traverses from originating station to destination station; and

3.   The transmission time of all the bits in the packet over the links of the ring.

We shall derive expressions for the average transfer delay suffered by packets on token ring networks for multiple-token, single-token, and single-packet operations.

To simplify the analysis. we make the following assumptions:

- The arrival process to each station is an independent Poisson process with a rate of λ packets/second;

- The average distance between the sending and receiving stations is one-half of the distance around the ring;

- There are M stations which are spaced so that the propagation delay between two consecutive stations is equal to τ_r/M, where τ_r is the ring latency;

- The packet length distribution is the same for each station, with mean $\bar{L}$ bits and second moment $\bar{L^2}$ (bits)^2, where the random variable L denotes the length of an arbitrary packet. The channel capacity is C bps so that the packet transmission time has the same distribution as L with mean $\bar{X} = \bar{L}/C$ seconds and second moment $\bar{X^2} = \bar{L^2}/C^2$ (seconds)^2; and

- Service at each station is of the exhaustive type.

To determine the average transfer delay for token ring networks, note the similarity between the operation of the token ring and the hub polling networks, as discussed in Chapter 8 (see also Figure 8-2(b) and Figure 9-1). The fact that hub polling typically requires a central controller, whereas token ring does not, does not make any difference in performance analysis. Operation of both networks is essentially the same.

The average time between the instant a station finishes its transmission of packets and the instant the next station receives a free token corresponds to the average polling interval. Thus, we adapt the average waiting time in a polling network, given by (8-9), for the token ring network. This equation is rewritten here, with $V = \bar{V}$ = constant and $\sigma_V^2 = 0$

$$W = \frac{M\lambda\bar{X^2}}{2(1-S)} + \frac{(M-S)\bar{V}}{2(1-S)} \qquad (9\text{-}3)$$

As given in (8-16), the average transfer delay for polling networks is

$$T = W + \bar{X} + \tau_{ave} \qquad (9\text{-}4)$$

To adapt this expression for token ring networks, the average delay corresponding to τ_{ave} in the polling network is $\tau_r/2$, one half of the ring latency, where τ_r is given by (9-2). Thus, the average transfer delay for a ring network can be expressed as

$$T = W + \bar{X} + \frac{\tau_r}{2} \qquad (9\text{-}5)$$

Furthermore, for ring networks, the service time can include any time interval during which the ring is inactive in the course of completing one packet transmission and becoming available for another transmission. This service time is implicitly present in the network throughput S of the expression of W. Specifically, the service time X in (9-3) must be modified and is replaced by the total time required in processing a packet and in becoming ready to process the next packet. Let this service time be X_e, called the effective service time. Thus, S in (9-3) is replaced by

$$S_e = M\lambda\bar{X}_e \qquad (9\text{-}6)$$

and the quantity $M\bar{V}$ corresponds to the ring latency τ_r.

The modifications to the average transfer delay for polling networks given by (9-5) result in the following expression for the average transfer delay for token ring networks:

$$T = \frac{M\lambda\overline{X_e^2}}{2(1 - S_e)} + \frac{(1 - S_e/M)\tau_r}{2(1 - S_e)} + \bar{X} + \frac{\tau_r}{2} \qquad (9\text{-}7)$$

9-4-1. Average Transfer Delay for Multiple-Token Operation. This mode of operation assumes that a new free token is generated immediately after the transmission of the last bit of the last packet. Thus, the effective service time and the packet transmission time are the same; that is, $\overline{X_e} = \bar{X}$ and $\overline{X_e^2} = \overline{X^2}$. In addition, the quantity $M\bar{V}$ corresponds to the ring latency τ_r. Equation (9-7) then gives the average transfer delay for multiple-token operation as

$$T = \frac{M\lambda \overline{X^2}}{2(1-S)} + \frac{(1-S/M)\tau_r}{2(1-S)} + \overline{X} + \frac{\tau_r}{2} \, , \, \tau_r \leq \overline{X} \qquad (9\text{-}8)$$

where $S = M\lambda\overline{X}$. From (9-8), we see that if $\tau_r << \overline{X}$, then the third term $\overline{X}$ on the right-hand side has the dominating effect on the average transfer delay T.

9-4-2. Average Transfer Delay for Single-Token Operation. In this mode, a new free token is generated only when the transmitting station has received its own busy token after a round trip. Thus, there can be only one token on the ring at any one time.

If the packet transmission time $\overline{X}$ is greater than the ring latency τ_r, the station is still transmitting when its busy token returns after a round trip. Thus, it cannot generate a new free token until it has finished the transmitting packet, and the operation is completely identical to that of multiple-token operation.

On the other hand, if the packet transmission time is less than the ring latency, single-token operation can be different from multiple-token operation because the effective service time X_e is greater than the packet transmission time X due to periods when the transmitting station is waiting for the busy token to return. During these periods the whole network is tied up by the busy token.

For single-token operation, the relationship between the packet transmission time X and the ring latency τ_r is of central importance. We shall consider the cases of fixed packet transmission time and exponentially distributed packet transmission time separately.

For fixed packet transmission time, $X = \text{constant} = \overline{X}$. Thus, as noted previously, if $X = \overline{X} \geq \tau_r$, the single-token operation is exactly the same as the multiple-token operation, and hence the average transfer delay T is given by (9-8) with $\overline{X^2} = \overline{X}^2$. If $X = \overline{X} < \tau_r$, however, the single-token operation becomes different from the multiple-token operation since the effective service time $X_e = \tau_r > \overline{X}$. Thus, $\overline{X}_e = \tau_r$ and $\overline{X_e^2} = \overline{X}_e^2 = \tau_r^2$, for the ring cannot be used to transmit a new packet until the busy token of the previous packet has circulated completely around the ring.

With $\bar{X}_e = \tau_r$, (9-6) becomes

$$S_e = M\lambda\tau_r = \frac{\tau_r}{\bar{X}}S = \alpha S \qquad (9\text{-}9)$$

where $\alpha = \tau_r/\bar{X} > 1$.

Substitutions of $\overline{X_e^2} = \tau_r^2$ and S_e in (9-9) into (9-7) yield the average transfer delay for single-token operation as

$$T = \frac{M\lambda\tau_r^2}{2(1 - \alpha S)} + \frac{(1 - \alpha S/M)\tau_r}{2(1 - \alpha S)} + \bar{X} + \frac{\tau_r}{2} \;, \; \alpha > 1 \qquad (9\text{-}10)$$

For exponentially distributed packet transmission times with mean $\bar{X}$, the distribution function is given by

$$F_X(x) = \begin{cases} 0 & , \; x < 0 \\ 1 - e^{-x/\bar{X}} & , \; x \geq 0 \end{cases} \qquad (9\text{-}11)$$

It follows that the effective service time X_e has the distribution function

$$F_{X_e}(x) = \begin{cases} 0 & , \; x < \tau_r \\ 1 - e^{-x/\bar{X}} & , \; x \geq \tau_r \end{cases} \qquad (9\text{-}12a)$$

the density function

$$f_{X_e}(x) = \begin{cases} 0 & , \; x < \tau_r^- \\ (1 - e^{-x/\bar{X}})\delta(x - \tau_r) & , \; \tau_r^- < x < \tau_r^+ \\ \dfrac{1}{\bar{X}}e^{-x/\bar{X}} & , \; x > \tau_r^+ \end{cases} \qquad (9\text{-}12b)$$

the first moment

$$\overline{X_e} = \tau_r + \overline{X}e^{-\alpha} \tag{9-13}$$

and the second moment

$$\overline{X_e^2} = \tau_r^2 + 2(1 + \alpha)\overline{X^2}e^{-\alpha} \tag{9-14}$$

The results of (9-13) and (9-14) together with S_e as given by (9-9) can be substituted in (9-7) to obtain the average transfer delay:

$$T = \frac{[\alpha^2 + 2(1 + \alpha)e^{-\alpha}]S\overline{X}}{2[1 - (\alpha + e^{-\alpha})S]} + \frac{[1 - (\alpha + e^{-\alpha})S/M]\tau_r}{2[1 - (\alpha + e^{-\alpha})S]} +$$

$$\overline{X} + \frac{\tau_r}{2} \,, \alpha > 1 \tag{9-15}$$

9-4-3. Average Transfer Delay for Single-Packet Operation. In the single-packet mode, the sending station generates a new free token only when it has received and removed all of the packets it has transmitted. Thus, there is never more than one packet circulating on the ring. In this case, there is always a delay of τ_r seconds after the complete transmission of each packet so that the effective service time is $X + \tau_r$. The first and second moments of this effective service time are, respectively,

$$\overline{X_e} = \overline{X} + \tau_r \tag{9-16}$$

and

$$\overline{X_e^2} = \overline{X^2} + 2\tau_r\overline{X} + \tau_r^2 \tag{9-17}$$

Using (9-16) and (9-17) together with (9-6) in (9-7) yields an expression for the average transfer delay for single-packet operation as

$$T = \frac{M\lambda(\overline{X^2} + 2\tau_r\overline{X} + \tau_r^2)}{2[1 - (1 + \alpha)S]} + \frac{[1 - (1 + \alpha)S/M]\tau_r}{2[1 - (1 + \alpha)S]} + \overline{X} + \frac{\tau_r}{2} \quad (9\text{-}18)$$

Example 9-1. Consider a single-token ring that has the following given parameters:

- One hundred stations on the ring;
- Ring length of 1 km;
- Ring bit rate of 10 Mbps;
- Mean packet length of 1000 bits;
- Independent Poisson arrival process to each station with a rate of 10 packets/second;
- Station latency of 2 bits; and
- Propagation delay of the ring of 5 μs per km;

(a) Determine the mean transfer delay for both constant and exponential packet lengths.

(b) If the number of stations on the ring is increased to 200 and the Poisson arrival rate is increased to 25 packets per second, calculate the mean transfer delay. All the other network parameters are unchanged.

(c) Repeat calculations in (b) for station latency of 5 bits.

Solution

(a) For constant packet length, $\overline{X^2} = \overline{X}^2$,

$\overline{X} = \overline{L}/C = 100$ μs , $\overline{X^2} = 10^4$ (μs)^2

$M = 100, \lambda = 10$ packets per second,

$\tau_r = \tau_p + M\tau_s = 25$ μs ,

$\alpha = \tau_r/\overline{X} = 0.25$,

$S = M\lambda\overline{X} = 0.1$

Since $\alpha < 1$, using (9-8) yields

$$T = 5.56 + 13.89 + 100 + 12.5 = 131.95 \ \mu s \ or \ 0.132 \ ms$$

For exponential packet length, $\overline{X^2} = 2\overline{X}^2$,

$$T = 11.12 + 13.89 + 100 + 12.5 = 137.51 \ \mu s \ or \ 0.1375 \ ms$$

(b)   Now, $M = 200$ and $\lambda = 25$ packets/second. Then

$$\tau_r = 45 \ \mu s \quad , \quad \alpha = 0.45 \quad , \quad S = 0.5$$

For constant packet length, (9-8) yields

$$T = 50 + 45 + 100 + 22.5 = 217.5 \ \mu s \ or \ 0.2175 \ ms$$

For exponential packet length, (9-8) yields

$$T = 100 + 45 + 100 + 22.5 = 267.5 \ \mu s \ or \ 0.2675 \ ms$$

(c)   Since $\tau_s = 5$ bits, the ring latency becomes

$$\tau_r = 5 + \frac{200 \times 5}{10} = 105 \ \mu s$$

$$\alpha = 1.05 > 1 \ .$$

Using (9-10) for constant packet length, we find

$$T = 58.03 + 110.24 + 100 + 52.5 = 320.77 \ \mu s \ or \ 0.3208 \ ms$$

For exponential packet length, using (9-15) yields

$$T = 212.07 + 173.79 + 100 + 52.5 = 538.36 \ \mu s \ or \ 0.5384 \ ms$$

9-5. OPERATION OF SLOTTED RING NETWORKS

Suppose that traffic on a ring is uniformly distributed between different source-destination pairs. It can be conceived that, on average, a packet need be transmitted to only half of the links of a ring network. Since, in the token ring network, a packet travels on every link, we see that half the network's transmission capacity is wasted. It is therefore possible to achieve twice the throughput, at least in principle, by using a different control strategy. Slotted rings and register insertion rings are such strategies that allow higher throughput.

The slotted ring control strategy is based on the empty slot principle. In the slotted ring, a number of fixed length slots circulate continuously on the ring, each slot containing a leading bit to indicate the slot as empty or full (Figure 9-4). This strategy was first proposed by J. Pierce and is referred to as the Pierce loop. Most of the development work on this technique was done at the University of Cambridge in England and it is known as the Cambridge ring.

All slots are initially marked empty. A station with data to send must break the data up into fixed-length packets called minipackets. It then waits until an empty slot arrives, marks the slot full, and inserts a minipacket of data as the empty slot goes by. To prevent hogging of the ring by one station (a phenomenon that a station may use more than its fair share of the empty slots for transmission), the sending station is not allowed to transmit another minipacket until its full slot makes a complete round trip, to be marked empty again by the sending station.

To illustrate the operation of a slotted ring, let us consider the four-station slotted ring in Figure 9-5. One of the stations, say, station 2, is the monitor station which is used to set up and maintain ring packetizing and to perform error checks. Each station has a latency of one bit. Propagation delay is ignored and the 4-bit latency is used to support one slot on the ring. The first bit of the 4-bit slot is used as the empty/full indicator. The operation is illustrated in Table 9-2 for two operating conditions: when only one station has data to transmit and when two stations have data to transmit.

When only station 1 has data to transmit, it captures the empty slot at $t = 0$, marks the slot full at $t = 1$ and inserts a minipacket of data, and then releases the ring by marking the full slot to empty at $t = 5$. Since the

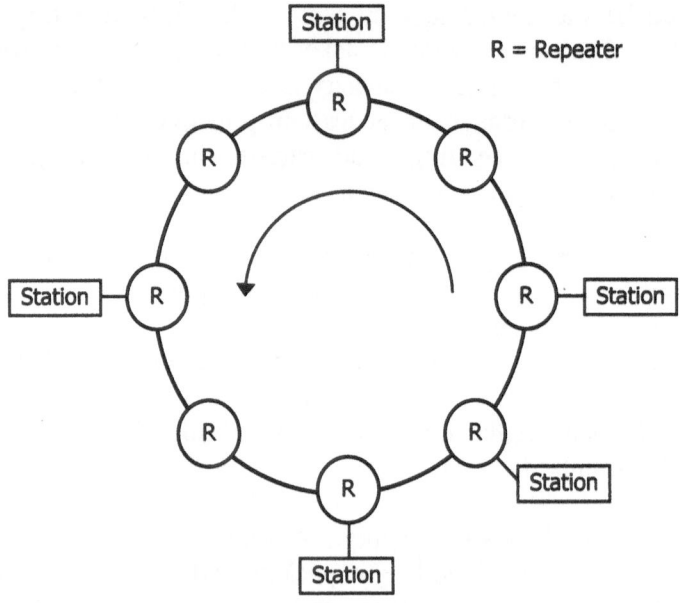

Figure 9-4. Slotted ring

other three stations do not have data to transmit, station 1 can send another minipacket only until $t = 8$. In this case, station 1 can make use of the ring only one-half of the time. This can be seen from the station 1 column in the table.

When stations 1 and 3 have data to transmit, the operation of the slotted ring is shown on the right-hand half of the table. Unlike the previous case, the outputs of all four stations at any one time are never empty. This can be observed by examining the rows in the table.

For slotted ring networks, the ring latency determines the total number of bits in the circulated slots. This fact has an important influence on the design of a slotted ring. For example, consider a 10 Mbps channel. The bit time is 0.1 μs. A typical channel propagation delay is 5 μs per kilometer so that one kilometer of cable can support 50 bits. If each station introduces a latency of 1 bit, then either a large number of stations or a long cable is required. Suppose that the ring has 100 stations connected by 2 kilometers of cable. The ring latency is equal to $2 \times 50 + 100 \times 1 = 200$ bits. This ring can then support four slots of 50 bits, ten slots of 20 bits, etc.

9-6. PERFORMANCE ANALYSIS FOR SLOTTED RING NETWORKS

To derive an expression for the average transfer delay in a slotted ring network, we make the following assumptions:

- Input data traffic to each station is an independent Poisson arrival process with a rate of λ packets/second;

- There are M stations on the network. Each station has a latency of τ_s seconds;

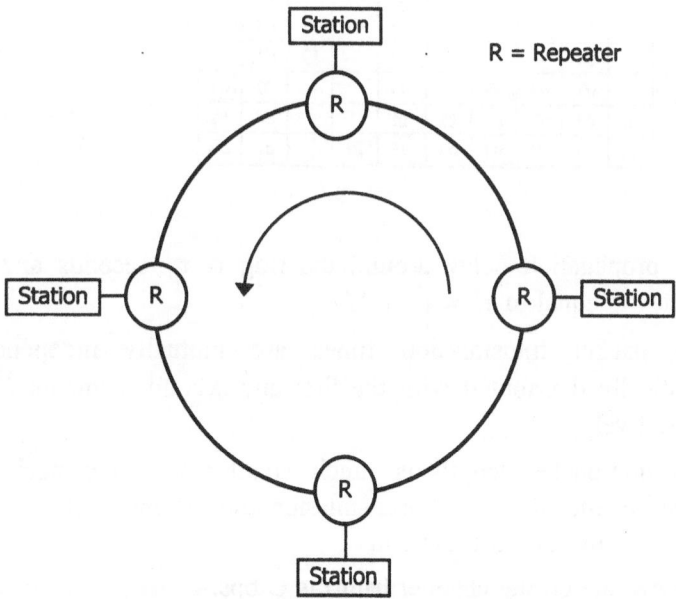

Figure 9-5. Four-station slotted ring

Table 9-2. Two operating conditions for a four-station slotted ring

Discrete time	Only #1 transmitting					#1 and #3 transmitting				
	In #1	Out #1	Out #2	Out #3	Out #4	In #1	Out #1	Out #2	Out #3	Out #4
0	□				□	□				□
		■					■			
2		d1	■				d1	■		
		d1	d1	■			d1	d1	■	
4	■	d1	d1	d1	■	■	d1	d1	d1	■
	d1	□	d1	d1	d1	d1	□	d1	d1	d1
6	d1		□	d1	d1	d1		□	d1	d1
	d1			□	d1	d1			■	d1
8	□				□	■			d3	■
		■				d3	■		d3	d3
10		d1	■			d3	d3	■	d3	d3
		d1	d1	■		d3	d3	d3	□	d3
12	■	d1	d1	d1	■	□	d3	d3		□
	d1	□	d1	d1	d1	■	d3			
14	d1		□	d1	d1		d1	■		
	d1			□	d1		d1	d1	■	
16	□				□	■	d1	d1	d1	■
		■				d1	□	d1	d1	d1
18		d1	■			d1		□	d1.	d1
		d1	d1	■		d1.			■	d1
20	■	d1	d1	d1	■	■			d3	■
	d1	□	d1	d1	d1	d3	■		d3	d3

Empty slot

■ = Full monitor bit
□ = Empty monitor bit
d1 = Data bit from station 1
d3 = Data bit from station 3

Full slot

- The propagation delay around the ring is τ_p seconds and the ring latency is equal to $\tau_r = \tau_p + M\tau_s$;

- The packet transmission times are mutually independent and identically distributed with the first and second moments $\bar{X}$ and $\overline{X^2}$, respectively;

- The minipacket length is much smaller than the packet length; likewise, the minipacket transmission time is much smaller than the packet transmission time; and

- The bit rate on the channel (ring) is C bps.

Since the time for a slot to circulate around the ring is very small (in comparison to a packet transmission time), it will be ignored in the analysis. Based on the above assumptions, the slotted ring network can be modeled as a distributed M/G/1 queueing system with arrival rate $M\lambda$ and general service time with first moment $\bar{X}$ and second moment $\overline{X^2}$.

We modify (9-5) to represent the average transfer delay for the slotted ring network and rewrite it as

$$T = W + \bar{X} + \frac{\tau_r}{2} \qquad (9\text{-}19)$$

where the average waiting time W is given by the Pollaczek-Khinchin formula (4-61)

$$W = \frac{\rho\bar{X}}{2(1-\rho)}(1 + \frac{\sigma_X^2}{\bar{X}^2}) \qquad (9\text{-}20)$$

Since the throughput is given by

$$S = M\lambda\bar{X} = \rho \qquad (9\text{-}21)$$

we replace ρ by S in (9-20) to yield

$$W = \frac{S\bar{X}}{2(1-S)}(1 + \frac{\sigma_X^2}{\bar{X}^2}) \qquad (9\text{-}22)$$

To take into account the large amount of overhead used in each minipacket for the slotted ring network operation, we let L_h be the number of header bits contained in each slot and L_d be the number of data bits in each slot. Then $L_h + L_d$ is the length of each slot in bits. Define the overhead factor

$$h = L_h/L_d \qquad (9\text{-}23)$$

From this, we see that the effective packet transmission time is increased to $(1 + h)\bar{X}$ seconds in a slotted ring network. It follows that the effective throughput is also increased to $(1 + h)S$.

Introducing these changes to (9-19) and (9-22) then gives the average transfer delay for the slotted ring network as

$$T = \frac{(1 + h)^2 S\bar{X}}{2[1 - (1 + h)S]}(1 + \frac{\sigma_X^2}{\bar{X}^2}) + (1 + h)\bar{X} + \frac{\tau_r}{2} \qquad (9\text{-}24)$$

As illustrated in Figure 9-5, for the actual operation of the slotted ring network, on average, a packet only needs to traverse half of the ring, since the effective channel bit rate as seen by any particular station is only half of the links of the ring. This implies that the packet transmission time $\bar{X}$ is effectively $2\bar{X}$. With this modification, we write (9-24) as

$$T = \frac{(1 + h)^2 S\bar{X}}{1 - (1 + h)S}(1 + \frac{\sigma_X^2}{\bar{X}^2}) + 2(1 + h)\bar{X} + \frac{\tau_r}{2} \qquad (9\text{-}25)$$

It is important to note that for a slotted ring, the number of slots on the ring m, the gap time g in seconds, the minipacket size $L_h + L_d$ in bits, the channel bit rate C in bps, the propagation delay around the ring τ_p in seconds, the number of stations M, the station latency τ_s in seconds, and the ring latency τ_r in seconds are related by the following expression

$$\tau_r = \frac{m(L_h + L_d)}{C} + g = \tau_p + M\tau_s \qquad (9\text{-}26)$$

In the design of a slotted ring network, the number of slots m is usually adjusted so that the gap g is minimized. If the station latencies are not sufficiently large to provide the required ring latency τ_r, additional artificial delays can be added in the station latency τ_s. Some of the ideas are illustrated by the following example.

Example 9-2. Consider a slotted ring of 1 kilometer in length with 50 stations. Each station has a latency of 1 bit. The slot size is 6 bytes with 3 bytes data. The channel bit rate is 10 Mbps and the propagation delay on the ring is 5 μs per kilometer.

(a) How many slots can the ring hold without adding any artificial delays? What is the gap time? If the packet size is a constant of 1200 bits long, what is the average transfer delay when the packet arrival rate at each station is (i) 10 packets per second, (ii) 40 packets per second?

(b) If the number of stations is increased to 100, how many slots can the ring hold without adding artificial delays? What is the gap time? What is the average transfer delay for each of the two arrival rates in (a) with the same constant packet length of 1200 bits?

Solution

(a) Using (9-26), the ring latency is given by

$$\tau_r = \tau_p + M\tau_s$$

$$= 5 \times 10^{-6} + 50 \times 1/(10 \times 10^6) \; seconds$$

$$= 10 \; \mu s$$

and for the slot size of 6 bytes,

$$\tau_r = m\,(6 \times 8)/(10 \times 10^6) + g$$

We see that the largest integer that makes τ_r less than or equal to 10 μs is $m = 2$. As a result, the gap time $g = 10 - 2(6 \times 8)/(10 \times 10^6) = 0.4\mu s$. The overhead factor is given by

$$h = L_h/L_d = (6 - 3)8/(3 \times 8) = 1$$

The packet transmission time is

$$\bar{X} = \bar{L}/C = 1200/(10 \times 10^6) = 120 \; \mu s$$

(i) For $\lambda = 10$ packets/second,
$$S = M\lambda\bar{X} = 50 \times 10 \times 120 \times 10^{-6} = 0.06$$

Since the packet length is constant, $\sigma_X^2 = 0$, using (9-25) gives the average transfer delay

$$T = \frac{(1+h)^2 S\bar{X}}{1-(1+h)S} + 2(1+h)\bar{X} + \frac{\tau_r}{2}$$

$$= 32.73 + 480 + 5$$

$$= 517.73 \ \mu s$$

(ii)   For $\lambda = 40$ packets/second,

$$S = 50 \times 40 \times 120 \times 10^{-6} = 0.24$$

and

$$T = 221.54 + 480 + 5 = 706.54 \ \mu s$$

(b)   For 100 stations, the ring latency is increased to
$$\tau_r = 5 \times 10^{-6} + 100 \times 1/(10 \times 10^6) = 15 \ \mu s \ .$$

Now the number of slots m on the ring can be three with a gap time of 0.6 μs .

(i)   For $\lambda = 10$ packets/second,

$$S = 100 \times 10 \times 120 \times 10^{-6} = 0.12$$

and

$$T = 75.79 + 480 + 5 = 560.79 \ \mu s$$

(ii)   For $\lambda = 40$ packets/second,

$$S = 100 \times 40 \times 120 \times 10^{-6} = 0.48$$

and

$$T = 5760 + 480 + 5 = 6245 \ \mu s$$

Observe that from (9-25), the maximum throughput is

$$S_{max} = \frac{1}{1+h} = 0.5$$

The average transfer delay will be very large when S is close to S_{max}. Under light and medium load conditions, say, $S \leq 0.3$, the dominant factor in the average transfer delay is the second term, $(1 + h)\bar{X}$, on the right-hand side of (9-25). When $S > 0.3$, the first term becomes dominant.

9-7. REGISTER INSERTION RINGS

The register insertion ring derives its name from the shift register in the interface of each station on the ring. The shift register, equal in size to the maximum packet length, is used for temporarily holding packets that circulate past the station. In addition, the station also has a transmit buffer for storing locally generated packets and a receive buffer for storing the input traffic of the station.

The distinctive structure of the interface is shown in Figure 9-6. The addition of the shift registers into the ring effectively increases the capacity and hence the throughput of the ring in terms of the total number of bits it can hold at any one time.

9-8. OPERATION OF REGISTER INSERTION RING NETWORKS

Operation of the register insertion ring can be described by the interface shown in Figure 9-7. First, consider the case in which the station has no data to send and the receive and transmit switches in Figure 9-6 are in position A. Bits circulate around the ring through the insertion buffer in the same way as for a slotted ring. When the insertion buffer is empty, the input pointer is at the rightmost position. When bits arrive from the ring, they are inserted bit by bit in the insertion buffer, with the input pointer shifting left for each bit. As soon as the full address field is in the insertion buffer, the station can determine if it is the addressee. If not, the packet is forwarded to the ring until the packet is gone. Now, if no further packets arrive, the input pointer will return to its initial rightmost position.

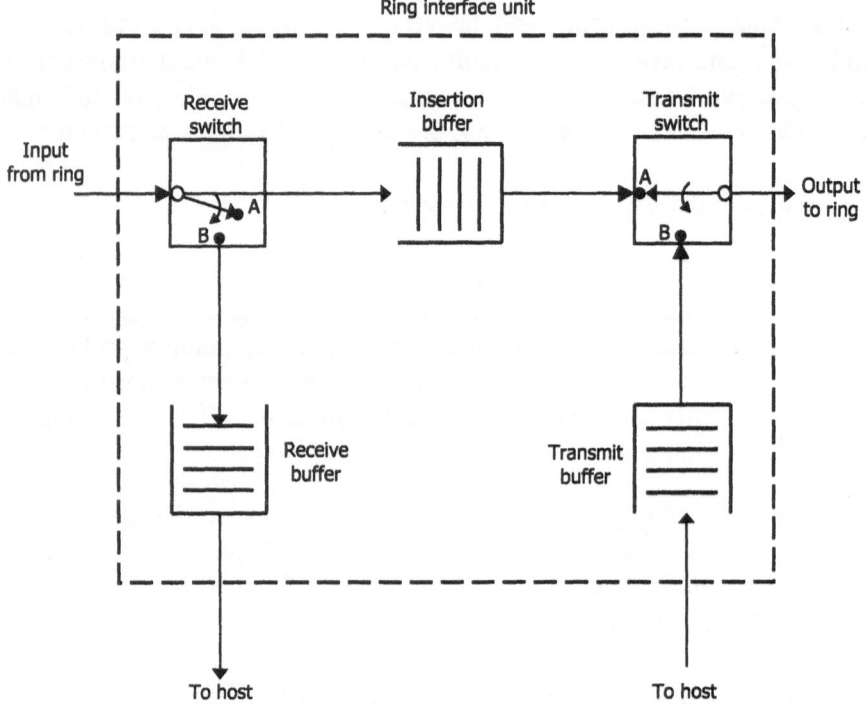

Figure 9-6. Register insertion ring interface

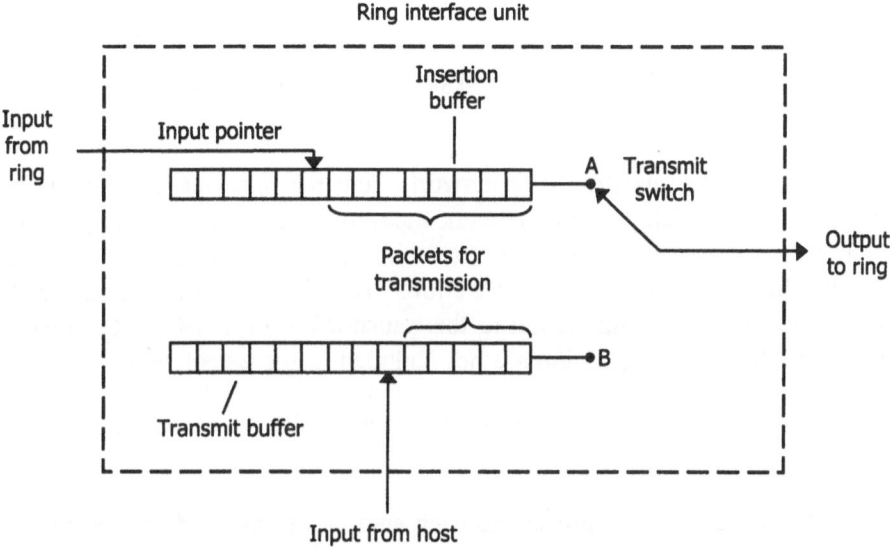

Figure 9-7. Store-and-forward buffering of packets in the ring interface

If an arriving packet is addressed to the station in question, the station has two choices. First, it can change the receive switch to position B, store the remainder of the packet in the receive buffer, and erase the address bits from the insertion buffer, thus removing the packet from the ring. In this case, some means of acknowledgement must be employed. The second alternative is to retransmit the packet as before, while copying it to the local station.

We see that there is always a delay at each station, whose minimum value is the length of the address field and whose maximum value is the length of the insertion buffer.

When the transmit switch is at position B, bits arriving from the ring will be stored in the insertion buffer. When the insertion buffer has spare capacity and the station has packets to send, selection of the transmit switch position depends on whether the ring uses station priority or ring priority. Station priority allows the transmit buffer to transmit first, whereas ring priority gives preference to the insertion buffer.

The principal advantage of the register insertion technique is high throughput by allowing multiple packets on the ring. However, this requires the recognition of an address prior to removal of a packet. If a packet's address field is damaged, it could circulate the ring indefinitely.

9-9. PERFORMANCE ANALYSIS OF REGISTER INSERTION RING NETWORKS

For register insertion ring networks, the performance criterion is the average transfer delay on the network measured from the time the packet is stored in the transmit buffer to the time the packet is received by the receive buffer of the destination station. The times required to move the packet from the sending station to the transmit buffer and from the receive buffer to the receiving station are not included in the calculation.

To develop a model for performance analysis, we make the following assumptions:

- The packet arrival process to each station's transmit buffer is Poisson with rate λ_t;

- The packet transmission times are mutually independent and identically distributed random variables with first and second moments $\bar{X}$ and $\overline{X^2}$, respectively;

- There are M stations on the network. Each station has a constant latency of τ_s seconds;

- All stations have the same traffic pattern: a station transmits to another station with k-separation (measured by the number of links between them) with probability q_k, $k = 1, 2, \cdots, M - 1$; this assumption implies that the probability of transmission between any two stations depends only on the separation of the two stations;

- The capacity of all buffers is infinite; and

- The packet interarrival times and service times are mutually independent; thus, each station can be modeled independently.

Figure 9-8 depicts a queueing model for the interface of Figure 9-6. A simplified model is shown in Figure 9-9 subject to the above six assumptions.

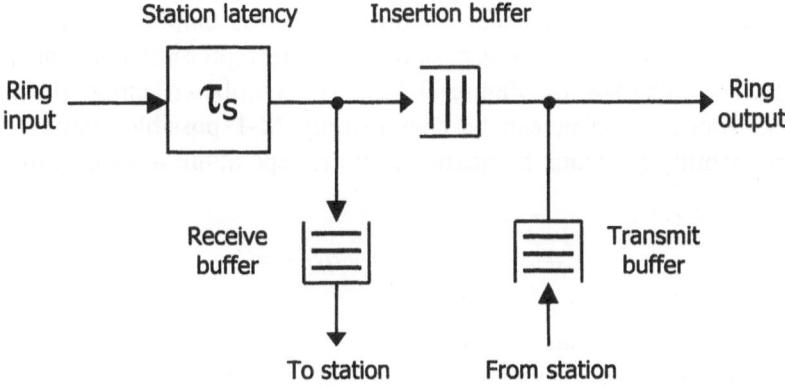

Figure 9-8. Queueing model for Figure 9-6

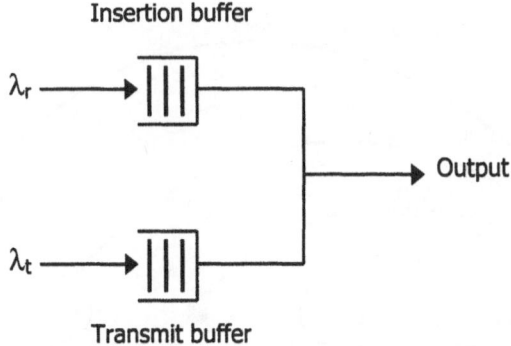

Figure 9-9. Simplified model for Figure 9-8

In the simplified model it is assumed that the input process to the insertion buffer is Poisson with rate λ_r which remains to be determined. Intuitively, it can be observed that λ_r depends on λ_t and q_j because λ_r is the sum of a portion of the generated traffic λ_t from the other M-1 stations.

First, we observe that a packet may wait in only one (transmit) buffer of the source station and does not go through the insertion buffer of either the source station or the destination station. Second, consider a test station, say, station 1 for convenience. According to assumption 4, the traffic transmission pattern of this test station can be represented by the traffic transmission diagram of Figure 9-10. The numbered circle denotes a station. Since a packet can be sent to only M-1 possible stations, each with probability q_j which is measured by the separation, it follows that

$$q_1 + q_2 + \cdots + q_{M-1} = 1 \qquad (9-27)$$

Note that the traffic transmission pattern shown in Figure 9-10 is valid and is the same for each station, although the order of the stations may be different.

Now let us calculate the total transit traffic at station M, contributed by all

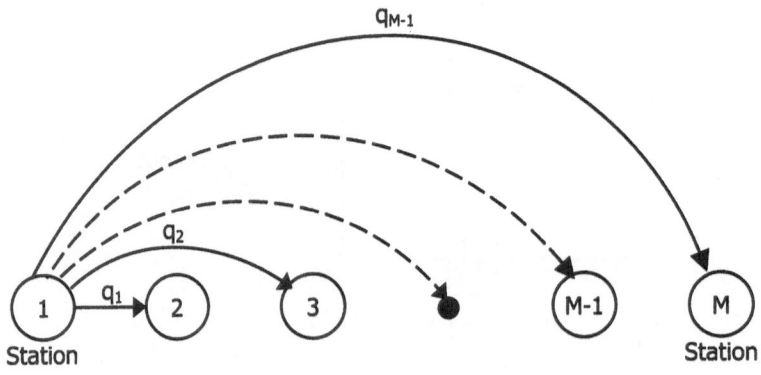

Figure 9-10. Traffic transmission diagram

the other M-1 stations. Figure 9-11 shows the contribution of transit traffic from each of the other M-1 stations, where only the transit traffic of each station (except station M) is depicted.

Like traffic transmission, the transit traffic shown in Figure 9-11 is the same for each station. By summing all the transit traffic at station M, we obtain

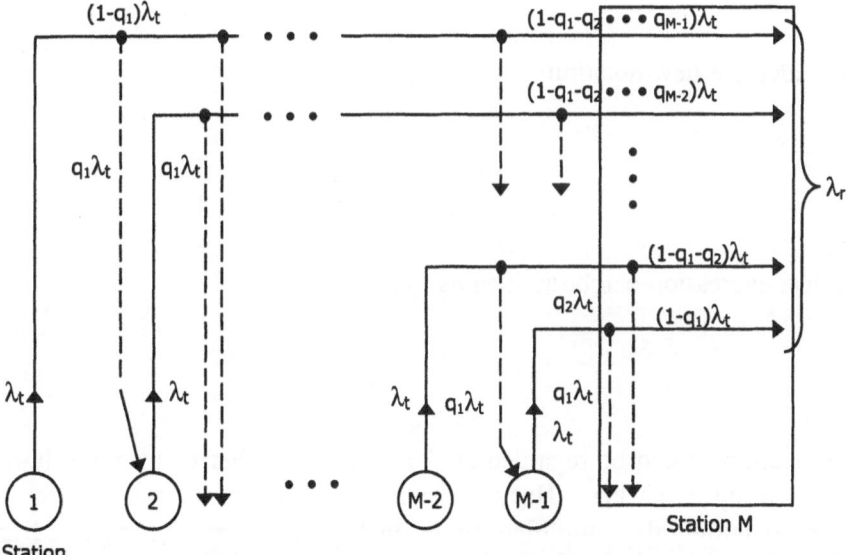

Figure 9-11. Transit traffic at station M

$$\lambda_r = (1 - q_1)\lambda_t + (1 - q_1 - q_2)\lambda_t + \cdots + (1 - q_1 - q_2 - \cdots - q_{M-1})\lambda_t$$

$$= [(M - 1) - (M - 1)q_1 - (M - 2)q_2 - \cdots - q_{M-1}]\lambda_t$$

$$= [q_2 + 2q_3 + \cdots + (M - 2)q_{M-1}]\lambda_t$$

$$= [\sum_{j=2}^{M-1} (j - 1)q_j]\lambda_t$$

$$= [\sum_{j=1}^{M-1} (j - 1)q_j]\lambda_t$$

$$= (\sum_{j=1}^{M-1} jq_j - 1)\lambda_t$$

where, in the third step, (9-27) has been used in the first term on the right-hand side.

Introducing a new notation

$$\alpha = \sum_{j=1}^{M-1} jq_j - 1$$

the last expression can be written as

$$\lambda_r = \alpha\lambda_t \tag{9-28}$$

The quantity α can be regarded as the average number of insertion buffers traversed by a packet. Having determined the arrival rate λ_r for the insertion buffer, the simplified model in Figure 9-9 can be regarded as a head-of-the-line nonpreemptive priority M/G/1 queueing system. It should be noted that service in the simplified model and in the queueing model of Figure 9-8, is different from the normal queue in that service is defined to occur only in the receive buffer of the destination station.

The simplified model in Figure 9-9 can now be used to determine the average waiting times W_t in the transmit queues and W_r in the insertion queues, under both ring and station priority conditions. The results obtained in (4-88) and (4-89) of Chapter 4 can be used to calculate W_r and W_t depending on which queue is given priority.

Case 1. Ring Priority (priority to insertion buffer)

For this case, we replace quantities with subscripts 1 and 2 by quantities with subscripts r and t in (4-88) and (4-89), respectively, to give

$$W_r = \frac{(\lambda_r + \lambda_t)\overline{X^2}}{2(1 - \rho_r)} \tag{9-29}$$

and

$$W_t = \frac{(\lambda_r + \lambda_t)\overline{X^2}}{2(1 - \rho_r - \rho_t)(1 - \rho_r)} \tag{9-30}$$

where

$$\overline{X_1} = \overline{X_2} = \overline{X}$$

$$\overline{X_1^2} = \overline{X_2^2} = \overline{X^2}$$

$$\rho_t = \lambda_t \overline{X}$$

and

$$\rho_r = \lambda_r \overline{X} = \alpha\lambda_t \overline{X} = \alpha\rho_t$$

Case 2. Station Priority (priority to transmit buffer)

In this case, the expressions (9-29) and (9-30) are interchanged to account for the interchange in priority, to give

$$W_t = \frac{(\lambda_r + \lambda_t)\overline{X^2}}{2(1 - \rho_t)} \tag{9-31}$$

and

$$W_r = \frac{(\lambda_r + \lambda_t)\overline{X^2}}{2(1 - \rho_r - \rho_t)(1 - \rho_t)} \tag{9-32}$$

Now consider the average packet transfer delay for the register insertion ring. We observe that the average delay consists of four components: the average delay (waiting time) in the transmit buffer W_t, the average packet transmission time $\overline{X}$, the average station latency traversed by a packet $(\alpha + 1)\tau_s$, and the average waiting delay due to insertion buffers αW_r. Adding all these four components gives the average transfer delay as

$$T = W_t + \overline{X} + (\alpha + 1)\tau_s + \alpha W_r \tag{9-33}$$

To simplify this expression, we calculate the quantity $W_t + \alpha W_r$ by defining the notation

$$\rho = \lambda_t \overline{X} = \rho_t$$

Then

$$\rho_r = \alpha \rho_t = \alpha \rho$$

Case 1. Ring Priority

Using (9-28), (9-29), and (9-30), $W_t + \alpha W_r$ can be expressed as

$$W_t + \alpha W_r = \frac{\lambda_t (1 + \alpha)^2 \overline{X^2}}{2(1 - \rho - \alpha\rho)} \tag{9-34}$$

Case 2. Station Priority

A similar calculation using (9-28), (9-31) and (9-32) yields

$$W_t + \alpha W_r = \frac{\lambda_t (1 + \alpha)^2 \overline{X^2}}{2(1 - \rho - \alpha\rho)} \tag{9-35}$$

a result identical to (9-34) for ring priority.

Thus, for both priority schemes, the average transfer delays are the same and equal to

$$T = \frac{\lambda_t (1 + \alpha)^2 \overline{X^2}}{2(1 - \rho - \alpha\rho)} + \overline{X} + (1 + \alpha)\tau_s \tag{9-36}$$

As in other networks, the throughput S can be defined as

$$S = M\rho \tag{9-37}$$

and the average transfer delay can be expressed in terms of S as

$$T = \frac{\lambda_t (1 + \alpha)^2 \overline{X^2}}{2[1 - (1 + \alpha)S/M]} + \overline{X} + (1 + \alpha)\tau_s \tag{9-38}$$

From this expression we see that the maximum network throughput is $M/(1 + \alpha)$, which can be greater than 1. This is a special feature of the register insertion ring network.

Example 9-3. A register insertion ring has M stations connected. All stations have independent Poisson traffic patterns and characteristics with the following parameters:

$$\lambda_t = 1 \text{ packet per second per station},$$

$$\tau_s = 16 \text{ bits},$$

$$C = 1 \text{ Mbps},$$

$$q_k = \frac{1}{M-1}.$$

The packet length has the first moment of 200 bits and second moment of 10100 (bits)^2. All buffers are assumed to have infinite capacity.

(a)   What is the maximum possible number of stations the register insertion ring can support for stable operation?

(b)   Determine the mean transfer delay for M = 400 stations.

Solution

(a)   Since the probability of transmission between any two stations is $q_k = 1/(M-1)$, then

$$\alpha = \sum_{k=1}^{M-1} kq_k - 1 = \frac{1}{M-1} \sum_{k=1}^{M-1} k - 1 = \frac{M}{2} - 1.$$

Using (9-37) and (9-38), we find the maximum throughput for stable operation

$$S_{max} = \frac{M}{1+\alpha} = M\lambda_t \bar{X}$$

or

$$M_{max} = \frac{2}{\lambda_t \bar{X}}$$

Hence

$$M_{max} = \frac{2}{1 \times 200/10^6} = 10^4 \; stations$$

(b) The mean packet transmission time is

$$\bar{X} = \bar{L}/C = 2 \times 10^{-3} \; seconds$$

and the second moment of the packet transmission time is

$$\overline{X^2} = \overline{L^2}/C^2 = 0.0101 \times 10^{-6}$$

For M = 400, we have

$$S = 400 \times 2 \times 10^{-3}$$

$$= 0.8$$

and

$$1 + \alpha = M/2 = 200$$

Using (9-38), we find the mean transfer delay

$$T = \frac{1 \times (200)^2 \times 0.0101 \times 10^{-6}}{2[1 - 200 \times 0.8/400]} + 2 \times 10^{-3} + 200 \times 16 \; 10^{-6}$$

$$= 0.3367 \times 10^{-3} + 3.2 \times 10^{-3}$$

$$= 5.5367 \; ms$$

9-10. SUMMARY

This chapter deals with the operation and performance analysis of local area networks with a ring geometry. Three types of ring networks are treated in detail: token rings, slotted rings, and register insertion rings. Depending on how the token is generated, operation of token rings can be further classified into single token, multiple token, and single packet operations. We derive expressions for the average transfer delay of all three types of token rings.

For slotted rings the overhead must be taken into account. This is expressed by the overhead factor. The actual operation of register insertion rings involves a number of queueing delays in series due to the tandem connection of a number of station interfaces between the transmitting and the receiving stations. A general model for such a connection requires a tandem connection of queues for which an analytical treatment is intractable. Thus, we develop a simplified model and derive an approximate expression for the average transfer delay.

Examples are given for illustration. Emphasis on modeling and analysis has been made throughout the chapter. Using the simplified model for the analysis of register insertion rings, it is shown that the average transfer delay expressions for both the ring priority and station priority schemes are the same.

REFERENCES

[1]   Hammond, J.L. and O'Reilly, P.J.P., Performance Analysis of Local Computer Networks, Reading, Mass.: Addison-Wesley, 1986.

[2]   Bertsekas, D. and Gallager, R., Data Networks, 2nd. ed., Englewood, Cliffs, N.J.: Prentice Hall, 1992.

[3]   Penney, B.K. and Baghdadi, A.A., "Survey of Computer Communications Loop Networks: Part 1 and 2", Computer Communications, Vol. 2, 1979, pp. 165-180, pp. 224-241.

[4]   Bux, W., "Local Area Subnetworks: A Performance Comparison", IEEE Trans. Communications, Vol. COM-29, No. 10, October 1981, pp. 1465-1473.

[5]   Bux, W. and Schlatter, "An Approximate Method for the Performance Analysis of Buffer Insertion Rings", IEEE Trans. Communications, Vol. COM-31, No. 1, January 1983, pp. 50-55.

[6] Ferguson, M.J. and Aminetzah, Y., "Exact Results for Non-Symmetric Token Ring Systems", IEEE Trans. Communications, Vol. Com-33, No. 3, March 1985, pp. 223-231.

PROBLEMS

9-1.   Consider the token ring network with 50 stations. The characteristics of the network are as follows:

- Channel bit rate = 10^6 bps;

- Station latency = 1 bit;

- Length of ring = 5 km;

- Constant packet length = 75 bits;

- Average arrival rate per station = 50 packets per second; and

- Propagation delay = 5 μs per km.

Calculate the average transfer delay for (a) multiple token operation, (b) single token operation, and (c) single packet operation.

9-2.   Consider the slotted ring network with the following features:

- Number of stations = 50;

- Channel bit rate = 5 Mbps;

- Length of ring = 1 km;

- Length of header = 24 bits;

- Number of slots in the ring = 4;

- Constant packet length = 2024 bits;

- Station latency = 2 bits;

- Packet arrival rate to each station = 5 packets per second; and

- Propagation delay = 5 μs /km.

Calculate (a) the number of data bits in each slot, (b) the gap time, and (c) the average transfer delay.

9-3.   A token ring network has 100 stations and a channel capacity of 1 Mbps. Input traffic to each station is independent Poisson with an average rate of 10 packets per second. All packets are 500 bits long. The station latency is 2 bits and the propagation delay for the complete ring is 10 μs. Calculate the average transfer delay for (a) multiple token operation, (b) single token operation, and (c) single packet operation.

9-4.   One hundred stations are connected to a local area ring network with a channel bit rate of 10 Mbps. Each station has a Poisson arrival process with an average rate of λ packets per second. All packets are of 500 bits long. Propagation delay for the complete ring is 50 μs and the station latency is 2 bits. What is the largest possible value of λ for stable operation for (a) token ring, (b) slotted ring with unity overhead factor, and (c) register insertion ring with an average number of insertion buffers traversed by a packet equal to fifty?

9-5.   A token ring network 10 km long is to be designed to connect M stations. The connections are to be made with 96 Kbps lines. Each station serves 50 terminals, each of which generates Poisson traffic with an average rate of 10 messages per minute. Propagation delay is 5 μs /km. Packets are of a fixed length of 240 bits. The station latency is 2 bits.

(a) Express the average transfer delay in terms of M and plot the result for multiple token operation.

(b) How many stations can be served if the average transfer delay is at most 0.1 second?

9-6.   A ring network has 100 stations to be connected with 1 Mbps lines. Each station generates Poisson traffic with an average rate of 1 packet per second. The packet length is a constant of 1000 bits. The propagation delay of the ring is 50 μs. Calculate the average transfer delay for (a) single token operation with 2 bits station latency, (b) slotted ring network with an overhead factor equal to 1, and (c) register insertion ring with equal probability of transmission to each of the other stations.

9-7.   A slotted ring network has 100 stations connected with 1 Mbps lines 20 km long. Each station has a latency of 2 bits. If the minipacket length must not exceed 50 bits, find two possible minipacket lengths and keep the gap as small as possible.

9-8.   A register insertion ring has 10 stations connected. All stations have identical traffic characteristics. The traffic transmission pattern is as follows.

Table 9-3. Traffic transmission pattern

Destination Source	1	2	3	4	5	6	7	8	9	10
1	--	1/6	1/6	1/8	1/8	1/8	1/12	1/12	1/16	1/16
2	1/16	--	1/6	1/6	1/8	1/8	1/8	1/12	1/12	1/16
3	1/16	1/16	--	1/6	1/6	1/8	1/8	1/8	1/12	1/12
4	1/12	1/16	1/16	--	1/6	1/6	1/8	1/8	1/8	1/12
5	1/12	1/12	1/16	1/16	--	1/6	1/6	1/8	1/8	1/8
6	1/8	1/12	1/12	1/16	1/16	--	1/6	1/6	1/8	1/8
7	1/8	1/8	1/12	1/12	1/16	1/16	--	1/6	1/6	1/8
8	1/8	1/8	1/8	1/12	1/12	1/16	1/16	--	1/6	1/6
9	1/6	1/8	1/8	1/8	1/12	1/12	1/16	1/16	--	1/6
10	1/6	1/6	1/8	1/8	1/8	1/12	1/12	1/16	1/16	--

Determine the maximum throughput.

CHAPTER 10

RANDOM ACCESS NETWORKS

10-1. INTRODUCTION

In Chapter 7, Section 7-5, we introduced the concept of random access in the discussion of pure ALOHA and slotted ALOHA networks. Random access networks are characterized by the absence of a channel access control mechanism. The access technique used in the ALOHA network is random in the sense that there is no predictable or scheduled time for any station to access the communication channel that is shared by all of the stations attached to the network. When the network has a bus-type geometry, the signal received at one station depends on the signals transmitted by all of the other stations. Typically, such a received signal is the sum of the signals transmitted from all other stations. This type of channel (medium) is called a multi-access channel.

Conceptually, a multi-access communication system can be regarded as a queueing system. Each station has a queue of packets to be transmitted with the multi-access channel as a common server. Ideally, the common server should view all of the waiting packets as being in one combined queue to be served by some appropriate queue discipline. Unfortunately, the server does not know which stations contain packets. Similarly, a station is unaware of packets at other stations.

In Chapter 7, we also obtained expressions for the throughput and the average transfer delay for both pure ALOHA and slotted ALOHA networks with Poisson input under steady-state conditions. However, this steady-state assumption is often not justified, thus, the performance results obtained in Chapter 7 apply only for a short interval of time.

In this chapter, we shall investigate the effect of the throughput S, the offered traffic G, and the backoff delay on the average transfer delay. We shall show that the expression (7-27) is valid when the set of integers for the backoff delay in a slotted ALOHA network approaches infinity; that is, when $K \to \infty$. Furthermore, we shall obtain a more accurate expression than (7-27) for S in terms of G. We shall derive a Markovian model for the slotted ALOHA network with a finite number of stations, and calculate the state transition probabilities in Section 10-3. Using the determined stationary state probabilities, we shall calculate the throughput and the average transfer delay.

Section 10-4 presents the concept of equilibrium point for performance analysis of slotted ALOHA. Section 10-5 explores the carrier-sense multiple access (CSMA) protocols. Section 10-6 investigates the carrier-sense multiple access with collision detection (CSMA/CD) protocols, where the throughput analysis is based on the concept of a cycle consisting of a busy period and an idle period.

10-2. PERFORMANCE ANALYSIS OF SLOTTED ALOHA WITH POISSON INPUT

Recall that the slotted ALOHA procedure requires that the time axis be divided into slots of fixed length P, which is the sum of the fixed packet transmission time $\bar{X}$, the processing time of the packet, and the maximum round-trip propagation delay. For local area networks, we assume that P equals $\bar{X}$ and we neglect the processing time and the propagation delay. Every transmitted packet must fit into one of these slots by beginning and ending in precise synchronization with the slot.

For the slotted ALOHA operation after its arrival, a ready packet is transmitted at the beginning of the next slot. If the packet is successfully received, then a positive acknowledgement is sent over a separate error-free channel by the receiver. After a delay determined by the maximum two-way propagation time of the channel and the processing time in the receiver, the sending station receives the acknowledgement. If a collision occurs, then no acknowledgement is sent. Thus, if the sending station does not receive an acknowledgement after a fixed time-out equal to the sum of the maximum two-way propagation time (quantized to slots) and the processing time in the receiver, the sending station selects a backoff time for retransmitting the packet. A flow chart for the slotted ALOHA protocol is given in Figure 10-1.

To obtain a more accurate expression than (7-30) for the average transfer delay, we first determine the average number of retransmissions per successful transmission by defining the following conditional probabilities q_n and q_r:

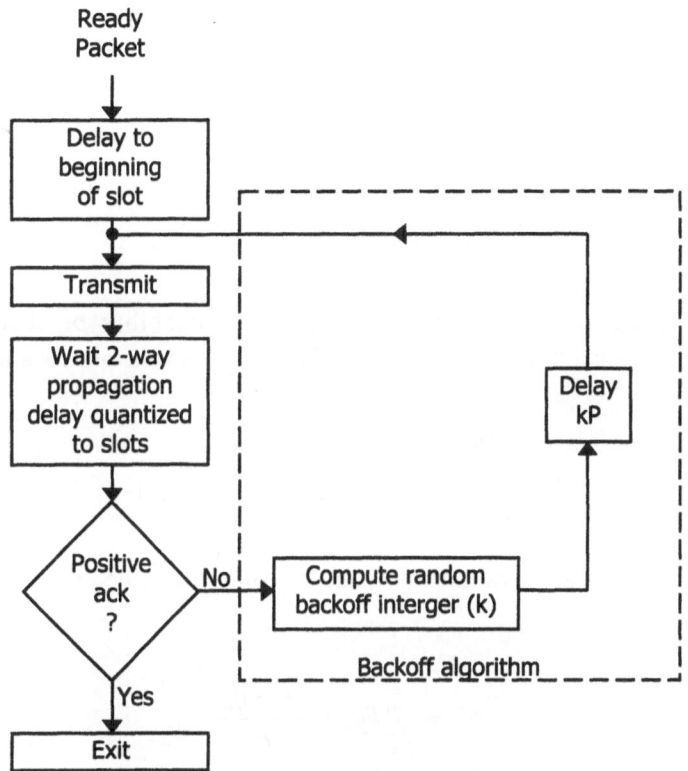

Figure 10-1. Flow chart for slotted ALOHA

q_n = The probability of a successful transmission given that the transmission is with a new packet; and

q_r = The probability of a successful transmission given that the transmission is with a retransmitted packet.

We derive the following expressions for q_n and q_r in (10A-12) and (10A-13) respectively in Appendix A at the end of this chapter as:

$$q_n = (e^{-G/K} + \frac{G}{K}e^{-G})^K e^{-S} \tag{10-1}$$

and

$$q_r = (\frac{e^{-G/K} - e^{-G}}{1 - e^{-G}})(e^{-G/K} + \frac{G}{K}e^{-G})^{K-1}e^{-S} \qquad (10\text{-}2)$$

We also show in Appendix B that

$$\lim_{K\to\infty} \frac{S}{G} = \lim_{K\to\infty} q_n = \lim_{K\to\infty} q_r = e^{-G} \qquad (10\text{-}3)$$

In terms of q_n and q_r, we can calculate the probability p_k that a given successfully transmitted packet requires exactly k retransmissions as

$$p_k = \begin{cases} q_n \,, k = 0 \\ (1 - q_n)(1 - q_r)^{k-1}q_r \,, k \geq 1 \end{cases} \qquad (10\text{-}4)$$

The average number of retransmissions N_r can then be determined from p_k as follows:

$$N_r = \sum_{k=0}^{\infty} kp_k = \frac{1 - q_n}{q_r} \qquad (10\text{-}5)$$

Substitution of N_r from (10-5) into (7-30) and neglecting τ_p and rP gives an expression for the average transfer delay

$$T = \frac{3P}{2} + (\frac{1 - q_n}{q_r})(\frac{K + 1}{2})P \qquad (10\text{-}6)$$

Thus, the normalized average transfer delay becomes

$$\hat{T} = 1.5 + (\frac{1 - q_n}{q_r})(\frac{K + 1}{2}) \qquad (10\text{-}7)$$

Furthermore, since G/S is the average number of transmissions per successful transmission, it follows from the definition of N_r that

$$G/S = N_r + 1 \tag{10-8}$$

Using (10-5) in (10-8) and solving for S gives

$$S = \frac{q_r G}{1 + q_r - q_n} \tag{10-9}$$

It is interesting to note that (10-6) and (10-9) are more accurate expressions than (7-30) and (7-27), respectively. Moreover, using (10-3) as $K \to \infty$, we see that both (10-6) and (10-9) reduce to (7-30) and (7-27), respectively.

The throughput S, as a function of G for several values of K, is plotted in Figure 10-2.

In the curves of Figure 10-2, note that the maximum throughput occurs at $G = 1$ for all values of K. The maximum throughput increases as K increases, reaching a peak value of $S_{max} = 1/e$ as $K \to \infty$. Most of the

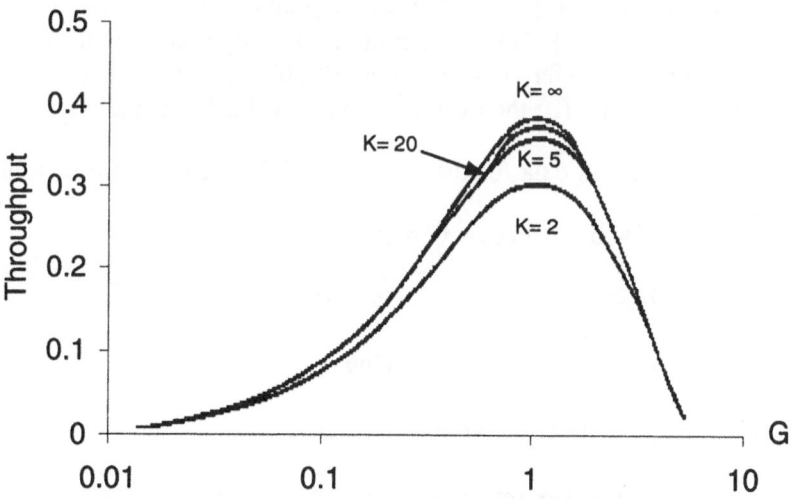

Figure 10-2. Throughput versus offered traffic for slotted ALOHA

variation with K occurs for $K < 20$.

Since the backoff delay is equivalent to the effective waiting time in a queue, the average number of packets in the backlog can be determined by Little's formula. Thus the average number of packets in the backlog is given by

$$N_b = M\lambda(T - P)$$

$$= S[(\frac{1 - q_n}{q_r})(\frac{K + 1}{2}) + \frac{1}{2}]$$

(10-10)

We see from (10-7) that the normalized average transfer delay of slotted ALOHA depends principally on the average retransmission delay of the packet. This fact has also been indicated by simulated studies.

Example 10-1. A slotted ALOHA network uses 1 Mbps lines to connect 100 stations. Each station has a Poisson input with an average rate of 0.1 packets per second. Each packet is 1000 bits long. The propagation delay and the collision detection time are negligible. The backoff algorithm randomly selects an integer k with equal probability from the set of 10 integers $\{ 0,1,2, \ldots 9 \}$. Under maximum throughput conditions, calculate (a) the average transfer delay using (7-30) and the more accurate expression (10-6), and (b) the average number of backlogs for both cases.

Solution. Given data for the slotted ALOHA network are as follows:

$M = 100$, $\lambda = 0.1$ packet/s, $C = 10^6$ bps,

$L = 1000$ bits, $K = 10$, $G = 1$, $S = 0.368$

Then $P = \bar{X} = 10^3 / 10^6 = 1\ ms$

The average number of retransmissions is

$N_r = \frac{G}{S} - 1 = 1/0.368 - 1 = 1.7174$

(a)   Neglecting the propagation delay and the collision detection time, we obtain from (7-30):

$$T = (1.5 + 1.7174 \times \frac{10 + 1}{2}) \times 1 = 8.7174 \ ms$$

Now, using (10-1) and (10-2), we find

$$q_n = (e^{-1/10} + \frac{1}{10} e^{-1})^{10} e^{-0.368} = 0.3792$$

and

$$q_r = \frac{(e^{-1/10} - e^{-1})}{1 - e^{-1}} (e^{-1/10} - \frac{1}{10} e^{-1})^{10-1} e^{-0.368} = 0.342$$

Using (10-6), we obtain

$$T = (1.5 + \frac{1 - 0.3792}{0.342} \times \frac{10 + 1}{2}) \times 1 = 11.4836 \ ms$$

(b)   From (7-30), we find the average number of backlogs:

$$N_b = S(\hat{T} - 1) = 0.368 (8.7174 - 1) = 2.84 \ \text{packets}$$

and the more accurate average number of backlogs is

$$N_b = 0.368(11.4836 - 1) \times 10^{-3} = 3.858 \ \text{packets}$$

10-3. PERFORMANCE ANALYSIS OF SLOTTED ALOHA WITH A FINITE NUMBER OF STATIONS

Because the slotted ALOHA system with Poisson input is always unstable, the performance analysis in the last section cannot treat the stability problem. In this section, we formulate a Markovian model for the slotted ALOHA system and obtain the stationary state probability distribution by calculating its state transition probabilities.

To develop a Markovian model for the slotted ALOHA system, we make the following assumptions:

● There are M stations or users, with each user generating single-packet messages;

● Each station is in one of two modes: the think mode and the retransmit mode. A station in the think mode is idle and has no packet for transmission. Thus it generates and transmits a new packet at the beginning of a time slot with probability σ. A station is said to be in the retransmit mode if it has a collided packet waiting for retransmission. A collided packet is transmitted in the current time slot with probability p. A station in the retransmit mode cannot generate a new packet;

• The channel propagation delay is negligible or zero.

We adopt the zero channel propagation delay assumption here for simplification of the modeling. However, the channel propagation delay is the most important factor characterizing multiple-access broadcast communication in a distributed environment. In reality, we cannot ignore the influence of the channel propagation delay in the modeling and analysis. We shall discuss this problem again later. With the above assumptions, we depict a model for the system in Figure 10-3.

In the model, a station in either the think mode or the retransmit mode will enter the think mode in the next slot if its packet transmission is successful in the current slot. If the transmission is unsuccessful, it will enter the retransmit mode in the next slot.

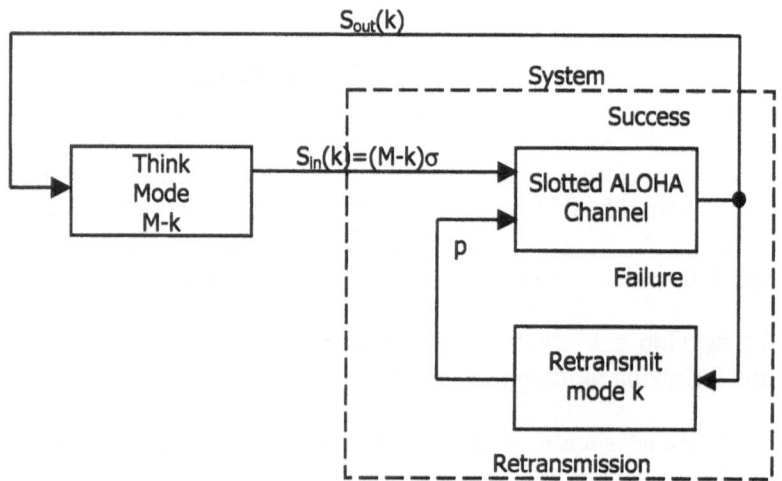

Figure 10-3. Model for slotted ALOHA system

Now let $N_R(t)$ be a random variable denoting the number of stations in the retransmit mode at time t. The system is said to be in state k at time t if $N_R(t) = k$. Thus, $N_R(t)$ is a discrete-time Markov chain with the discrete state space consisting of the set of integers $\{0, 1, 2, \cdots, M\}$. Let the one-step state transition probabilities be

$$p_{ij} = P\{N_R(t+1) = j \,|\, N_R(t) = i\}\,, i,$$

$$(10\text{-}11)$$

$$j = 0, 1, 2, \cdots, M$$

To calculate the state transition probabilities, we consider the following cases:

(a) For $j \le i - 2$, $p_{ij} = 0$. Since there can be only one successful transmission in a time slot, the value of $N_R(t)$ can decrease at most by one.

(b) For $j = i - 1$, the transition from state i to state j occurs when no new packet is generated and a retransmitted packet is successful. Thus,

$$p_{ij} = \binom{i}{1} p(1-p)^{i-1}(1-\sigma)^{M-i}\,, j = i-1$$

(c) For $j = i$, the transition from state i to state j occurs under two conditions: (i) when there is only one new transmission and there is no retransmission, or (ii) when there are no new transmissions and zero or at least two retransmissions. Thus

$$p_{ii} = \binom{M-i}{1}\sigma(1-\sigma)^{M-i-1}(1-p)^i + [1 - \binom{i}{1}p(1-p)^{i-1}](1-\sigma)^{M-i}$$

(d) For $j = i + 1$, the transition from state i to state j occurs when there is one new transmission and at least one retransmission. Thus,

$$p_{ij} = \begin{bmatrix} M - i \\ 1 \end{bmatrix} \sigma(1 - \sigma)^{M-i-1}[1 - (1 - p)^i], \, j = i + 1$$

(e) For $j \geq i + 2$, the transition from state i to state j occurs when there are $j - i \geq 2$ new transmissions. Thus,

$$p_{ij} = \begin{bmatrix} M - i \\ j - i \end{bmatrix} \sigma^{j - i} (1 - \sigma)^{M-j}, \, j \geq i + 2$$

The stationary state probability distribution $\{p_j\}$, $j = 0, 1, \ldots, M$ for the Markov chain, where

$$p_j = \lim_{t \to \infty} P\{N_R(t) = j\} = P\{N_R = j\} , \, j = 0, 1, 2, \ldots, M$$

can be computed by solving the following set of linear simultaneous equations:

$$p_j = \sum_{i=0}^{M} p_{ij} \, p_i, \, j = 0, 1, 2, \ldots, M \qquad (10\text{-}12)$$

and

$$\sum_{j=0}^{M} p_j = 1 \qquad (10\text{-}13)$$

Note that one of the equations in (10-12) is linearly dependent on the others with the normalization condition (10-13).

10-3-1. Throughput Analysis. We shall now calculate the throughput S that is defined as the average number of successfully transmitted packets per transmission time of a packet. Denote by $S_{out}(k)$ the conditional throughput, given that $N_R = k$. From the formula of total probability, we calculate the mean throughput

$$S = \sum_{k=0}^{M} S_{out}(k) \, p_k \qquad (10\text{-}14)$$

Since p_k are known, it remains to determine the conditional throughput $S_{out}(k)$ for all k. Given that $N_R = k$, $S_{out}(k)$ is equal to the conditional probability of exactly one (successful) packet transmission in a time slot. Thus, we have

$$S_{out}(k) = (1-p)^k(M-k)\sigma(1-\sigma)^{M-k-1} +$$

$$\qquad\qquad (10\text{-}15)$$

$$kp(1-p)^{k-1}(1-\sigma)^{M-k}$$

Therefore, substitution of $S_{out}(k)$ into (10-14) yields the mean throughput S.

10-3-2. Average Transfer Delay. Since the average channel input rate is

$$S_{in} = \sum_{k=0}^{M} S_{in}(k)p_k = (M - \bar{N}_R)\sigma \qquad (10\text{-}16)$$

where

$$\bar{N}_R = \sum_{k=0}^{M} kp_k \qquad \text{and} \qquad S_{in}(k) = (M-k)\sigma \qquad (10\text{-}17)$$

then, by applying Little's formula to the model in Figure 10-3, we obtain

$$T = \frac{\bar{N}_R}{\lambda_{in}} + P = \frac{\bar{N}_R P}{S_{in}} + P \qquad (10\text{-}18)$$

where $S_{in} = \lambda_{in} P$.

The second term P on the right-hand side of (10-18) represents the first transmission time of the packet. Under steady-state conditions,

$$S_{in} = S \tag{10-19}$$

From (10-16) to (10-19), we finally obtain the average transfer delay

$$T = \frac{MP}{S} - \frac{P}{\sigma} + P \tag{10-20}$$

Dividing (10-20) by P yields the normalized average transfer delay:

$$\hat{T} = \frac{M}{S} - \frac{1}{\sigma} + 1 \tag{10-21}$$

As seen from the above analysis, the retransmission probability p is implicitly contained in the expression for S and $\hat{T}$. We can investigate the effect of the values of p on the system performance by plotting the throughput S and the normalized average transfer delay $\hat{T}$ versus p for fixed values of M and σ. A typical plot is shown in Figure 10-4 for $M = 100$ and $\sigma = 0.0042$.

We see from Fig. 10-4 that as p increases to a certain threshold value at about 0.043, the throughput decreases and the average transfer delay increases very rapidly with p. This indicates that the system becomes unstable.

We have shown that the Markov chain model can be used to study the dynamic behavior and the performance analysis of multiple access protocols. However, it is very difficult to extend the Markov chain methods to more complicated protocols which are modeled as multi-dimensional Markov chains.

Example 10-2. In all of the analyses, we have assumed that the slotted ALOHA network is in statistical equilibrium. However, this assumption is not always satisfied. The major factors in determining stability of the slotted ALOHA are the number of stations M, the average

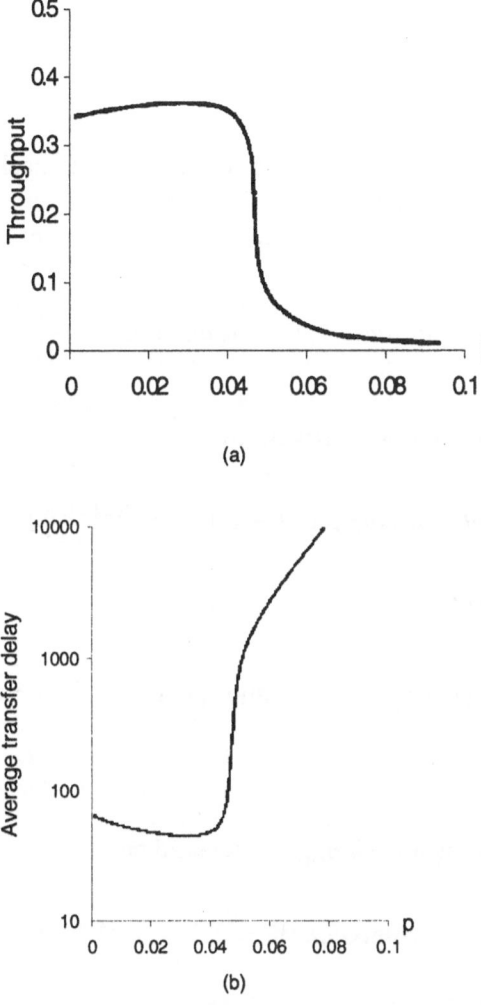

Figure 10-4. (a) Throughput versus p for slotted ALOHA
(b) Normalized average transfer delay versus
p for slotted ALOHA

backoff time $\bar{B}$, and the average number of stations in the retransmission mode, $\bar{N}_R$.

For simplicity, suppose that $\sigma = 1/Me$, $p = 1/M$, and M is large. (a) Assume that the system is in state k; that is, there are k stations in the retransmit mode. Determine the probability that a given packet transmission is successful, and calculate the throughput $S_{out}(k)$. (b) Determine the equilibrium point and the average transfer delay if the packet length is 1000 bits long, $M = 100$ stations, and the channel transmission rate is 1 Mbps.

Solution. (a) Since the probability that a given packet transmission is successful equals the probability that no other packets are transmitted in the same slot, then from (7-16) we obtain (when the system is in state k)

$$P \{ a \ successful \ ransmission \} = \frac{S(k)}{G(k)}$$

If the given packet is a new arrival, then

$$P \{ a \ successful \ transmission \} = (1 - \sigma)^{M-k-1} (1 - p)^k \approx e^{-G(k)}$$

where $G(k) = (M - k) \sigma + kp$ and the approximations for small x,

$$(1-x)^y \approx e^{-xy} \quad and \quad (1-x)^{y-1} \approx e^{-xy}$$

have been used.

If the given packet is a backlogged packet, then

$$P \{ a \ successful \ transmission \} = (1 - \sigma)^{M-k} (1 - p)^{k-1} \approx e^{-G(k)}$$

which has the same probability as the new arrival.

The throughput is equal to

$$S_{out}(k) = G(k) e^{-G(k)}$$

(b) When the system is in state k, the throughput to the system is

$$S_{in}(k) = (M - k)\sigma = \left(1 - \frac{k}{M}\right)\frac{1}{e}$$

or

$$\frac{k}{M} = 1 - e\, S_{in}(k)$$

Now we write

$$G(k) = 1 + (1 - e)\, S_{in}(k)$$

or

$$S_{in}(k) = \frac{1 - G(k)}{e - 1}$$

At equilibrium, $k = k_e$, $S_{in}(k) = S_{out}(k) = S$, $G(k) = G$.

We then have

$$G\, e^{-G} = \frac{1 - G}{e - 1}$$

Solving for G numerically, we find that

$$G = 0.4862$$

and

$$S = \frac{1 - G}{e - 1} = 0.3$$

Since $P = \bar{X} = 1\ ms$ and $\sigma = 1/Me$, we determine from (10-20) that

$$T = \frac{100}{0.3} - \frac{1}{100 \times 2.7183} + 1 = 334.3296\ ms$$

It is interesting to note from (10-20) that if the number of stations M increases to infinity, there will be no finite average transfer delay.

When studying the infinite-station case, a multi-access system is often defined as being stable for a given arrival rate if the average transfer delay per packet is finite. In this sense, the ordinary slotted ALOHA with a Poisson input is unstable.

10-4. PERFORMANCE ANALYSIS OF SLOTTED ALOHA BY EQUILIBRIUM POINT

In order to analyze more complicated protocols in slotted ALOHA, an analytical technique called the equilibrium point analysis may be used. The equilibrium point analysis assumes that the system is always at an equilibrium point. Therefore, the equilibrium point analysis does not involve the calculation of state transition probabilities.

Consider again the model in Figure 10-3. Let $N_R(t)$ be a random variable (called the channel backlog) denoting the number of stations in the retransmit mode at time t. The channel input rate at time t is

$$S(t) = [M - N_R(t)]\sigma \qquad (10\text{-}22)$$

Let $[N_R(t), S(t)]$ denote the channel state vector. From (10-15) and (10-17), we have

$$S_{out}(k) = (1 - p)^k (M - k)\sigma(1 - \sigma)^{M-k-1}$$
$$+ kp(1 - p)^{k-1}(1 - \sigma)^{M-k} \qquad (10\text{-}23)$$

and

$$S_{in}(k) = (M - k)\sigma \qquad (10\text{-}24)$$

An equilibrium point is defined as a point k_e at which the channel input rate is equal to the channel output rate. By equating (10-23) and (10-24), we write

$$S_{in}(k_e) = S_{out}(k_e) \tag{10-25}$$

Thus, the equilibrium point is a solution of (10-25). In solving this equation, k_e is assumed to be a real number in spite of the fact that the variable k is an integer. If the number of solutions is one, then the system is said to be stable. Otherwise, it is said to be unstable.

We plot $S_{in}(k)$ and $S_{out}(k)$ versus k in Figure 10-5, where an intersection of the straight line (the channel load line) representing $S_{in}(k)$ and the curve for $S_{out}(k)$ corresponds to an equilibrium point.

Figure 10-5(a) and Figure 10-5(b) represent a stable system because they have only one equilibrium point. The system corresponding to Figure 10-5(c) is unstable. It is important to note that the system corresponding to Figure 10-5(b) is stable but overloaded. Therefore, by a stable system we usually mean the one in Figure 10-5(a) rather than Figure 10-5(b).

From (10-22), we see that $N_R(t)$ and $S(t)$ are constrained to lie on the channel load line (10-24). In Figure 10-5, arrows on the channel load lines indicate directions of increasing backlog size k if $S_{in}(k) > S_{out}(k)$ and decreasing backlog size k if $S_{in}(k) < S_{out}(k)$. As shown in Figure 10-5(c), on an unstable equilibrium point such as the middle one, a small perturbation will drift $N_R(t)$ away from the equilibrium point. An unstable system always has two locally stable equilibrium points and one unstable equilibrium point. In that sense, an unstable system is also called a bistable system. Generally, a stable system stays near the stable equilibrium point all the time and a bistable system goes back and forth between the two locally stable equilibrium points.

Note that the equilibrium point has two interesting properties.

(a) The steady-state throughput-delay performance of a stable system is closely approximated by its stable equilibrium point;

(b) In an unstable system, the throughput-delay performance at a locally stable equilibrium point can be achieved only for some finite time period.

In equilibrium point analysis, the throughput S is approximated by the conditional expectation of the number of successful packet transmissions per packet transmission time given that the system is in state

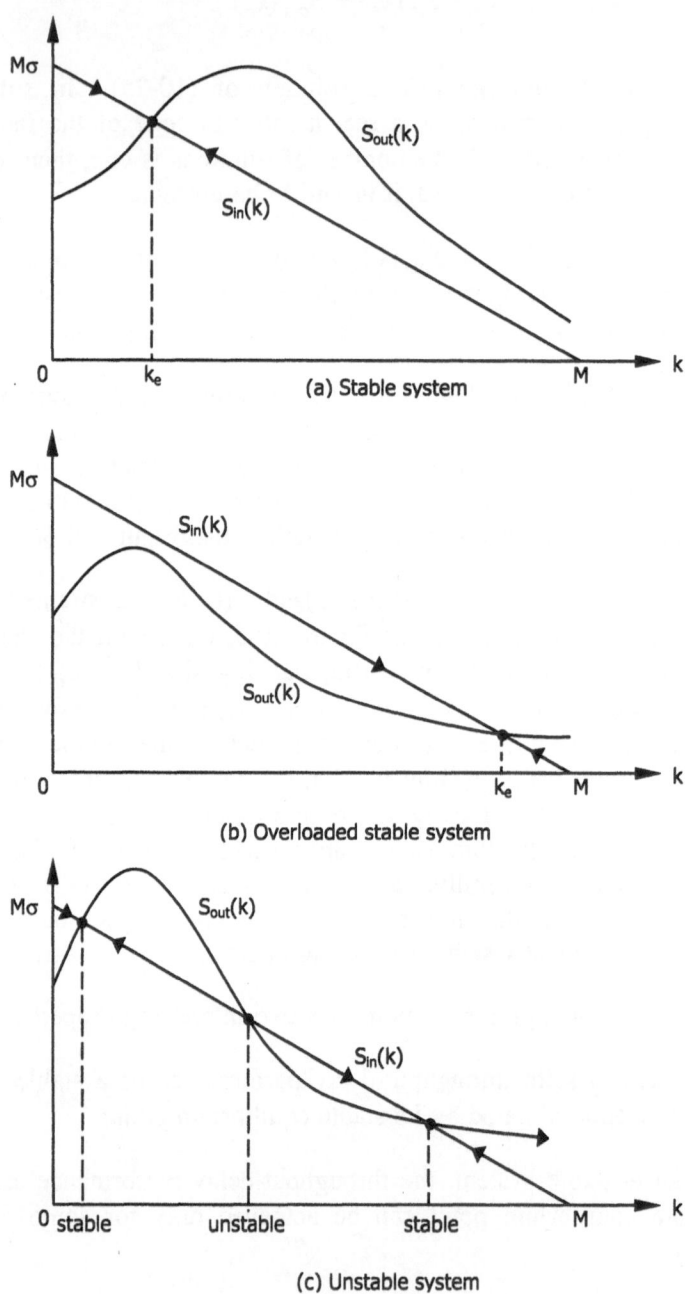

Figure 10-5. Stability of equilibrium points

k; that is,

$$S = E[S(k)] = S(k_e) \qquad (10\text{-}26)$$

Using $S(k_e)$, we can calculate the average transfer delay. For example, in the slotted ALOHA of Figure 10-3, we have from (10-21).

$$\hat{T} = \frac{M}{S(k_e)} - \frac{1}{\sigma} + 1 \qquad (10\text{-}27)$$

The Modified Model. When we apply the equilibrium point analysis to a Markovian model such as that of Figure 10-3, we often modify the model in order to simplify the analysis. We modify the model in Figure 10-3 by merging the two inputs to the slotted ALOHA channel into one. Since we usually assume bursty users, we consider the modification under the condition $\sigma < p$. We show the modified model in Figure 10-6, where the think mode in Figure 10-3 has been decomposed into two tandem phases, the think mode and the transmit mode, and where the retransmit mode in Figure 10-3 has become a part of the transmit mode in Figure 10-6.

A station which has just successfully transmitted a packet moves into either the think mode with probability $1 - \sigma/p$ or the transmit mode with probability σ/p. A station which has just had a collision enters the transmit mode with a probability of 1. A station in the think mode will move into the transmit mode at the next slot with probability σ, and a station in the transmit mode transmits a packet (that is, moves out from the transmit mode) with probability p. The modified model in Figure 10-6 is equivalent to the model in Figure 10-3 from the viewpoint of the probabilistic behavior.

Now we let $N_T(t)$ be a random variable denoting the number of stations in the transmit mode at time t. Thus, $N_T(t)$ is a Markov chain. From the modified model, the conditional throughput $S(k)$ in state k is given by

$$S(k) = kp(1-p)^{k-1}, \; k = 0, 1, 2, \cdots, M \qquad (10\text{-}28)$$

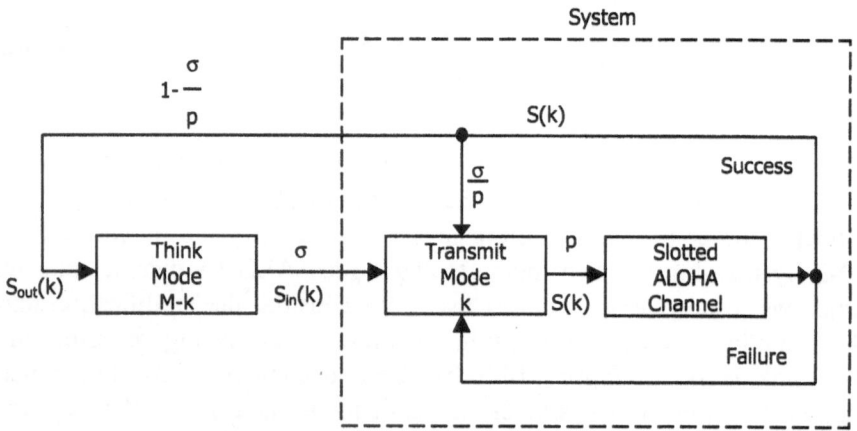

Figure 10-6. Modified model for the system in Figure 10-3 for $\sigma < p$.

and

$$S_{out}(k) = (1 - \frac{\sigma}{p}) S(k)$$

which has a simpler form than that of (10-15). Since

$$S_{in}(k) = (M - k)\sigma$$

then by applying the procedure for obtaining the equilibrium point to the modified model, we have

$$(M - k)\sigma - (1 - \frac{\sigma}{p})S(k) = 0 \qquad (10\text{-}29)$$

Solving this equation numerically for k under the condition that k is a nonnegative real number yields an equilibrium point k_e. Thus, the throughput S and the normalized average transfer delay $\hat{T}$ can be expressed as

$$S = S(k_e) = k_e p (1 - p)^{k_e - 1} \tag{10-30}$$

and

$$\hat{T} = \frac{M}{S(k_e)} - \frac{1}{\sigma} + 1 \tag{10-31}$$

respectively.

10-5. CARRIER-SENSE MULTIPLE ACCESS (CSMA) PROTOCOLS

For local area networks with stations located close together, propagation delays are usually much smaller than the packet transmission time. For such networks, it is feasible for a station that has a packet to transmit to listen to the channel to determine if it is busy before a transmission is attempted. If the channel is sensed busy, the station can defer its transmission until the channel becomes idle. This process is called carrier sensing. Networks with carrier sensing are called carrier-sense multiple access (CSMA) networks.

For satellite channels, carrier sensing is useless because the propagation delays are much greater than the packet transmission time. For networks with propagation delays much shorter than the packet transmission time, CSMA-type protocols can have much smaller average transfer delays and higher throughput than the ALOHA protocols, since the carrier sensing can reduce the number of collisions and the length of collision intervals.

The carrier sensing information can be used to reduce the number of collisions. There are several ways to make use of this information. We may divide CSMA protocols into two general classes: non-persistent and p-persistent. Operations for each of these classes can be slotted or nonslotted.

For slotted operation, the time axis is divided into slots of τ_p seconds, where τ_p is the end-to-end propagation delay of a bus or the maximum round-trip for a station on a broadband tree network. Like the slotted ALOHA, all stations for slotted CSMA operation must be synchronized and can start transmission only at the beginning of a slot.

Persistent and Nonpersistent CSMA Protocols. We describe CSMA protocols with the flow chart of Figure 10-7(a), which is an extension of that in Figure 10-1 to include a carrier-sense box. For slotted operation, all time delays are quantized to slots of τ_p seconds.

Note that the flow chart of Figure 10-7(a) reduces to that of Figure 10-1 for slotted ALOHA if the carrier-sense box is replaced by a direct connection. The CSMA protocols differ from the slotted ALOHA in having the decisions based on carrier sensing. If, at a certain instant, a station has a packet ready for transmission, the packet is called a ready packet and the station is called a ready station, irrespective of whether the packet is a new or a retransmitted one.

Nonpersistent CSMA. With this protocol, collisions are reduced because a station that finds the channel busy always defers the transmission. When a station becomes ready, it senses the channel and carries out the following procedure:

1. If the station senses that the channel is idle, it transmits the packet;

2. If the channel is sensed busy, the station uses the backoff algorithm to reschedule the packet to a later time. After then, the station senses the channel again, and the procedure is repeated.

Persistent CSMA. Persistent CSMA protocols have two different versions: 1-persistent CSMA and p-persistent CSMA. The 1-persistent CSMA never allows the channel to go unused when there is a ready station. The operation is as follows. A ready station senses the channel and then carries out the following procedure:

1. If the station senses that the channel is idle, it transmits the packet;

2. If the station senses the channel to be busy, it keeps on sensing the channel until the channel becomes idle and then transmits the packet.

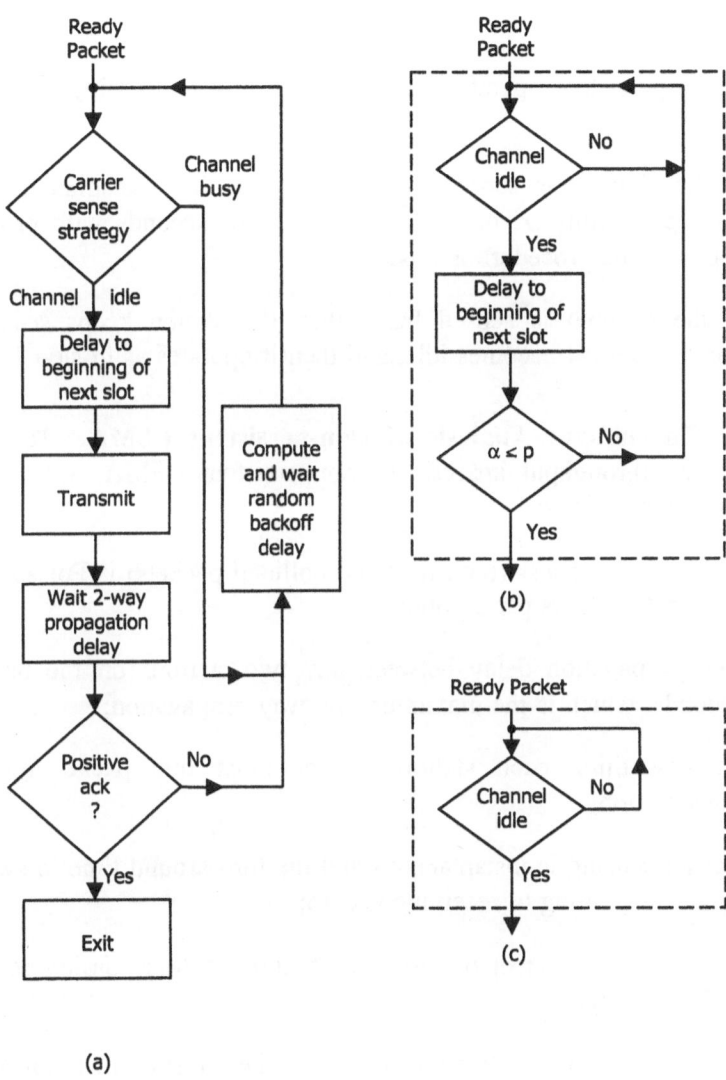

(a)

(b)

(c)

Figure 10-7 (a) Flow diagram for non-persistent CSMA protocols
(b) Carrier sense box for p-persistent CSMA
(c) Carrier sense box for non-persistent CSMA

When two or more ready stations sense the channel to be busy, with 1-persistent CSMA they all wait until the channel becomes idle and then transmit. In this case, a collision must occur (with probability 1). If we reduce this probability to p, $0 < p < 1$, the chance of a collision is reduced.

For p-persistent operation, a ready station after sensing the channel, operates as follows:

1. If the channel is sensed idle, then the station transmits the packet with probability p, or the station waits τ_p seconds with probability 1-p, and the procedure is repeated;

2. If the channel is sensed busy, then the station keeps sensing the channel until it becomes idle, and then it operates as in step 1.

10-5-1. Throughput Analysis of Non-persistent CSMA. To give an approximate throughput analysis of nonpersistent CSMA, we make the following assumptions:

•   The arrival process (both new and collided packets) is Poisson with a rate of Λ packets per second;

•   The propagation delay between any two stations on the bus is τ_p seconds, which is the maximum one-way propagation;

•   At any time, each station has at most one packet ready for transmission;

•   Carrier sensing is instantaneous and the turn-around time in switching from transmitting to receiving is zero;

•   The channel is error-free so that failure of transmission is due to collision only;

•   The overlap of any fraction of two packets results in retransmission of both packets.

Let $G = \Lambda P$ be the offered traffic and S be the throughput, which is defined as the number of successful transmissions per packet transmission time. A model of the system is shown in Figure 10-8.

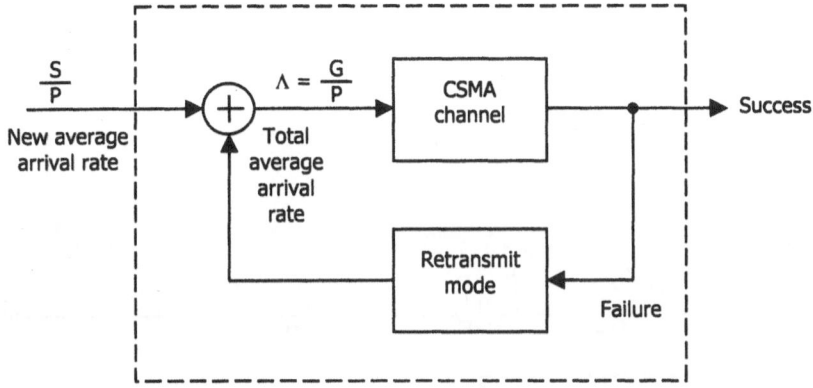

Figure 10-8. Model of CSMA channel

Suppose that at time t, the channel is idle and packet 0 arrives as shown in Figure 10-9. This packet is transmitted immediately with a transmission time of P seconds. Since the propagation time is τ_p, the other stations do not sense packet 0 before time $t + \tau_p$. If these stations have arrivals in the time interval $[t, t + \tau_p)$, they transmit the packets immediately. Packet n is the last packet that arrives at a random time Y in $[t, t + \tau_p)$. After time $t + \tau_p$, all stations have sensed the channel to be busy. According to the nonpersistent CSMA protocols, further arrivals such as packet $n + 1$ will be rescheduled for transmission at a later time. Transmission of packet n is completed at time $t + Y + P$. All stations sense the end of packet n at time $t + Y + P + \tau_p$, but up to this time, all arrivals after packet n are rescheduled for transmission at a later time.

After time $t + Y + P + \tau_p$, all stations sense the channel to be idle. The next transmission at t_1 initiates another cycle. The intervals $[t, t + Y + P + \tau_p)$ and $[t + Y + P + \tau_p, t_1)$ are called a busy period and an idle period, respectively. A cycle consists of a busy period and an idle period.

The busy period shown in Figure 10-9 is an unsuccessful busy period because there are more than one arrival in $[t, t + \tau_p)$. On the other

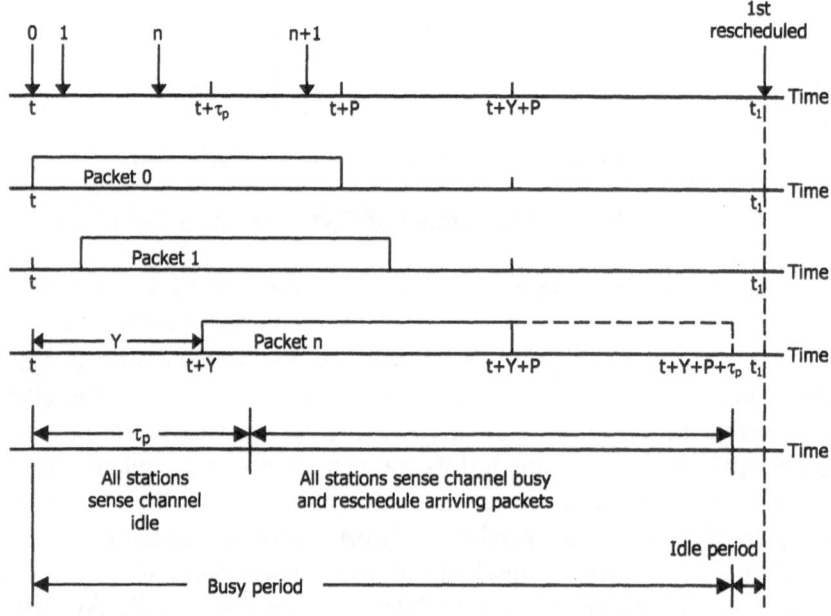

Figure 10-9. A cycle containing an unsuccessful busy period
for non-persistent CSMA

hand, if packet 0 is the only arrival in $[\, t, t + \tau_p \,)$, then the busy period is
a successful busy period.

From this observation, we see that the vulnerable period for CSMA
protocols is the maximum one-way propagation delay τ_p. This is an
improvement when compared with vulnerable periods of P and $2P$ for the

slotted ALOHA and the pure ALOHA, respectively, provided that $\tau_p < P$.

Using the notion of busy-idle period (or cycle) just described, we can calculate the throughput for the non-persistent CSMA channel. Let V denote the length of time for successful transmissions in a cycle, and let $\bar{V}$ denote its average value. Since the average length of a cycle is $\bar{B} + \bar{I}$, where $\bar{B}$ and $\bar{I}$ are the average values of the busy period and the idle period, respectively, the throughput can be expressed as

$$S = \frac{\bar{V}}{\bar{B} + \bar{I}} \tag{10-32}$$

This expression follows from the fact that under steady-state conditions, the throughput (the average number of successful transmissions) S is equal to the ratio of the average time for transmitting successful packets to the average cycle time. It remains to find $\bar{V}$, $\bar{B}$, and $\bar{I}$.

Since the average time for successful transmissions in a cycle equals the product of the probability of successful transmissions and the packet transmission time, we write

$$\bar{V} = P\{packet\ 0\ is\ successfully\ transmitted\} \times P \tag{10-33}$$

Furthermore, since packet arrivals are Poisson with an average rate of G/P packets per second, and since the probability that packet 0 is successfully transmitted is equal to the probability that there are no other arrivals in $[t, t + \tau_p)$, we have

$$P_s = P\{packet\ 0\ is\ successfully\ transmitted\} = e^{-G\tau_p/P} \tag{10-34}$$

and hence

$$\bar{V} = Pe^{-G\tau_p/P} \tag{10-35}$$

Referring to Figure 10-9, note that the busy period B has the length

$$B = P + \tau_p + Y \qquad\qquad (10\text{-}36)$$

where Y is a random variable. It is not necessary to differentiate between a successful and unsuccessful transmission or collision in (10-36) because the busy period B here includes successful transmissions as a special case; that is, $Y = 0$. Thus,

$$\bar{B} = P + \tau_p + \bar{Y} \qquad\qquad (10\text{-}37)$$

To find the distribution function of Y, we note that $t + Y$ is the last packet arrival time in $[t, t + \tau_p)$, so that $P\{Y \leq y\}$ is the probability of no arrivals in $[y, \tau_p)$. Thus,

$$P\{Y \leq y\} = e^{-G(\tau_p - y)/P}, \quad 0 \leq y \leq \tau_p \qquad\qquad (10\text{-}38)$$

This distribution function has a jump of size $e^{-G\tau_p/P}$ at $y = 0$.

The average value of Y is given by

$$\bar{Y} = \int_0^{\tau_p} y\,dP\{Y \leq y\} = \tau_p - \frac{P}{G}(1 - e^{-G\tau_p/P}) \qquad\qquad (10\text{-}39)$$

and from (10-37), the average busy period is

$$\bar{B} = P + 2\tau_p - \frac{P}{G}(1 - e^{-G\tau_p/P}) \qquad\qquad (10\text{-}40)$$

To compute the average idle period $\bar{I}$, we recall that a Poisson arrival process with mean rate G/P and an exponential interarrival time with mean P/G are equivalent. The idle period is simply the time interval between the end of a busy period and the next arrival epoch. Because of the Markov property (or memoryless property) of the exponential interarrival time, the average idle period and the mean interarrival time are equal. Thus,

$$\bar{I} = P/G \tag{10-41}$$

Substituting $\bar{V}$, $\bar{B}$, and $\bar{I}$ into (10-32) yields

$$S = \frac{Pe^{-G\tau_p/P}}{P + 2\tau_p - \dfrac{P}{G}(1 - e^{-G\tau_p/P}) + \dfrac{P}{G}} \tag{10-42}$$

or

$$S = \frac{Ge^{-\alpha G}}{G(1 + 2\alpha) + e^{-\alpha G}} \tag{10-43}$$

where

$$\alpha = \tau_p/P \tag{10-44}$$

In particular, if $\alpha = 0$ or $\tau_p = 0$, then

$$S = G/(1 + G) \tag{10-45}$$

From (10-43), we see that as α becomes small, S approaches this limiting value.

Curves showing the relationship between S and G from (10-43) are given in Figure 10-10 for several values of α.

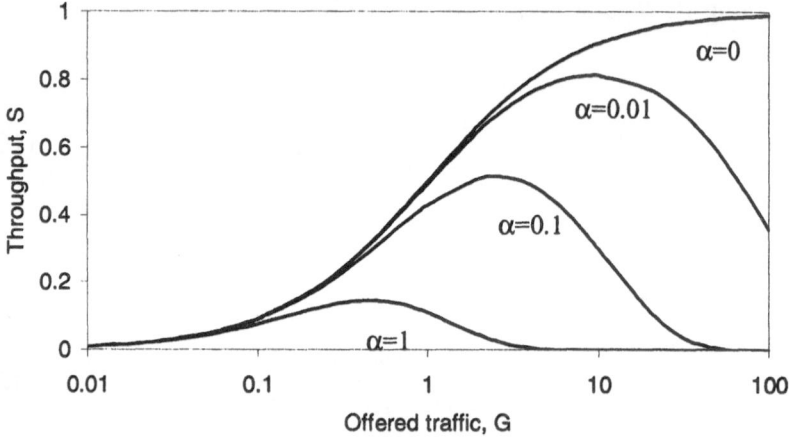

Figure 10-10. S versus G for non-persistent CSMA

10-5-2. Throughput Analysis of Slotted Non-persistent CSMA. For an approximate analysis of the slotted nonpersistent CSMA, in addition to the assumptions for the non-persistent CSMA, we assume that the time axis is divided into slots of equal length τ_p. A diagram for the slotted non-persistent CSMA, similar to Figure 10-9, is shown in Figure 10-11.

Suppose that the channel is idle at time 0 and there is at least one arrival in the first slot. Packet 0 is transmitted at time 1. If this is the only packet arrival in the first slot, then it is a successful transmission. Otherwise, the transmission is unsuccessful. In either case, the busy period is exactly $\tau_p + P$ seconds.

We shall compute S in terms of $\overline{V}$, $\overline{I}$ and $\overline{B}$ using (10-32). First, consider the events occurring in the first slot (0, τ_p) and their associated probabilities. Note that

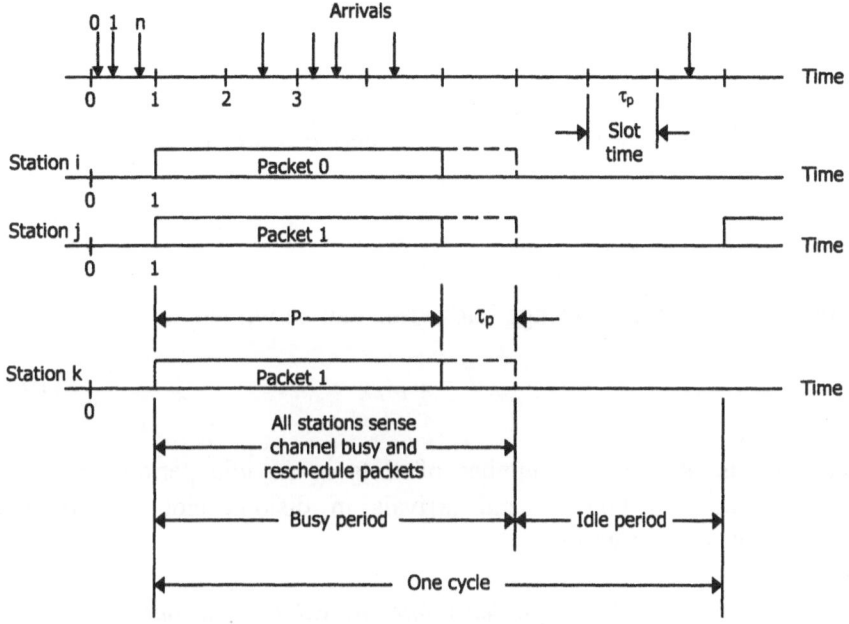

Figure 10-11. A cycle containing a busy period for slotted non-persistent CSMA

$$P_s = P \{packet \; 0 \; is \; a \; successful \; transmission\}$$

$$= P \{one \; arrival \; in \; (\; 0, \; \tau_p \;) \; given \; that \; some \; arrivals \; in \; (\; 0, \; \tau_p \;)\}$$

$$= \frac{P \{one \; arrival \; in \; (0, \; \tau_{p)} \; and \; some \; arrivals \; in \; (0, \; \tau_{p})\}}{P \{some \; arrivals \; in \; (0, \; \tau_{p})\}} \qquad (10\text{-}46)$$

$$= \frac{P \{exactly \; one \; arrival \; in \; [\; 0, \; \tau_p \;)\}}{P \{some \; arrivals \; in \; [\; 0, \; \tau_p \;) \;\}}$$

$$= \frac{\alpha G e^{-\alpha G}}{1 - e^{-\alpha G}}$$

Thus, using (10-33), we obtain

$$\bar{V} = \frac{P \alpha G e^{-\alpha G}}{1 - e^{-\alpha G}} \qquad (10\text{-}47)$$

From Figure 10-11, we see that the busy period is a deterministic quantity:

$$\bar{B} = P + \tau_p \qquad (10\text{-}48)$$

In the slotted case, the average idle period can be written as

$$\bar{I} = \bar{N}_I \tau_p \qquad (10\text{-}49)$$

where $\bar{N}_I$ is the average number of slots in an idle period. Since the arrival process is Poisson and arrivals in disjoint slots are mutually independent, we can write

$$P\{N_I = 0\} = P\{at\ least\ one\ arrival\ in\ \tau_p\ seconds\}$$
$$= 1 - e^{-\alpha G} = q \qquad (10\text{-}50)$$

and

$$P\{N_I = 1\} = P\{no\ arrival\ in\ the\ first\ slot\} \times$$

$$P\{at\ least\ one\ arrival\ in\ the\ second\ slot\} \qquad (10\text{-}51)$$

$$= (1 - q)q$$

In general, we obtain

$$P\{N_I = k\} = (1 - q)^k q\ ,\ k = 0, 1, \cdots \qquad (10\text{-}52)$$

Therefore, the mean value of N_I is

$$\bar{N}_I = \sum_{k=0}^{\infty} k(1-q)^k \quad q = \frac{1-q}{q} \tag{10-53}$$

Using (10-49), (10-50), and (10-53), we obtain

$$\bar{I} = \frac{\tau_p e^{-\alpha G}}{1 - e^{-\alpha G}} \tag{10-54}$$

Finally, substituting expressions (10-47), (10-48), and (10-54) for $\bar{V}$, $\bar{B}$ and $\bar{I}$, respectively in (10-32), we obtain

$$S = \frac{\alpha G e^{-\alpha G}}{1 + \alpha - e^{-\alpha G}} \tag{10-55}$$

Graphs of S versus G for several values of α are plotted in Figure 10-12.

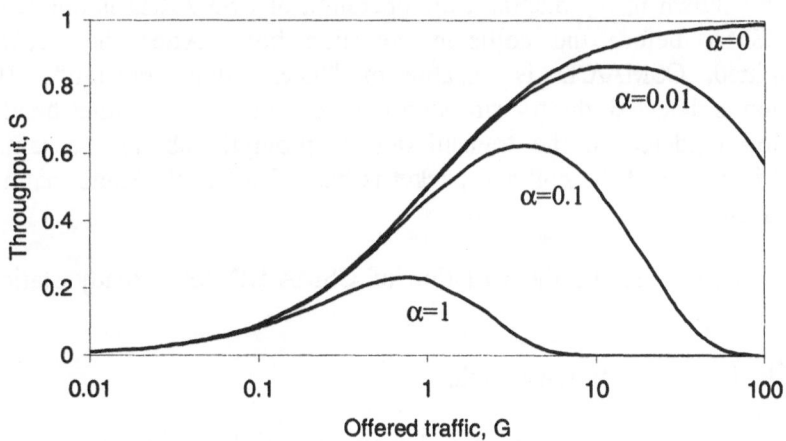

Figure. 10-12. S versus G for slotted non-persistent CSMA

10-6. CARRIER-SENSE MULTIPLE ACCESS WITH COLLISION DETECTION (CSMA/CD) PROTOCOLS

A further improvement on the carrier-sense multiple access protocols can be obtained by using collision detection. Collision detection, also called "listen while transmit", means to detect a collision shortly after it occurs. Thus, unsuccessful transmissions can be aborted promptly, minimizing the length of unsuccessful periods.

To implement the collision detection, the transceiver must monitor the channel before transmitting and also while transmitting. Like the CSMA protocols, the CSMA/CD protocols can have variations such as slotted and unslotted, and non-persistent and p-persistent. For all of these variations, the use of collision detection provides improved performances for local area networks with a short bus such that the normalized propagation delay is small. If a local area network does not have a small normalized propagation delay, CSMA/CD is not always the logical choice. Nevertheless, CSMA/CD is perhaps the most commonly used random access protocol for local area networks.

Basic Operation. The basic operation of the CSMA/CD can be described by the flow chart in Figure 10-13.

As shown in the diagram, the operation of CSMA/CD is the same as for CSMA before the collision detection box. After the packet is transmitted, CSMA/CD is capable of "listen while transmit". If no collision is detected, the transmission is successful. On the other hand, if a collision is detected, the transmission is promptly aborted, a jamming signal is sent, and the collided packet is backed off in the same manner as for CSMA.

We can describe the variation of CSMA/CD for a ready station as follows.

1.   If the channel is sensed idle:

*   for the nonpersistent and 1-persistent CSMA/CD, the packet is transmitted;

*   for p-persistent CSMA/CD, the packet is transmitted with probability p, and is delayed by τ_p seconds with probability $1 - p$;

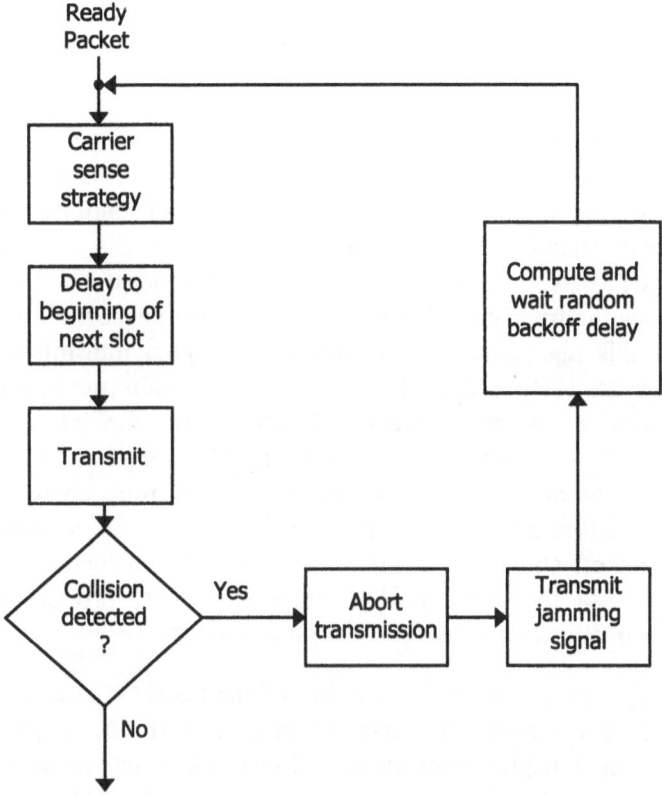

Figure 10-13. Flow chart for CSMA/CD

2. If the channel is sensed busy:

- for nonpersistent CSMA/CD, the packet is backed off and the procedure is repeated;

- for 1-persistent CSMA/CD, the packet is deferred until the channel is sensed idle and then transmitted; and

- for p-persistent CSMA/CD, the packet is deferred until the channel is sensed idle and then transmitted with probability p, and delayed by τ_p seconds with probability $1 - p$. In the latter case, the procedure is repeated after the τ_p seconds delay.

3.   If a collision is detected while transmitting, the transmission is aborted and a jamming signal is sent, and all stations involved in a collision backoff; and

4.   After the packet backs off, the procedure is repeated.

The jamming signal is not an essential feature of CSMA/CD. However, it is commonly used by local area networks such as Ethernet. The jamming signal is used to ensure that all other stations know of the collision and backoff. Commercial local area networks often use the 1-persistent CSMA/CD protocol with a backoff algorithm. For example, Ethernet uses the truncated binary exponential type backoff algorithm. One version of this algorithm allows only 16 attempted transmissions. After then, the packet is discarded. For each retransmission, the backoff time is $2r\tau_p$, where r is an integer selected from the set of integers $(0, 1, \cdots, 2^k - 1)$ with $k = $ min $(\ 10,$ number of attempted transmissions to date). In this case, the backoff time changes with the number of collisions. Thus, the backoff strategy is a dynamic one. Dynamic backoff strategy is sometimes called a linear incremental backoff strategy because the mean backoff time is linearly proportional to the number of collisions experienced by a given packet.

To illustrate the detailed operation of the CSMA/CD using jamming, consider the two-dimensional diagram in Figure 10-14. Suppose that at time 0, station A begins transmission. Since other stations will not detect the transmission until after the propagation delay of τ_p seconds, CSMA/CD has a vulnerable period of τ_p seconds. For most analyses, τ_p is taken to be the maximum end-to-end propagation delay of the channel. If two specific stations are considered as in Figure 10-14, τ_p is denoted by τ_{AB}. A station requires ε seconds to detect a collision.

Now consider that before the signal from station A arrives at station B, station B initiates a transmission at time Y, where Y is a random variable. Shortly after the transmission of station B, the signal from station A arrives, and ε seconds later, station B detects the collision and then sends out a jamming signal for J seconds. The jamming signal from station B reaches station A at time $Y + \tau_{AB} + \varepsilon$. Station A then sends out a jamming signal for J seconds. Station B will receive the complete jamming signal at time $Y + J + 2\tau_{AB} + \varepsilon$. During this period, the channel is always busy. After time $Y + J + 2\tau_{AB} + \varepsilon$, the channel becomes idle.

We can define busy period and idle period for the CSMA/CD channel in the same manner as for the CSMA channel. An idle period is a time interval during which the channel is not used by any station. A busy period is an unsuccessful busy (collision) period if more than one station transmits in the period. A busy period is a successful busy period if only one station transmits in the period. In Figure 10-14, the busy period $[0, Y + J + 2\tau_{AB} + \varepsilon]$ is an unsuccessful busy period.

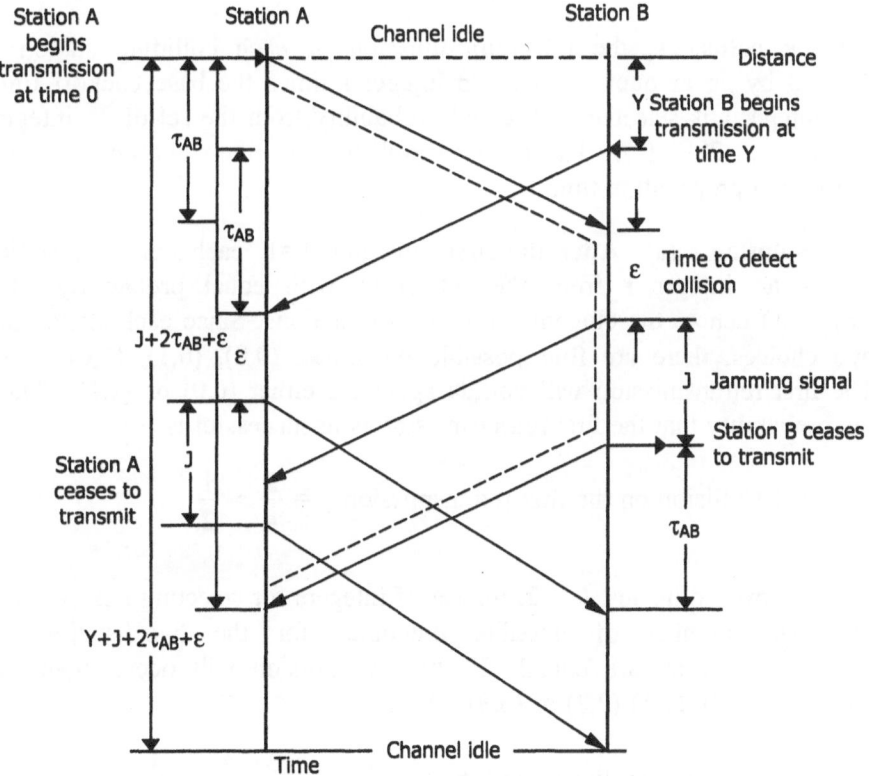

Figure 10-14. Time diagram for CSMA/CD

Example 10-3. Consider the transmission of two stations on a CSMA/CD network. Suppose that the backoff algorithm used is the binary exponential algorithm, where the propagation delay between the colliding stations is neglected. Find

(a) The probability that the first retransmission is unsuccessful;

(b) The probability that the second retransmission is unsuccessful, and the probability that the nth retransmission is unsuccessful; and

(c) The probability that the first successful transmission occurs on the nth retransmission.

The binary exponential backoff algorithm is described as follows:

After k collisions, the kth retransmission of each colliding station is delayed by an amount equal to an integer r times the base backoff time. The integer r is selected with equal probability from the set of 2^k integers $\{0, 1, \ldots, 2^k - 1\}$. The base backoff time is often chosen twice the end-to-end propagation time.

Solution. (a) After the first collision, k=1, each colliding station selects an integer r from the set $\{0,1\}$ with equal probability. Let (r_1, r_2) denote the outcomes for the two stations. Since each station has two choices, there are four possible outcomes: (0,0), (0,1), (1,0), (1,1). The first retransmission will collide again for either (0,0) or (1,1). Thus, the probability that the first retransmission is unsuccessful is

$$P \{ \text{Collision on the first retransmission} \} = \frac{2}{4} = \frac{1}{2}$$

(b) after two collisions, k = 2, the set of integers for selecting r is $\{0, 1, 2, 3\}$. The number of possible outcomes for the two stations is $2^2 \times 2^2 = 16$; all are equally likely. A collision will occur again for outcomes (0,0), (1,1) (2,2) or (3,3). Thus,

$$P \{ \text{Collision on the second retransmission} \} = \frac{4}{16} = \frac{1}{2^2}$$

In general, for k = n, the set of integers for selecting r is $\{0, 1, \ldots, 2^n - 1\}$ which has 2^n integers in total. There are $2^n \times 2^n = 2^{2n}$ possible choices of

(r_1 , r_2), of which 2^n cause another collision; that is,

$$(0,0) , (1,1) , \ldots , (2^n - 1 , 2^n - 1)$$

Thus, we obtain

$$P \{ \text{Collision on the nth retransmission} \} = \frac{2^n}{2^n \times 2^n} = \frac{1}{2^n}$$

(c) In order that the first successful transmission occurs on the nth retransmission, the first (n-1) retransmissions must result in collisions, and the nth retransmission (or equivalently the (n+1)th attempt) must be a successful transmission. Thus, we have

P { First successful transmission on the nth retransmission }

= P { Collision on each of first (n-1) retransmissions and no collision on the nth retransmission }

$$= \prod_{k=1}^{n-1} P \{ \text{Collision on the kth retransmission} \} \times P \{ \text{no collision on}$$
the nth retransmission }

$$= \frac{1}{2} \frac{1}{2^2} \cdots \frac{1}{2^{n-1}} \left(1 - \frac{1}{2^n} \right)$$

$$= \frac{1}{2^{(n-1)n/2}} \left(1 - \frac{1}{2^n} \right)$$

Throughput Analysis for Non-persistent CSMA/CD. To give a throughput analysis for non-persistent CSMA/CD, we make the following assumptions in addition to those given in Section 10-5:

- The time ε required to detect a collision is negligible; and

- The jamming time J is the same for all stations.

We use the same approach in (10-32) to calculate the throughput S in terms of $\bar{V}$, $\bar{B}$ and $\bar{I}$. Consider first the average busy period $\bar{B}$. Since a busy period can be either a successful transmission period with probability

P_s and length $P + \tau_p$, or a collision period (unsuccessful period) with probability $1 - P_s$ and average length $\bar{C}$, we write

$$\bar{B} = P_s(P + \tau_p) + (1 - P_s)\bar{C} \qquad (10\text{-}56)$$

where P_s is the probability of a successful transmission, and C is the length of a contention period. Note that P_s is equal to the probability of having no arrivals in the vulnerable period $[0, \tau_p)$. Thus,

$$P_s = e^{-G\tau_p/P} = e^{-\alpha G} \qquad (10\text{-}57)$$

It remains to compute the average length of a collision period $\bar{C}$. To simplify the calculation, we let τ_p be the propagation delay between every pair of stations. This assumption implies that all adjacent stations are equally spaced. Figure 10-15 shows three colliding transmissions.

In the figure, suppose that station 1 begins a transmission at time 0, while stations 2 and 3 begin transmission at times X and Y respectively, with X being less than Y. Observe that the instant at which the first packet collides with the packet of station 1 determines the length of the collision period C. Thus,

$$C = J + 2\tau_p + X \qquad (10\text{-}58)$$

and

$$\bar{C} = J + 2\tau_p + \bar{X} \qquad (10\text{-}59)$$

To compute the distribution function of X, it is useful to think of X as the first arrival time in $[0, \tau_p)$, given that there is at least one arrival in this interval. Thus, $P\{X > x\}$ is the conditional probability of no arrivals in $[0, x)$, given that at least one arrival occurs in $[0, \tau_p)$. We can write

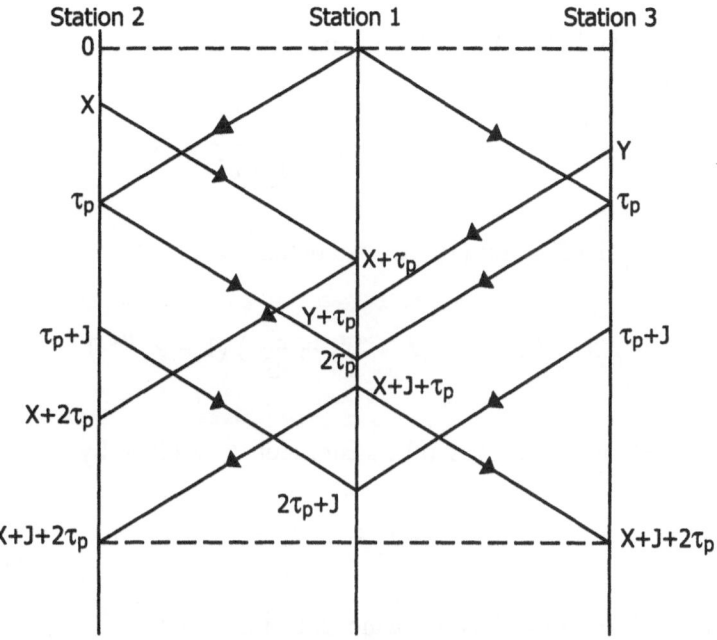

Figure 10-15. Three colliding transmissions in a collision period for CSMA/CD channel.

$$P\{X>x\} = \frac{P\{no\ arrivals\ in\ [0,x)\} \times P\{at\ least\ one\ arrival\ in\ [x,\tau_p)\}}{P\{at\ least\ one\ arrival\ in\ [0,\tau_p)\}}$$

$$(10\text{-}60)$$

$$= \frac{e^{-Gx/P}\ [\ 1 - e^{-G(\tau_p - x)/P}\ ]}{1 - e^{-\alpha G}}, \qquad 0 \leq x \leq \tau_p$$

by using the fact that Poisson arrivals in nonoverlaped time intervals are independent.

The average value of $\bar{X}$ is given by

$$\bar{X} = - \int_0^{\tau_p} x dP \{X > x\} = \left[\frac{1}{\alpha G} - \frac{e^{-\alpha G}}{1 - e^{-\alpha G}} \right] \tau_p \qquad (10\text{-}61)$$

Using (10-59) and (10-61), $\bar{C}$ can be expressed as

$$\bar{C} = J + (2 + \frac{1}{\alpha G} - \frac{e^{-\alpha G}}{1 - e^{-\alpha G}}) \tau_p \qquad (10\text{-}62)$$

From (10-56), (10-57) and (10-62), we obtain

$$\bar{B} = Pe^{-\alpha G} + (J + 2\tau_p + \frac{P}{G})(1 - e^{-\alpha G}) \qquad (10\text{-}63)$$

The average time of a successful transmission $\bar{V}$ is given by

$$\bar{V} = P_s P = Pe^{-\alpha G} \qquad (10\text{-}64)$$

The average idle period $\bar{I}$ is the same as that of (10-41):

$$\bar{I} = P/G \qquad (10\text{-}65)$$

Using these results in (10-32) yields

$$S = \frac{Ge^{-\alpha G}}{(G - 1)e^{-\alpha G} + \alpha G (2 + J/\tau_p)(1 - e^{-\alpha G}) + 2} \qquad (10\text{-}66)$$

Throughput Analysis for Slotted Non-persistent CSMA/CD. For slotted non-persistent CSMA/CD, the throughput analysis is closely parallel to that for the slotted non-persistent CSMA presented in Section 10-5.

From (10-47), we have the average length of time for a successful transmission:

$$\bar{V} = \frac{P\alpha Ge^{-\alpha G}}{1 - e^{-\alpha G}} \tag{10-67}$$

Since the length of the idle period does not depend on whether or not the CSMA protocol has collision detection, we obtain from (10-54)

$$\bar{I} = \frac{\alpha Pe^{-\alpha G}}{1 - e^{-\alpha G}} \tag{10-68}$$

To compute $\bar{B}$ using (10-56), we require the probability of a successful transmission P_s in (10-46),

$$P_s = \frac{\alpha Ge^{-\alpha G}}{1 - e^{-\alpha G}} \tag{10-69}$$

We also need the average length of the collision period $\bar{C}$. Consider two stations, both transmitting a packet at time 0 (at the beginning of a slot). These transmissions certainly lead to a collision. Of course, τ_p is the slot time equal to the end-to-end propagation delay. For slotted operation, the jamming time J is also an integral number of slots. Figure 10-16 shows such a collision period for slotted CSMA/CD.

Examination of the figure indicates that the collision period has a constant length of $2\tau_p + J$. Thus, the average length of the collision period is

$$\bar{C} = 2\tau_p + J \tag{10-70}$$

and from (10-56), (10-69) and (10-70), we find the average length of the busy period:

$$\bar{B} = \frac{\alpha Ge^{-\alpha G}}{1 - e^{-\alpha G}}(P + \tau_p) + \left[1 - \frac{\alpha Ge^{-\alpha G}}{1 - e^{-\alpha G}}\right](2\tau_p + J) \tag{10-71}$$

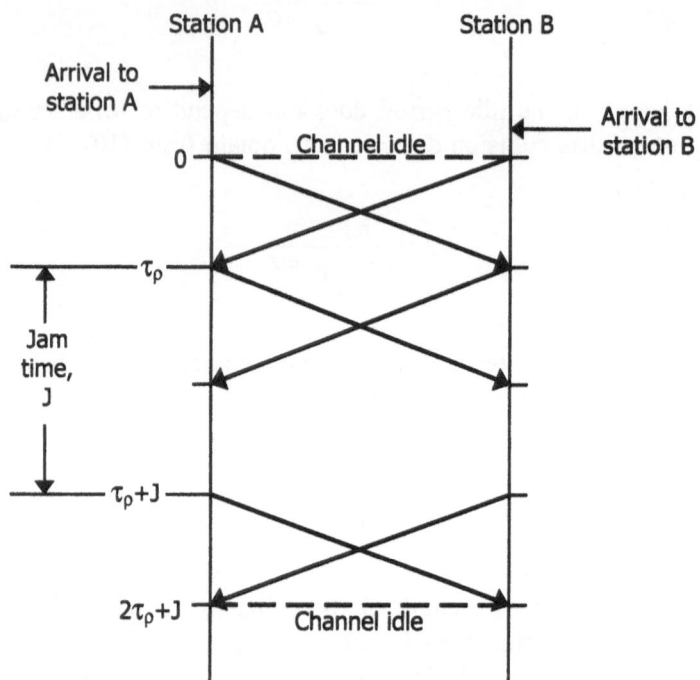

Figure 10-16. Collision period for slotted CSMA/CD

Using the results for $\bar{V}, \bar{B}$ and $\bar{I}$ in (10-32) yields

$$S = \frac{\alpha G e^{-\alpha G}}{\alpha G e^{-\alpha G} + \alpha + \alpha(1 + \gamma)(1 - e^{-\alpha G} - \alpha G e^{-\alpha G})} \quad (10\text{-}72)$$

where $\gamma = J/\tau_p$.

It is interesting to note that if

$$P = \tau_p + J$$

then

$$\gamma = \frac{J}{\tau_p} = \frac{1}{\alpha} - 1$$

Introducing $1/\alpha - 1$ for γ in (10-72) yields

$$S = \frac{\alpha G e^{-\alpha G}}{1 + \alpha - e^{-\alpha G}}$$

which is identical to the throughput in (10-55) for slotted non-persistent CSMA.

Note that the throughput for non-persistent CSMA/CD does not reduce to the result for non-persistent CSMA. This is because the length of the collision period for CSMA/CD depends on the arrival time of the first colliding packet, while that for CSMA depends on the last colliding packet.

10-7. SUMMARY

This chapter deals with random access networks, which do not have a channel access control mechanism for controlling when a station can transmit. The basic random access method was discussed in Section 7-5 for using the pure ALOHA network with Poisson input. The chapter begins with the slotted ALOHA with Poisson input. Unlike the pure ALOHA, the slotted ALOHA allows packets to be transmitted only at the beginning of the next slot after arrival. With this restriction, the slotted ALOHA method is not a completely random access method in the strictest sense; it also requires that all stations maintain slot synchronization. A drawback of the slotted ALOHA system with Possion input is that the system is always unstable.

For stability considerations, the chapter treats a slotted ALOHA system with a finite number of stations. Generally, the throughput as well as the average transfer delay depends on the retransmission probability. Instability can occur when the retransmission probability exceeds a certain threshold value. By assuming the existence of an equilibrium point in the system, it is possible to apply the Markov chain method for stability considerations and performance analysis.

If hardware at each station is available for listening to the channel and determining whether the channel is free before transmission, then carrier-sense methods can be employed. These methods are carrier-sense multiple access (CSMA) and carrier-sense multiple access with collision detection (CSMA/CD). Both methods can be further classified into nonpersistent and p-persistent, and slotted and nonslotted. By means of CSMA or CSMA/CD, the maximum throughput can be increased.

By adding still another hardware feature of monitoring the channel before transmitting and also while transmitting to a station, further improvement in maximum throughput over the CSMA is possible. Generally speaking, CSMA/CD performs better than CSMA. Application of both CSMA and CSMA/CD requires that the normalized propagation delay over the channel be small. Small normalized propagation delays are typical for local area networks. Thus, significant improvements in maximum throughput can be obtained.

REFERENCES

[1]   Lam, S.S., "Packet Switching in a Multiple-Access Broadcast Channel with Application to Satellite Communications in a Computer Network", Ph.D. dissertation, Computer Science Department, University of California, Los Angeles, March 1974.

[2]   Kleinrock, L., Queueing Systems, Vol. 2, Computer Applications, New York: John Wiley & Sons, 1976.

[3]   Kleinrock, L. and Lam, S.S., "Packet Switching in a Multiaccess Broadcast Channel: Performance Evaluation", IEEE Trans. Communications, Vol. COM-23, No. 4, April 1975, pp. 410-423.

[4]   Lam, S.S. and Kleinrock, L., "Packet Switching in a Multiaccess Broadcast Channel: Dynamic Control Procedures", IEEE Trans. Communications, Vol. COM-23, No. 9, September 1975, pp. 891-904.

[5]   Lam, S.S. and Tobagi, F.A., "Packet Switching in Radio Channels: Part I - Carrier Sense Multiple Access Methods and Their Throughput - Delay Characteristics", IEEE Trans. Communications, Vol. COM-23, No. 12, December 1975, pp. 1400-1416.

[6] Tobagi, F.A. and Kleinrock, L., "Packet Switching in Radio Channels: Part IV - Stability Considerations and Dynamic Control in Carrier Sense Multiple Access", IEEE Trans. Communications, Vol. COM-25, No. 10, October 1977, pp. 1103-1119.

[7] Hammond, J.L. and O'Reilly, P.J.P., Performance Analysis of Local Computer Networks, Reading, Mass.: Addison-Wesley, 1986.

[8] Tobagi, F.A. and Hunt, V.B., "Performance Analysis of Carrier Sense Multiple Access with Collison Detection", Computer Networks, Vol. 4, 1980, pp. 245-259.

[9] Tasaka, S., Performance Analysis of Multiple Access Protocols, Cambridge, Mass.: MIT Press, 1986.

[10] Massey, J.L., "Some New Approaches to Random-Access Communications", in Multiple Access Communications, ed. Abramson, N., Pitscataway, Mass.: IEEE, 1993.

[11] Capetanakis, J.I., Tree Algorithms for Packet Broadcast Channels", IEEE Trans. Inform. Theory, Vol. IT-25, No. 5, September 1979, pp. 505-515.

[12] Tsybakov, B.S. and Bakirov, V.L., "Packet Transmission in Radio Networks", in Multiple Access Communications, ed. Abramson, N., Piscataway, Mass.: IEEE, 1993.

[13] Bertsekas, D. and Gallager, R., Data Networks, 2nd ed., Englewood Cliff, N.J.: Prentice Hall, 1992.

PROBLEMS

10-1. A slotted ALOHA multi-access channel connects M nodes. Assume that when a node is in the transmit mode, it can accept a packet with probability σ, and when in the retransmit mode, it cannot accept a packet. The slot size is one time unit. Let

$N_R(k)$ = the number of nodes in the retransmit mode at the beginning of the kth slot; and

$$\bar{N}_R = \sum_{j=0}^{M} jp_j, \text{ where } p_j = P\{N_R(k) = j\}.$$

Note that $\bar{N}_R$ depends on the particular manner in which collisions are resolved, but we assume that $\bar{N}_R$ is given here.

(a)   Find the average number of accepted arrivals per slot N_a as a function of $\bar{N}_R$, M, and σ.

(b)   Find the average number of packets in the system N_{syst}, which is the average number of backlogs plus the average number of new arrivals during the previous slot.

(c)   Obtain an expression for the average transfer delay T.

10-2.   Assume, for simplicity, that each transmitted packet in a slotted ALOHA system is successful with some fixed probability γ. New packets are assumed to arrive at the beginning of a slot and are transmitted immediately. If a packet transmission is unsuccessful, it is retransmitted with probability p in each successive slot until successfully transmitted.

(a)   Find the normalized average transfer delay $\hat{T}$.

(b)   Suppose that the number of nodes M is large, and that σ and p are small. Show that in state k, the probability γ that a given packet transmission is successful is approximately equal to $e^{-G(k)}$, where $G(k) = (M - k)\,\sigma + kp$, and σ and p are defined in Section 10-3.

10-3.   An approximate expression between S and G for slotted ALOHA networks is $S = Ge^{-G}$.

(a)   Determine an expression for the average transfer delay per successful packet transmission in terms of G, P, and τ_p. The backoff algorithm is the binary exponential backoff that is used in Ethernet. This backoff algorithm can be described as follows: If a packet has been transmitted unsuccessfully k times, then the probability of transmission in successive slots is $p_k = 2^{-k}$. When a packet initially arrives at a station, it is transmitted immediately in the next slot.

(b)   Suppose that the network has the following parameter values:

- Channel bit rate = 1 Mbps;

- Average packet length = 1000 bits;

- $\tau_p = 25$ μs, S = 0.3;

- The detection delay for determining whether or not a transmission is successful is negligible.

  Determine the average transfer delay in slots or packet times.

10-4.   A slotted ALOHA network has M stations, each with an infinitely large buffer and a Poisson arrival of average rate λ packets per second. New arrivals are considered as backlogs immediately rather than transmitted in the next slot. When a station contains one or more packets, it independently transmits one packet in each slot with probability p. Assume that any given transmission is successful with probability p_S.

(a)   Show that the average time from the beginning of a backlogged slot until the completion of the first successful transmission at a given station is $\dfrac{1}{pp_S}$ and the second moment of this time is $\dfrac{(2 - pp_S)}{(pp_S)^2}$.

(b)   Note that the assumption of a constant probability for a successful transmission allows each station to be considered independently. Use the service time results of part (a) to show that the average transfer delay is

$$T = \frac{1}{pp_S (1 - \rho)} + \frac{1 - 2\rho}{2(1 - \rho)} , \rho = \frac{\lambda}{pp_S}$$

(c)   Assume that $p_S = 1$ (this assumption corresponds to very light loading and yields a smaller T than any other value of p_S). Calculate T for $p = \dfrac{1}{M}$ and compare it with the average transfer delay for TDMA if M is large.

10-5.   A group of personal computers uses a pure ALOHA access scheme to communicate over a bus network. All packets are 50 bytes long and the channel bit rate is 120 Kbps. If each computer generates 10 packets per minute on average, how many computers can the network support?

10-6.   Consider the transmission of two stations on an Ethernet that uses CSMA/CD (carrier sensing multiple access with collision detection) and a truncated binary exponential type of backoff algorithm which allows an initial attempt plus 15 retransmissions, each delayed by an integer r times the base backoff time. The interger r is selected randomly from a set of integers $\{0,1, \ldots, 2^{k-1}\}$, where k = min { number of retransmissions to date, 10}. After 16 attempts, the packet is discarded. Find the probability that a packet is discarded, and determine the average number of retransmissions per packet.

10-7.   A slotted CSMA/CD network has M stations operating with slots of identical length equal to twice the end-to-end propagation delay; that is, $2\tau_p$. Under heavy load conditions, assume that each station always has a packet in queue. The probability of a station initiating a transmission in a slot (at the beginning of the slot boundary) is σ.

(a)   Calculate the probability P_S of some station having a successful transmission in a given slot. What value of σ maximizes P_S?

(b)   What is the average number of slots wasted before some station has a successful transmission? Express the average contention time $\bar{C}$ in terms of σ, M, and τ_p.

(c)   Let P be the transmission time of a successful packet. Derive an expression for the throughput S in terms of P, σ, M, and τ_p, where $\tau_p \ll P$. Show that S is maximized with respect to σ by the same maximized value P_S as in part (a). Find an expression for the maximum throughput.

(d)   For a channel bit rate of 3 Mbps, packets of length 1024 bits, and propagation delay of 8 μs, calculate the maximum throughput when M = 4, 64, and 256. What limit does the maximum throughput approach as M tends to infinity?

APPENDIX A - Determination of q_n and q_r

Equation (10-5) expresses the average number of retransmissions N_r in terms of the probabilities q_n and q_r of (10-1) and (10-2), respectively.

To derive q_n and q_r, we first recall that after transmission of a packet, a station must wait r slots to determine if the transmission is successful or has collided. Consider the event of a successful transmission which takes place in the current slot, labelled slot C in Figure A-1. If the successfully transmitted packet in slot C is a retransmitted packet, it must have had a collision in an earlier slot, such as slot A. Thus, the collision must have taken place in one of the K earlier slots such that the current slot is $(r + k)$ slots after the transmission of the collided packet, where $0 \le k \le K-1$ and k is chosen at random from the set of integers, $\{0, 1, 2, ..., K - 1\}$.

For the retransmission to be successful, all of the following three events must occur:

1. No new packets are generated in slot C;

2. No other packets that collided in slot A are retransmitted in slot C; and

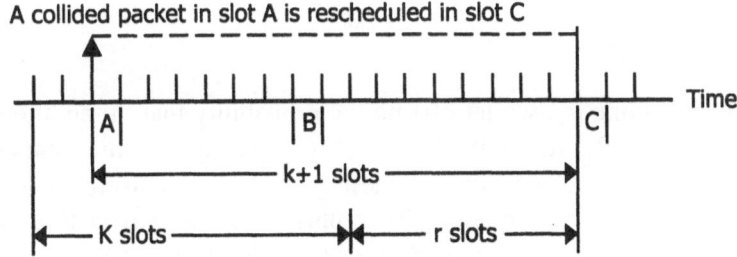

Figure 10 A-1. Retransmission of a packet in slot C from a collision in slot A

3.   No packets that collided in one of the other K-1 slots, other than slot A, are rescheduled for slot C.

Assume that the above three events are independent. Let

$$q_i = P\{Event\ i\ occurs\ \},\quad i=1,\ 2,\ 3.$$

Using these probabilities, we have

$$q_r = q_1\ q_2\ q_3^{K-1} \tag{10A-1}$$

If the packet that is successfully transmitted in slot C is a new packet, two independent events must take place:

(a)   No other new packets are generated in slot C; and

(b)   No retransmissions occur in slot C from collisions in any earlier slots.

It follows that

$$q_n = q_1\ q_3^{K} \tag{10A-2}$$

It now remains to determine the probabilities q_i for $i = 1,\ 2,\ 3$. Since the arrival process of new packets is Poisson with a rate of λ packets per slot and there are M stations, we have

$$q_1 = e^{-M\lambda P} = e^{-S} \tag{10A-3}$$

To determine q_2, we let q(j) be the probability that, in addition to the packet successfully transmitted in slot C, exactly j other packets are transmitted in slot A and none of these packets is retransmitted in slot C. Since the packet arrival process (new plus retransmitted) is Poisson with rate Λ, the probability of exactly j arrivals in slot A is given by

$$\frac{e^{-G}G^j}{j!},\ \text{where}\ G = \Lambda P \tag{10A-4}$$

Furthermore, the probability of not transmitting a collided packet in slot C is the probability of not choosing one value of the back off integer k, which results in retransmission in slot C. This probability is equal to $1 - \frac{1}{K}$. Assuming all events are independent, $q(j)$ can be expressed as

$$q(j) = e^{-G} \frac{G^j}{j!} \left(1 - \frac{1}{K}\right)^j \tag{10A-5}$$

Let us define the following two events:

X = In addition to the successfully transmitted packet in slot C, at least one other packet is transmitted in slot A and none of these packets is transmitted in slot C,

and

Y = A collision occurs in slot A (or at least one other packet is generated in slot A).

Then we have

$$P(X,Y) = \sum_{j=1}^{\infty} q(j) \tag{10A-6}$$

Since the packet arrival process is Poisson with rate Λ, then

$$P(Y) = \sum_{j=1}^{\infty} \frac{e^{-G} G^j}{j!} = 1 - e^{-G}, \ G = \Lambda P \tag{10A-7}$$

The probability q_2 is the probability $P(X|Y)$. Therefore,

$$q_2 = \sum_{j=1}^{\infty} q(j)/(1 - e^{-G}) \tag{10A-8}$$

Substituting (10A-5) into (10A-8) yields

$$q_2 = \frac{e^{-G}}{1 - e^{-G}} \sum_{j=1}^{\infty} \frac{G^j}{j!} (1 - \frac{1}{K})^j = \frac{e^{-G/K} - e^{-G}}{1 - e^{-G}} \qquad \text{(10A-9)}$$

To calculate q_3, we consider the event that no retransmissions from collisions in a slot other than slot A (for example, slot B) occur in slot C. This event can occur in one of three mutually exclusive ways:

(a)   No arrivals and hence no transmission occurs in slot B;

(b)   One arrival and hence a successful transmission occurs in slot B so that no retransmission is needed in slot C; and

(c)   Two or more transmissions take place in slot B, but none is retransmitted in slot C.

The probabilities of the first two events are determined by the Poisson arrival process to be e^{-G} and Ge^{-G}, respectively. The last event is the union of the events with probability $q(j)$ for j from 2 to ∞. Thus, the probability of the last event is

$$\sum_{j=2}^{\infty} q(j) = \sum_{j=2}^{\infty} e^{-G} \frac{G^j}{j!} (1 - \frac{1}{K})^j$$

$$\qquad \text{(10A-10)}$$

$$= e^{-G}(e^{G-G/K} - 1 - G + \frac{G}{K})$$

Since q_3 is the probability of the union of the above three mutually exclusive events, we finally obtain

$$q_3 = e^{-G} + Ge^{-G} + \sum_{j=2}^{\infty} q(j) = e^{-G/K} + \frac{G}{K}e^{-G} \qquad \text{(10A-11)}$$

Substituting q_1, q_2, and q_3 into (10A-1) and (10A-2) yields the results in (10-1) and (10-2), respectively, as

$$q_n = (e^{-G/K} + \frac{G}{K} e^{-G})^K e^{-S} \qquad \text{(10A-12)}$$

and

$$q_r = \left[\frac{e^{-G/K} - e^{-G}}{1 - e^{-G}} \right] (e^{-G/K} + \frac{G}{K} e^{-G})^{K-1} e^{-S} \qquad \text{(10A-13)}$$

APPENDIX B - Determination of the limiting values of q_n and q_r

From (10-9) and (10A-1) to (10A-3), we have

$$\frac{S}{G} = \frac{q_r}{1 + q_r - q_n}$$

$$q_n = e^{-S} q_3^K$$

$$q_r = e^{-S} q_2 q_3^{K-1}$$

where

$$q_2 = \frac{e^{-G/K} - e^{-G}}{1 - e^{-G}}$$

$$q_3 = e^{-G/K} + \frac{G}{K} e^{-G}$$

Since G is a constant, $\lim_{K \to \infty} q_2 = \lim_{K \to \infty} q_3 = 1$, and $q_n = \frac{q_3}{q_2} q_r$,

then

$$\lim_{K \to \infty} q_n = \lim_{K \to \infty} q_r$$

It follows that

$$\lim_{K\to\infty} \frac{S}{G} = \lim_{K\to\infty} q_n = \lim_{K\to\infty} q_r$$

We now proceed to find this common limit, which we denote by β.

Since S is also a constant,

$$\lim_{K\to\infty} q_n = e^{-S} \lim_{K\to\infty} q_3^K$$

Note that as $K\to\infty$,

$$\lim_{K\to\infty} S = G \lim_{K\to\infty} \frac{S}{G} = \beta G$$

where

$$\beta = \lim_{K\to\infty} q_n = e^{-\beta G} \lim_{K\to\infty} q_3^{K.}$$

We need to calculate

$$\lim_{K\to\infty} q_3^K = \lim_{K\to\infty} (e^{-G/K} + \frac{G}{K} e^{-G})^K$$

$$= \lim_{K\to\infty} e^{-G}(1 + \frac{G}{K} e^{-[(K-1)/K]G})^K$$

$$= e^{-G(1 - e^{-G})}$$

Thus, we obtain

$$\beta = e^{-G(1 + \beta - e^{-G})}$$

A solution to this functional equation is

$$\beta = e^{-G}$$

To show that β is the unique solution, we assume that β' is another solution.

If $\beta' > e^{-G}$, then $1 + \beta' - e^{-G} > 1$. Thus,

$$\beta' = e^{-G(1 + \beta' - e^{-G})} < e^{-G}$$

which is a contradiction.

If $\beta' < e^{-G}$, then $1 + \beta' - e^{-G} < 1$. Hence,

$$\beta' = e^{-G(1 + \beta' - e^{-G})} > e^{-G}$$

which leads to a similar contradiction.

Therefore, we conclude that $\beta = e^{-G}$ is the unique solution.

ANSWERS TO SELECTED PROBLEMS

Chapter 1

1-4   4.9

1-5   (a) 20.83, (b) 20.41, (c) 0.82

Chapter 2

2-2   (a) $\frac{1}{2}(R\sqrt{\frac{C}{L}} + G\sqrt{\frac{L}{C}})$, (b) $\sqrt{\frac{L}{C}}$, (c) $\frac{R}{2}\sqrt{\frac{C}{L}}$

2-3   (a) $\sqrt{RG}$, (b) $\omega\sqrt{LC}$

2-4   (a) 0.35 nepers/km, (b) 141 Ω

2-5   103 Ω

2-6   4.56 mH/km

2-7   0.003 $\mu F/km$

2-8   7 km

2-9   4.6 km

Chapter 3

3-1   3 × 3; 52,488

3-2   52,488

3-3 7.6×10^{-11}

3-4 2.2×10^{-6}

3-5   0.02

Chapter 4

4-1   (a) 0.96 seconds, (b) 1.79 seconds

4-2   (a) 0.6, (b) 3.6 min.

4-3   (a) 0.4, (b) 1.33 min., (c) 6.9 min., (d) 1.8

4-4   (a) 110, (b) 0.3, (c) 0.06 seconds, (d) 0.42

4-5   (a) 300, (b) 0.12s

4-6   (a) 710 packets/sec., (b) 0.64 ms

4-7   (a) 163 packets/sec., (b) 0.025 sec., (c) 3.12, (d) 0.019s, (e) 0.88

4-8   (a) $\dfrac{\lambda}{\mu(\lambda + \mu)}$, (b) $\dfrac{1}{\mu} - \dfrac{1}{\lambda} (1 - e^{-\lambda/\mu})$

4-9   (a) $\lambda\tau$, (b) $\lambda^2(\sigma^2 + \tau^2) + \lambda\tau$, (c) $\lambda^2\sigma^2 + \lambda\tau$

4-10   (a) 0.78, (b) 0.14

4-11   (a) $\dfrac{dP_0(t)}{dt} = -\lambda P_0(t) + \mu P_1(t)$

$\qquad \dfrac{dP_1(t)}{dt} = -\mu P_1(t) + \lambda P_0(t)$

$\qquad$ (b) $P_0(t) = \dfrac{\mu}{\lambda + \mu} + [P_0(0) - \dfrac{\mu}{\lambda + \mu}] e^{-(\lambda+\mu)t}$

$\qquad\qquad P_1(t) = \dfrac{\lambda}{\lambda + \mu} + [P_1(0) - \dfrac{\lambda}{\lambda + \mu}] e^{-(\lambda+\mu)t}$

4-12   (a) 909 packets/sec., (b) 0.909 erlangs

4-13   (a) 24×10^{-4}, (b) 0.51s, (c) 10.2

4-14   (a) 0.1, (b) 1.6 ms

4-15   (a) $(k + 1)(\lambda/\mu)^k (1 - \dfrac{\lambda}{\mu})^2$, (b) $2a/(1 - a)$, $a = \lambda/\mu$

4-16   (a) 40, (b) 0.9, (c) 0.3 s, (d) 0.6

4-17   (a) $(1 - \sigma)\sigma^{k-1}$, (b) $\dfrac{1}{\sigma}(1 - \dfrac{\lambda}{1 - \sigma})(\alpha\sigma)^k$

4-18   (a) 1.63, (b) 7.09 s

4-19   (a) 0.67, (b) 1.33, (c) 2

4-20   (a) $\dfrac{(\lambda/\mu)^k}{k!}\, e^{-\lambda/\mu}$, (b) $\mu(1 - e^{-\lambda/\mu})$, (c) $1 - e^{-\lambda/\mu}$

4-21   (a) 100, (b) 0.4 s

4-22   (c) $\dfrac{(\lambda/\mu)^k}{k!}\, e^{-\lambda/\mu}$

4-23   (c) $\dfrac{1 - \mu_2/\mu_1}{1 - (\mu_2/\mu_1)^{M+1}}(\dfrac{\mu_2}{\mu_1})^k$

4-24   (a) $\dfrac{(1 - \rho)\rho^k}{1 - \rho^{n+1}}$, $k \leq n$, $\rho = \lambda/\mu$, (b) $\rho[\dfrac{1 - (n + 1)\rho^n + n\rho^{n+1}}{(1 - \rho)(1 - \rho^{n+1})}]$

       (c) $\dfrac{1}{\mu}\,\dfrac{1 - (n + 1)\rho^n + n\rho^{n+1}}{(1 - \rho)(1 - \rho^{n+1})}$

4-25   (a) $(1 - \rho)\dfrac{\lambda}{s + \lambda}\hat{H}(s) + \rho\hat{H}(s)$, (b) $\lambda e^{-\lambda t}$; $1 - e^{-\lambda t}$, $t \geq 0$,

       (c) $\rho\delta(t - T) + (1 - \rho)e^{-\lambda(t-T)}1(t - T)$;
       $1 - (1 - \rho)e^{-\lambda(t-T)}$, $t \geq T$

4-26 $\dfrac{(1 + 4\rho) - \sqrt{1 + 8\rho}}{2}$, $\rho = \lambda/\mu$

4-27   (a) $\sigma e^{-(1-\sigma)\mu t}$, (b) $\sigma/\mu(1 - \sigma)$

4-28   (a) 3/4, (b) $\dfrac{1}{4}(\dfrac{3}{4})^k$, (c) 1.5 s

4-29   (i) (a) 0.5665, (b) 1.087, (c) 0.217 s

       (ii) (a) 0.565, (b) 1.085, (c) 0.0217 s

4-30 $\dfrac{\lambda}{sr(2sr - \lambda s - \lambda)}[b^2 + b(s - 1) + \dfrac{(s - 1)(2s - 1)}{6}]$

Chapter 5

5-1   (a) 22, (b) 21, (c) 0.015

5-2   (a) 0.0267, (b) 0.02

5-3   0.75

5-4   (a) 1000 ccs, (b) 0.1 ccs per subscriber, (c) 0.0028, (d) 0.01, (e) 0.72

5-7   (a) $(n + 1)/s\mu$ $seconds$,

$$\text{(b)} \quad \frac{n+1}{s\mu} + \frac{1}{\mu} \sum_{k=1}^{s} \frac{1}{k}, \text{ (c) 1/s, (d) (s - 1)/2 s}$$

5-8   (a) 0.03; 3,

(b) 0.09; 2.7s; 3; 0.4s

5-9   (a) 44.37, (b) 0.27, (c) 44.73

5-10   (a) 8.29, (b) 13

5-11   (a) 1/3, 1/3, 1/3, (b) 6.67, (c) 8

5-12   (a) $\dfrac{a^k}{k!}e^{-a}$

5-13   (a) 0.025, (b) 0.02, (c) 0.11 min; 0.95 min

5-15   (a) 0.154; 0.4201, (b) 2.73, (c) 15.12; 22.27, (d) 2

5-16   (a) 9.63; 7.704; 0.8, (b) 0.37; 0.8134; 1.2

Chapter 6

6-1   (a) 15, (b) 10

6-2   (a) 0.0131, (b) 0.0066, (c) 0.013

6-3   (a) 0.0236, (b) 0.1816, (c) 2.38 s, (d) 0.0213

Chapter 7

7-1   (a) 40, (b) 0.1

7-2   0.8; 0.94

7-3   81.8; 86.72

7-4   (a) 58.33 ms, (b) 33.43 ms

7-5   (a) 50, (b) 93

7-6   (a) 1.33 ms, (b) 0.33 ms

7-7 $1 - \dfrac{(M-1)g}{C}; \dfrac{(M-1)g}{C-(M-1)g}$

7-8   (a) $1 - y$, where y is the overhead factor, (b) 1.98%

7-9   (a) 0.24, (b) 1.4 packets/s, (c) 0.6, (d) 7.3 ms

7-10   (b) 0.4862

Chapter 8

8-1   (a) 93, (b) 3571

8-2   (a) 1667 bps, (b) 4.26 s; 4.32 s; 4.3 s, (c) 7.66 s

8-3   1200 bps

8-4   (a) 52, (c) 80

Chapter 9

9-1   (a) 167.13 μs, (b) 167.13 μs, (c) 217.05 μs

9-2   (a) 35, (b) 2 μs, (c) 1.52 ms

9-3   (a) 1.06 ms, (b) 1.06 ms, (c) 1.4 ms

9-4   (a) 142 packets/s, (b) 1000 packets/s, (c) 392 packets/s

9-5   (b) 50

9-6   (a) 1.32 ms, (b) 2.38 ms, (c) 24.16 ms

9-7   50 bits; 30 bits

9-8   2.41

Chapter 10

10-1   (a) $(M - \bar{N}_R)\sigma$, (b) $\bar{N}_R + (M - \bar{N}_R)\sigma$, (c) $1 + \dfrac{\bar{N}_R}{(M - \bar{N}_R)\sigma}$

10-2   (a) $1 + \dfrac{1 - \gamma}{\gamma p}$, (b) $e^{-G(k)}$

10-3   (b) 2.42 ms

10-5   331

10-6 2.465×10^{-32}; 2.0283

10-7   (a) $M\sigma(1 - \sigma)^{M-1}$; $1/M$,

       (b) $(1 - P_s)/P_s$; $(1/P_s - 1)2\tau_p$,

       (c) $\dfrac{P}{P + [\dfrac{\dfrac{M}{(1 - \dfrac{1}{M})^{M-1}}}{} - 1]2\tau_p}$

       (d) 0.9255